The Complete Lover

The Complete Lover

*Eros, Nature, and Artifice
in the
Eighteenth-Century
French Novel*

Angelica Goodden

CLARENDON PRESS · OXFORD
1989

Oxford University Press, Walton Street, Oxford OX2 6DP

Oxford New York Toronto
Delhi Bombay Calcutta Madras Karachi
Petaling Jaya Singapore Hong Kong Tokyo
Nairobi Dar es Salaam Cape Town
Melbourne Auckland

and associated companies in
Berlin Ibadan

Oxford is a trade mark of Oxford University Press

Published in the United States
by Oxford University Press, New York

British Library Cataloguing in Publication Data
Goodden, Angelica
The complete lover: Eros, nature, and
artifice in the eighteenth-century French novel.
1. French literature, 1715–1789 — Critical studies
I. Title
840.9' 005
ISBN 0-19-815820-3

Library of Congress Cataloging in Publication Data
Data available

Phototypeset by Dobbie Typesetting Limited, Plymouth, Devon
Printed and bound in
Great Britain by Biddles Ltd
Guildford and King's Lynn

Contents

Abbreviations

The following abbreviations have been used in the footnotes and bibliography:

AUMLA   *Journal of the Australian Universities Modern Languages Association*
FMLS   *Forum for Modern Language Studies*
JES   *Journal of European Studies*
JHS   *Journal of Hellenic Studies*
MLR   *Modern Language Review*
PMLA   *Publications of the Modern Language Association of America*
RHLF   *Revue d'Histoire littéraire de la France*
RhM   *Rheinhessisches Museum*
RLC   *Revue de Littérature comparée*
RR   *Romanic Review*
RSH   *Revue des Sciences humaines*
TAPA   *Transactions of the American Philological Association*
ZfSL   *Zeitschrift für französische Sprache und Literatur*

Introduction

ALTHOUGH, according to Sade's *Idée sur les romans*, 'On appelle roman l'ouvrage *fabuleux* composé d'après les plus singulières aventures de la vie des hommes',[1] in eighteenth-century France novels began to deal more closely with the matters of ordinary life than had their seventeenth-century forebears. They wended an often rambling way through the landscape of feeling, pausing to analyse sentiment when they encountered it with an amplitude rarely possible in drama or poetry. But they retained enough of imaginative power still to offer their readers a form of escapism, in the shape of idealized emotions or passionate intensities which set their subjects above the everyday. Some novels of the period, admittedly, are barely concerned with this central emotion, but they are rare: perhaps the most celebrated is Montesquieu's *Lettres persanes*, a work of cultural reportage in which the love-interest is supplied by a sub-plot that seldom impinges on the author's discourse about social institutions and mores. But for Montesquieu's contemporary Marivaux, it went without saying that novels were specifically directed at moving their reader emotionally: 'puisque le roman n'est fait que pour le cœur, quand il le touche, doit-on s'en plaindre?'[2]

The types of erotic relationship described in eighteenth-century French fiction are varied. There is the liaison governed by intellect, of which Laclos paints an unforgettable portrait; the merely sensual union, whose nature is explored by Sade; the sentimental pairing, which Rousseau extols in lyrical prose. But few novelists of the day confine their investigation to a single mode of loving. Far commoner is the attempt to show in what proportions heart, mind, and senses combine in the individual who seeks erotic satisfaction with another human being; and in the more idealistic works of the time the author's concern is to show precisely how these elements may be admixed to form the complete lover. Such a being, it is true, is more often projected than actualized in the novel, and completeness remains a distant goal. But a clear lesson none the less emerges from the fiction

---

[1] *Idée sur les romans*, ed. Octave Uzanne (Paris, 1878), p. 3.
[2] *Les Aventures de* *** *ou les effets surprenants de la sympathie*, Œuvres de jeunesse, ed. Frédéric Deloffre (Paris, 1972), p. 5.

of the period. On the evidence of Crébillon *fils*'s *Les Egarements du cœur et de l'esprit*, Prévost's *Manon Lescaut*, Rousseau's *La Nouvelle Héloïse*, Laclos's *Les Liaisons dangereuses*, and other novels of the century, reason must guide sentiment, and sentiment reason, in the man or woman who aspires to a state of erotic fulfilment.

In order to be 'finished',[3] the lover must (whether involuntarily or by intent) undergo certain experiences. The effort to educate his character—a common theme—arises from the perception that eros is a social phenomenon whose manifestations ought not to go unchecked. The lover must learn to distinguish carnal from spiritual impulses, the material from the metaphysical. He (and especially she, for eighteenth-century woman is a weaker vessel than the male) must learn to understand the pathological effects of passion, to observe and where possible moderate its influence over body and soul. Environment, too, has its part to play in forming the complete lover: the hothouse of the town may polish him but promote dissipation of his moral being, while the calm of the rustic world, by contrast, can encourage the development of unadulterated feeling. Urban backgrounds also create quite different perceptions of the love-object's physical attractiveness from those known in country fastnesses: a sophisticated air is often preferred in the former, a natural one in the latter. It is incumbent on the lover, besides, to discover the best means of expressing his feelings to the loved one, as well as to learn what separates erotic relationships from other types which also involve the tender emotions, such as friendship. This book successively examines such matters, and tries to show how the eighteenth-century lover's response to them helps shape his erotic personality.

The lover's response, furthermore, is complemented by the reader's, for reading novels can have the effect of sharpening (or dulling) one's perception of eros. Rousseau's experience is instructive here. In the *Confessions* he describes the dangerous knowledge of the passions which he acquired at an early age through an obsessive reading of novels:

Ma mère avait laissé des romans. Nous nous mîmes à les lire après souper mon père et moi. Il n'était question d'abord que de m'exercer à la lecture

---

[3] See Laclos, *Les Liaisons dangereuses*, in *Œuvres complètes*, ed. Laurent Versini (Paris, 1979), p. 1182, note 4. 'Finir' was a traditional euphemism in 'galant' language for the act of consummating an affair.

par des livres amusants; mais bientôt l'intérêt devint si vif que nous lisions tour à tour sans relâche, et passions les nuits à cette occupation. Nous ne pouvions jamais quitter qu'à la fin du volume. Quelquefois mon père, entendant le matin les hirondelles, disait tout honteux: allons nous coucher; je suis plus enfant que toi.

En peu de temps j'acquis par cette dangereuse méthode, non seulement une extrême facilité à lire et à m'entendre, mais une intelligence unique à mon âge sur les passions. Je n'avais aucune idée des choses, que tous les sentiments m'étaient déjà connus. Je n'avais rien conçu; j'avais tout senti.[4]

Although this passage suggests that Rousseau's reading of fiction taught him unassailable and impressive truths about emotion, imaginative literature may evidently also falsify conceptions of ordinary human conduct and feeling. The author's artifice, as Diderot suggests in *Jacques le fataliste*, can be deployed to affect our rational and emotional view of the subject-matter presented. Sometimes, however, and especially over a period of time, his intended effect is altered by factors beyond his control. To modern readers, for instance, it may well appear that eighteenth-century literature is too loud in its protestations, that it cheapens passion by proclaiming it so insistently, and that the habitual exhibitionism of its characters puts their honesty of feeling in question. People as complacent as this about their emotional reactions, we may reflect, cannot be undergoing experience genuinely; they dramatize things too vaingloriously to be sincere. To have 'felt' everything through an acquaintance with eighteenth-century prose fiction, in other words, may be to have experienced unlikely intensities of passion—hatred, lust, joy, despair *in extremis*. In *Les Egarements du cœur et de l'esprit* Crébillon's Meilcour finds that when he attempts to sustain an advance towards the woman of his heart, his past reading of novels provides him with no reliable indication of how he should proceed.[5] If fiction is principally concerned with the eternal verities of human emotion, its themes may still be of an extraordinary kind, and of scant use in the conduct of ordinary life.

The naïve reader who takes novels to be final authorities on phenomena like human love and desire, then, may be saddling himself with an incomplete conception of their nature. Despite the

---

[4] *Les Confessions*, in *Œuvres complètes*, ed. Bernard Gagnebin and Marcel Raymond, 4 vols (Paris, 1959–69), i. 8.

[5] Crébillon *fils*, *Les Egarements du cœur et de l'esprit*, in *Romanciers du XVIIIᵉ siècle*, ed. Etiemble, 2 vols (Paris, 1965), ii. 54.

claims Rousseau makes for it in the *Confessions*, fiction is not necessarily a reliable guide to human feeling, because it need not suffer the constraint of verisimilitude. Besides, when we consider the prevalence of censorship in eighteenth-century France, to say nothing of the ingrained influence of Christian morality even in an age of growing secularization, we can understand why the contemporary novel should often exhibit carelessness or ignorance of certain emotional states. On the evidence of eighteenth-century fiction, it seems not to be the novel's task to provide a full account of sexual arousal, desire, pleasure, and many other aspects of erotic love. The 'sensible' climate of the period often means that sentiment informs the novel's detailing of sexual encounters to a degree which may surprise modern readers, or seem to them quaint; the physical nature of sex is rarely discussed. Delicacy naturally prohibits many narrators, even obsessive analysts like Marivaux's Marianne, from clinically examining the phenomena attending the arousal of passion. *Manon Lescaut* is frequently cited for its discreet handling of erotic encounters, but is no more reticent in this respect than many novels of the day; and even Laclos's scandalous *Liaisons dangereuses* refrains from discussing the practical aspects of love-making.

Mechanistic materialism enjoyed considerable favour among eighteenth-century intellectuals, but it is uncommon to find reflections of contemporary discoveries about man's physical functioning in the novel's depiction of eros. Where pleasure is discussed, it is more likely to be in metaphysical than in materialist terms. There is much evidence in the fiction of the age of a desire to see a governing spiritual impulse behind the facts of erotic attraction and desire. The old theory of sympathy, which ancient philosophers had seen as material in origin, was given a part-metaphysical explanation even by writers like Diderot who wanted to account for the world and its inhabitants in narrowly physical terms. Causal explanations were tempered by appeal to the mystical essence of the human soul, the 'je ne sais quoi' which drew people uncomprehendingly together.

The phenomenon of human love, to judge by the novels of Marivaux, Crébillon *fils*, Rousseau, Laclos, and Saint-Pierre, is not to be illuminated by an examination of and assimilation to animal sexuality. The central message of novels like *La Vie de Marianne*, *Les Egarements du cœur et de l'esprit*, and *La Nouvelle Héloïse* is that humans have ideas about love which are not available to other creatures, that people are self-conscious beings endowed with reason and gifted with a power of language in which the thoughts that are a central part of emotional response can be formulated. If amatory experience can be adumbrated by a consideration of human physiology—as Diderot half-facetiously suggests in his early novel *Les Bijoux indiscrets*—then there is little evidence of that fact in the eighteenth-century novel, aside from the works of Sade and like-minded writers. What novels of the period characteristically examine, by contrast, is the type of arousal that can be undergone by humans alone. In this respect, perhaps, we can sensibly talk of the novel as a rational enterprise, for even at its most 'feeling' it describes states which are the product of mental apprehension. But this is not to say that eighteenth-century novelists resorted to the explicitly intellectual in their presentation of a distinctively human love. Nothing, indeed, could be more at odds with most readers' experience of that period's fiction than such a notion.

All this suggests that love, mysterious as its origins may be, is also subject to voluntary control, and thus properly part of moral life. Rousseau's view is that man, although a fallen creature directed towards immorality by created social institutions, has the power to regain the virtue he originally possessed. *La Nouvelle Héloïse*, perhaps the most influential novel of the century, is an allegory constructed around these assumptions, being the story of a woman who achieves a state of exemplary virtue after yielding it up in an irregular love-relationship. Human passion, Rousseau implies, is not an ungovernable force, but one amenable to human will and intention. It is striking that this should be one lesson of a work often regarded as the epitome of eighteenth-century sensibility. However highly feeling may be valued, according to Rousseau, it has no lasting worth when divorced from will. True affection and love are necessarily informed by active intention; only when they lack that element can they be reductively described as merely sentimental. Love, Rousseau argues, does not consist in a burst of feeling which we are unable to govern (he opposes such a view, at least, until the final pages of the book), but is an attitude endorsed by part of our conscious being. Such a belief lies, at any rate in theory, behind the many eighteenth-century novels whose authors protest the morality of their enterprise, for morality has no place in a universe from which free will is absent.

Love, then, is no more forcibly imposed by supernatural agency than it is a merely physical phenomenon. It is a properly human emotion, but one which belongs to our incorporeal as well as our material selves. To say that it is in some sense subject to the will does not mean that we may simply love to order: we find it bizarre, if not sad, to read of someone who decides, like Constant's *Adolphe*, that it is time he experienced passion, and sets about procuring it for himself. The convergence of various philosophical and religious traditions in our culture means that we are usually inclined to value the spiritual in love more highly than the physical, while perhaps feeling that in an ideal partnership the two orders come together. Laclos describes the latter state near the end of *Les Liaisons dangereuses*, but is also careful to show the precariousness of such a union.

In the eighteenth-century novel, morality is still intimately connected with sexual continence. Therein lies the newness of *La Nouvelle Héloïse*, where Rousseau makes a woman who has once fallen and then regenerated herself the ultimate symbol of human virtue. In *Les Liaisons dangereuses*, which it greatly influenced, we witness the God-fearing Présidente de Tourvel adapting her rigid Christian morality in accordance with the radical change wrought in her by her adulterous love for Valmont: the ultimate sanction for her acts is now her lover's happiness rather than the demands of Christian virtue, which prohibits their relationship. In opening herself to the full force of passion, and in her eventual willingness to answer all its dictates, the Présidente acknowledges what many thinkers of the period argued, namely that morality is not a fixed absolute, but a variable. She modifies her formerly inflexible view that marriage represents the ultimate moral and social order, a perfect example of the 'mediocrity', or mean, which leads to solid happiness, and is equally distant from libertinism and asceticism. The bonds of love and philanthropy that tie the humanitarian to his fellows hold the Présidente to the man who now stands at the centre of her universe. Before yielding to her feelings for Valmont, however, Mme de Tourvel had seen the ideal relationship between the sexes (epitomized by marriage) as being not 'amour-passion' but 'amour-bienfaisance', in which the soul was purged of the feelings of temptation and violence often thought to be inseparable from eros. The pleasure derived from such a union was what the eighteenth-century called 'volupté', in the spiritual rather than carnal sense of the word[6]—not

---

[6] [Rémond le Grec], *Agathon. Dialogue sur la volupté*, in *Recueil de divers écrits sur l'amour et l'amitié etc.* (Brussels, 1736).

a joy of the flesh, but a state of bliss dependent on mind and taste, free from the brute urgings of instinct. Its signs were moderation, calmness of soul, and delicacy of heart and spirit.

This optimistic philanthropic view is ultimately belied both in Laclos's novel and in *La Nouvelle Héloïse*. But its insufficiency, which Rousseau at least is less than straightforward in admitting, should not be taken to indicate the primacy of destructive sexuality in passion. If the eighteenth century saw the sexual as dangerous, it did so less in a Racinian spirit than in terms of the (negative) influence of social convention. In *Clarissa*, a work which influenced many novelists in eighteenth-century France, Richardson implicitly criticizes the power of a society which, in effect, destroys a woman suspected of having lost her sexual purity. Like Rousseau after him, he wanted to show that chastity of mind is a greater thing than chastity of body, and that the ultimate fulfilment of happiness should not be dependent on the possession of an unruptured hymen. The vulgar world (the world that tolerates adulterous relationships after matrimony) condemns woman on a technicality; true humanitarianism finds saving virtue in a morality which is not that of the market-place.[7] Even after the rape, like Julie after her freely chosen submission to Saint-Preux, Clarissa remains inwardly spotless.

Other eighteenth-century novels likewise praise a morality of sentiment. Des Grieux's fickle mistress speaks more profoundly than she knows when she tells him, 'la fidélité que je souhaite de vous est celle du cœur.'[8] This is sentimentality in an elevated sense, and the libertine is its arch-enemy. He sees conventional protestations of virtue as mere posturing, a spur to his pride, an invitation to expose the hollowness of received morality. But when he encounters a character who, unlike Prévost's Manon, understands the possibility of uniting erotic desire with heartfelt love, he is given pause for thought. In the case of Laclos's Valmont, however, the pause is tragically brief, and when it is over the desire for power, reasserting itself, brings destruction in its train.

Erotic love has the quality of combining desire—the libertine's guiding impulse—with care for the other person. This is the mature

[7] See Christopher Hill, 'Clarissa Harlowe and her Times', in *Samuel Richardson: A Collection of Critical Essays*, ed. John Carroll (Englewood Cliffs, N. J., 1969), pp. 116 ff.

[8] Abbé Prévost, *Histoire du chevalier des Grieux et de Manon Lescaut*, ed. Frédéric Deloffre and Raymond Picard (Paris, 1965), p. 147.

union which des Grieux understands and Manon does not. To be erotically in love with someone is to identify oneself with his or her good, as des Grieux does with Manon's, and to want one's own good to be his or her own. Such is the mutuality which Valmont and the Présidente fleetingly experience. Usually erotic love runs a course (rarely traced in the novel, which habitually concerns itself only with the progress towards consummation) that leads, through a sense of the beloved's irreplaceable value, to the substitution for sexual desire of a love resting on companionship and trust. In *La Nouvelle Héloïse* Julie and her husband miss out on the first stage and settle into this phase immediately upon marriage, whereupon Julie systematically denigrates its antecedent in letters to her former lover Saint-Preux. 'Mature' partners like these are invulnerable to the hurts and uncertainties that accompany sexual yearning. But possibly, as Valmont tries to persuade Mme de Tourvel, they are the losers thereby. It is perhaps true, as a philosopher has recently argued,[9] that a metaphysical illusion is always at the heart of such desire; Laclos's Mme de Rosemonde, at least, tries to persuade the Présidente so. But those of romantic disposition, among whom many novel-readers may be numbered, will be unwilling to heed such warnings as the old woman utters.

The eighteenth-century novel describes the dignity, pathos, and exaltation of love. To call it 'sentimental' is not necessarily to denigrate it, despite the frequency with which the word is used as a term of abuse. The sentimentality of eighteenth-century fiction can make it vapid, mawkish, and winsome, and its modern reader may rightly object to the overtness of its appeal to his own feelings. It is often intellectually second-rate; but it is not necessarily facile, and the loudness of its proclamations may be in order. In emphasizing the relation between the capacity to feel love and the ability to do good it is inherently no more naïve than when it argues that literature can affect the reader's emotions positively. If the sentimental climate of the eighteenth century ruined much art, its products are often of considerable aesthetic and moral stature: it led to some false posturing and to a great deal of genuine feeling. The age of Rousseau, Diderot, and Laclos was an age both of sentiment and of reason, and the novel reflects that duality; as a supreme vehicle for the expression of love

---

[9] See Roger Scruton, *Sexual Desire* (London, 1986).

it can still absorb us, in its wayward lyricism no less than in its resolute control.

Selecting material relevant to an examination of love in the eighteenth-century French novel is perhaps like selecting a lover: personal inclination is decisive, even though some will find the choice it dictates hard to comprehend. In other words, I make no apology for omitting all reference to well-known works like Challe's *Illustres Françaises* or Lesage's *Gil Blas*, since I believe eclecticism to be as necessary to scholarship as to taste. It is a truth universally acknowledged, and one which many eighteenth-century novels confirm, that love has the power to surprise those who fall under its dominion. If my choice of texts sometimes appears similarly unexpected, the unexpectedness may perhaps be seen as a reflection of that principle. Except when, as with ancient writers, a corpus is essentially the sum of what has survived, the idea of a fixed body of relevant texts constituting a 'syllabus' simply means that someone's selection has been officially endorsed; capriciousness, on the other hand, is a necessary condition of imaginative criticism. There is another aspect of this book which possibly calls for justification. Analysing love as it appears in literature is both stimulating and alarming: I have sometimes felt like the fictional character who flung himself upon his horse and rode madly off in all directions,[10] and in response to this feeling I have severely limited both the range of topics discussed and the number of novels used to illustrate the discussion. An investigation into this subject could, evidently, have been as long as Rousseau's *La Nouvelle Héloïse* or Louvet's *Faublas*, while still leaving many matters untouched and many volumes unmentioned. But books, like lovers, can only be complete by being selective.

---

[10] Stephen Leacock, *Nonsense Novels* (London and New York, 1921), p. 59. The novel in question is *Gertrude the Governess*.

1

The Art (and Artfulness) of Loving

THERE is a famous scene in Prévost's *Manon Lescaut* in which the young hero des Grieux, imprisoned at Saint-Lazare for the acts of lawlessness he has committed under the influence of eros, debates with his Jesuit friend Tiberge the relative security attendant on loving God and loving mortals. Although des Grieux carries off the technical victory in his support of human passion, he is forced to acknowledge the chanciness it entails: the delights of love are transitory, even forbidden, and oblige man to suffer torment and privation in their pursuit.[1] The same connection between eros and mutability is emphasized by other eighteenth-century novelists, among them Marivaux (who in *La Vie de Marianne* shows his beautiful heroine being deserted by the unworthy Valville), Rousseau (whose unfinished sequel to *Emile*, *Emile et Sophie*, describes the young couple's moral collapse in the dangerous environment of the city, and Sophie's pregnancy by another man), and Laclos (whose *Liaisons dangereuses* depicts the abrupt termination by the rake Valmont of the love which had seemed to promise him salvation). According to these writers, human love is neither dependable nor predictable in its course, and its unsettling changes can bring deepest misfortune on the unwary.

The immutability of the Christian God had meant that the tactics of theological love were simple, and involved a comparatively straightforward psychology in which success was assured provided that the intention was right: God took the will for the act. With secular love we are back in the pagan world, where even the gods change, and Cupid is mercurial. Goodwill and devotion are ineffectual without a repertoire of gambits more suited to the card-table than to the prie-dieu. In this world, in which expertise in the theory of fantasies and foibles is more important than eternal verities, the focus shifts from intellect and will to the cunning art of diagnosing the passions of the body as well as the soul.

[1] pp. 91–3

In *Les Liaisons dangereuses* Valmont and the Marquise de Merteuil seek to master eros through intellect, subjecting it and those it rules to their 'méthodes' and 'projets'. Only near the end of the novel does this governance evidently falter, and the two protagonists get their melodramatic come-uppance; although before that time Valmont has half-conceded that he is 'maîtrisé', like a schoolboy, by his love for the Présidente de Tourvel.[2] Fiction of the age often concerned itself with similar, if less disturbing, attempts to master or confer stability on passional life. Such attempts sometimes involved the acquisition of an art of loving which sought to guarantee its practitioner lasting success in affairs of the heart (offspring of the Ovidian *Ars amatoria*): Crébillon's *Les Egarements du cœur et de l'esprit* and Rousseau's *Emile* describe educations of this kind. And sometimes the human ambition to control unruly emotion found expression in the creation of a special environment, such as the Clarens of Rousseau's *La Nouvelle Héloïse*, where goodwill and practical activity united (at least in intention) to snuff out erotic feeling. Alternatively, removal to a new environment—such as the Louisiana of *Manon Lescaut*—was shown to usher in a changed attitude to the impulses fostered by the old one, resulting in the effort to transform an irresponsible and hence unsettling love into a reflective one.

On the evidence of *Les Egarements du cœur et de l'esprit*, *Manon Lescaut*, Restif de la Bretonne's *Le Paysan perverti* and *La Paysanne pervertie*, *Les Liaisons dangereuses*, and Louvet's *Faublas*, learning about love in the 'civilized' world of eighteenth-century France is a long-drawn-out process, and one which can be aided by the guidance of a mentor (male or female). Acquaintance with the world of mere sexuality, on the other hand, comes easily, and the influence of more experienced parties here is as often nefarious as salutary. In Crébillon's novel the seventeen-year-old Meilcour enters the world of high society 'avec tous les avantages qui peuvent y faire remarquer';[3] despite the efforts of Mme de Lursay to save him for herself and decency, the hold of the roué Versac on the youth has begun to tell at the point where the novel breaks off. Restif's two peasants are young when they leave their village for the city, where their sexual innocence will rapidly be compromised before being lost altogether. The courtesan Juliette, sister of the virtuous Justine in Sade's *Les Infortunes de la vertu*, is fifteen when she is obliged to leave her Paris

---

[2] p. 288.   [3] p. 13.

convent and make her way in the world, asks a procuress for protection, and begins her novitiate in a house of easy virtue; and it is at the same age that Laclos's Cécile likewise quits the cloister to face the seductions of society.

That certain moral qualities suffer in all this is clear, at least to the individual who has only partially adapted himself to the social environment of the *haut monde*. Meilcour regrets the loss of sincerity entailed by the need to act a part in worldly circles. Listening to Versac's magisterial exposé of the way to social advancement, he feels disquiet at the older man's doctrine: 'je doute que je puisse jamais adopter un système qui m'obligerait à cacher les vertus que je puis avoir, pour me parer des vices que je n'aurais pas.'[4] Versac praises his protégé for his high principles, but assures him that morality and 'mondanité' are often incompatible with one another. The youth with aspirations to be 'formed' for the world, he implies, effectively has no choice: ethics have to be subordinated to social advancement. One must learn, as Versac has learnt, to disguise one's character perfectly.

Pensez-vous que je me sois condamné sans réflexion au tourment de me déguiser sans cesse? Entré de bonne heure dans le monde, j'en saisis aisément le faux. J'y vis les qualités solides proscrites, ou du moins ridiculisées, et les femmes, seuls juges de notre mérite, ne nous en trouver qu'autant que nous nous formions sur leurs idées . . . Je sacrifiai tout au frivole; je devins étourdi, pour paraître plus brillant; enfin, je me créai les vices dont j'avais besoin pour plaire: une conduite si ménagée me réussit.

Je suis né si différent de ce que je parais, que ce ne fut pas sans une peine extrême que je parvins à me gâter l'esprit.[5]

Firm intent is needed to acquire the skills of social intercourse in this rarefied atmosphere. Spontaneity, Meilcour learns, is discouraged if not forbidden, for it lays the individual open to the threat of chance. Life in the world, by contrast, is a willed construction, and love as much as other worldly skills is therefore practised according to a set of rules. The orderly, or apparently orderly, habits of society will prescribe the methods to be adopted by those who engage in amorous pursuits. Meilcour is struck by the fine distinctions dictating correct procedure in this regard, his awareness of which he owes to Mme de Lursay.

---

[4] p. 159.   [5] p. 153.

A l'égard de l'amour, reprit-elle, je vous ai, je pense, déjà répondu que ce n'était pas une excuse légitime [for a man's being indelicate]. Pour les bontés dont vous me parlez, je conviens que j'en ai pour vous, mais il en est de plus d'une espèce, et je crois que les miennes ne vous mettent en droit de rien . . . Elle ajouta à cela mille choses finement pensées, et me fit enfin entrevoir de quelle nécessite étaient les gradations. Ce mot, et l'idée qu'il renfermait, m'étaient totalement inconnus. Je pris la liberté de le dire à Madame de Lursay, qui, en souriant de ma simplicité, voulut bien prendre la peine de m'instruire.[6]

Saint-Preux describes this kind of delicacy as quintessentially French, and sarcastically contrasts it with the roughness of Swiss souls like Julie and himself.[7] But an 'art de plaire' such as is taught in French high society is not the same as a prudently regulated attitude to the matter of sentimental and sexual education. Most methods of social intercourse presuppose that an orderly approach to living which shows connections and interrelationships is needed to save humans from the threat of conflict and upheaval. Such a mode of controlling fate may be precarious, however, since—as Meilcour perceives—it often rests on the shifting sands of duplicity rather than the bedrock of a shared morality; and few but the naïve would suppose that any art could altogether eliminate contingency from emotional life.

In 1759 Caraccioli adverts to the demeaning frivolity of an age which pays scant attention to cultivating useful sciences, preferring to lose itself in the enjoyment of pointless arts. The word Caraccioli uses in *La Jouissance de soi-même* to designate the kind of art he has in mind makes it clear that the one which is most keenly studied is that of love: all things 'concour[ent] au triomphe de la volupté'.[8] His words pinpoint the sort of regularity by which people seek to confute chance and rule their existence:

Les gens du monde étudient aujourd'hui les plaisirs comme on étudiait autrefois la philosophie; ceux qui passent pour les plus sages sont ceux qui se font un système de la volupté, qui la goûtent avec réflexion, qui l'enseignent avec méthode, et qui la raffinent à force d'expériences.[9]

This elevated class is able to conceive and propagate values which have no necessary connection with those of other social groups.

---

[6] p. 84.   [7] *La Nouvelle Héloïse*, *Œuvres complètes*, ii. 249.
[8] *La Jouissance de soi-même* (Utrecht and Amsterdam, 1759), pp. ix–x.
[9] p. 189.

Meilcour's reaction to Versac's exposition of the way the *haut monde* behaves in matters of love, liking, and 'goût' is a salutary reminder of that. We have little sense, reading this novel, that moral values which should regulate relationships based on emotion have an eternally fixed position and can therefore be appealed to by people in general in their debates about what constitutes right action. Instead, there is the uncertainty that worries the philosopher, although not his interlocutor, in Diderot's *Le Neveu de Rameau*: what one man calls virtue another may call vice, and the other way about.[10] If this is true (and the many attempts in the eighteenth century to refute time-honoured Christian values indicated that it might be), there can be no such things as bounds of decency or propriety, in emotional affairs and elsewhere.

Dolmancé's stated opinion in Sade's *La Philosophie dans le boudoir* that the words 'vice' and 'virtue' have a purely local significance, that all depends on the mores of a particular people and the climate under which it lives,[11] is recognizably related to Versac's assertion that virtues are arbitrary concepts.[12] The idea of relativity which both men propose found much favour in the eighteenth century. When des Grieux's father makes it plain to his son that social convention allows him, even encourages him, to take a mistress, but demands that she be better than Manon, he is tacitly appealing to just such a notion. When the pious Mme de Tourvel finds that she can square her conscience with the fact of adulterously loving Valmont by adducing the overriding importance of procuring his 'bonheur', she is in effect adapting the concept of virtue to changed circumstances. The *summum bonum* is now Valmont's well-being, for he is her new God. But morality, whether relative to love or not, needs to be based on a deep human understanding about what represents a value. Moral norms do exist, and convention may be more than a local creation. This is not to deny the force of variable social and cultural assumptions in shaping the standards by which we apportion praise or blame for human actions: the law, a leitmotiv of *Les Liaisons dangereuses*, is an assemblage of such standards, and its existence betokens an attempt by humans to give stability to institutions and relationships in a world of flux. Without norms, humans live in a climate of rootlessness and disarray, as *La Philosophie dans le boudoir*

---

[10] *Le Neveu de Rameau*, ed. Jean Fabre (Geneva and Lille, 1950), p. 62.
[11] *La Philosophie dans le boudoir* (Paris, 1972), pp. 64–5.   [12] pp. 151–2.

makes distressingly apparent. Crébillon depicted a degree of such instability in *Les Egarements*, although even he, according to Melchior Grimm, imagined nothing to match the unsettled universe of *Les Liaisons dangereuses*.[13]

The temptation to deliver dogmatic judgements on propriety and impropriety is especially strong where love is concerned. The novels I have been discussing all try to show this temptation being avoided, generally in the interests of educating one of their characters, be it Meilcour in *Les Egarements*, or Cécile and Danceny in *Les Liaisons dangereuses*. At the same time, their authors want to show that this education is corrupting, that certain forms of loving carry morally compromising risks. The little world of high society they depict makes certain contingent interventions into the moral code which add up to transgressions. In so far as the reader supports such transgressions (sympathizing, for example, with attractive miscreants like des Grieux and Manon, or Valmont and Mme de Merteuil, despite disapproving of their actions), he is guilty of moral inconsistency. (The realization that this is the case may, of course, be part of the educative process that the author wishes him to undergo.)

The idea of educating people in emotion presupposes that emotion is responsive to teaching. It lies within human control, that is to say, and is not primarily a visceral spasm or another type of purely physical disturbance, but may be consciously guided, developed, suppressed, or submitted to any other kind of voluntary force. According to Chavannes, in an essay on education, all that is done by rule, or by reflection designed to give direction to what nature suggests, is an art. It is man's achievement exclusively, for animals, however regular their activity, operate by instinct, having no sense of rules to be applied, and knowing nothing of invention, of creating *ex nihilo*. The fact that man has transcended nature is his glory, and his alone.[14] Meilcour's education in love involves his exposure either to as varied (but still controlled) an 'environment of feeling' as possible, or to as many second-hand situations as is feasible; hence Mme de Lursay's significance as one who describes scenarios which Meilcour has not encountered before, and hence the importance of the literary leitmotive of novel and drama in the story.

[13] *Correspondance littéraire, philosophique et critique*, ed. Grimm, Diderot, Raynal, Meister etc., 16 vols (Paris, 1877–82), xiii. 108–11.

[14] A. C. Chavannes, *Essai sur l'éducation intellectuelle, avec le projet d'une science nouvelle* (Lausanne, 1787), p. 29.

The emotions can be 'schooled' because it is possible for teachers to modify their pupils' understanding of erotic matters.[15] Mme de Lursay teaches her young charge that certain reactions to perceived occurrences are mistaken, and that he should 'look before he leaps'— should consider the likelihood, for example, of a woman of her substance concealing a lover in her closet while guests are still in her house, as Meilcour at one point accuses her of doing. The neophyte learns to measure the appropriate response to different circumstances, although these may vary according to time and milieu: in the eighteenth century people (of the upper and middle classes at least) seem to have responded more fulsomely to most emotion-inducing situations than is now considered appropriate. Inasmuch as love is not primarily an instinct, but an art that can be learnt, it befits those concerned with upbringing to devote themselves to inculcating principles of correct behaviour in the young. Thus parents, like the excellent Mme de Meilcour, impart wisdom to their offspring before permitting their launch into society.[16] His mother could not save him from cutting a fatuous figure in the world, Meilcour tells us, but she effectively limited the damage he was able to do himself and others.

As is well known, however, eighteenth-century parents, particularly aristocratic ones, were often uninvolved in their children's education. Mme de Volanges in *Les Liaisons dangereuses* is a case in point. Novels featuring incest (such as Restif's *Anti-Justine*) or near-incest (like Mirabeau's *Le Rideau levé*) offer dubious examples of parental involvement. But at least one eighteenth-century novelist himself illustrates the responsible parental concern that might be met with in the bourgeoisie. In his *Mémoires pour Catherine II* Diderot remarks that the only omission in the education girls received in Catherine's 'maison' was a course on the human anatomy. He himself arranged lessons for his daughter on this subject with the former painter Mlle Biheron, who used wax models 'qui [avaient] la vérité de la nature sans en offrir le dégoût', and so taught her pupil 'ce que c'était que la pudeur, la bienséance et la nécessité de dérober aux hommes des parties dont la nudité, dans l'un et l'autre sexe, les aurait

---

[15] See Roger Scruton, 'Emotion, Practical Knowledge and Common Culture', in *Explaining Emotions*, ed. Amélie Oksenberg Rorty (Berkeley, Los Angeles, and London, 1980), p. 524.

[16] pp. 13, 107.

réciproquement menés au vice'.[17] Angélique Diderot was not the
only one to benefit from this instruction: twenty other girls of good
family and a hundred society women attended the classes, girls
receiving their lessons between the ages of sixteen and eighteen, or
a year or two before marriage.

The same procedure is recommended in Mme de Genlis's insipid
letter-novel on education, *Adèle et Théodore* (1782). This work
presents a remarkable instance of devotion by aristocrats to the
education of their children. In the first letter of the novel—written
at three o'clock in the morning, for reasons which are not divulged,
but suggest a state of high excitement in the writer—the Baron
d'Almane explains to his correspondent that he intends to leave Paris
with his wife and two children for a period of two years, because
'j'attends [d'Adèle et de Théodore] le bonheur de ma vie, et je me
consacre entièrement à leur éducation.'[18] This turns out to be not
quite true, as Adèle is educated by her mother, and Théodore alone
by the Baron. Such a division of effort was common in the eighteenth
century: rare were the instances of females being granted, as they
had apparently been in Sparta, equality of treatment with males.
From the educational treatises by Rollin and Fénelon, via the latter's
pedagogical novel *Télémaque*, through Rousseau's *Emile* to Saint-
Pierre's *Etudes de la nature*, the assumption was constantly made
that a girl's upbringing should be directed towards making her a good
wife, mother, and mistress of the household.[19] Although Sophie, in
Emile, is to be allowed to read Emile's *Spectator*, she is to do so
with an eye to learning wifely duties from it.[20] (Emile, meanwhile,
is to be lent her beloved *Télémaque*.) Throughout the section of his
novel that deals with Emile's search for a wife, Rousseau proclaims
that men neither want nor require a learned spouse, a view echoed
by his disciple Saint-Pierre in the *Etudes de la nature*.

The story of Abélard and Héloïse, together with the legend sur-
rounding their (initially tutorial) relationship, exerted a considerable

[17] *Mémoires pour Catherine II*, ed. Paul Vernière (Paris, 1966), pp. 86–7; also
Correspondance, ed. Georges Roth, 16 vols (Paris, 1955–70), viii. 211.

[18] *Adèle et Théodore, ou Lettres sur l'éducation*, 3 vols (Paris, 1782), i. 2.

[19] Charles Rollin, *De la manière d'enseigner et d'étudier les belles-lettres, par
rapport à l'esprit et au cœur*, 2 vols (Paris, 1740), i. 30; Fénelon, *Télémaque* (Paris,
1968), p. 210; Bernardin de Saint-Pierre, *Discours sur cette question: Comment
l'éducation des femmes pourrait-elle contribuer à rendre les hommes meilleurs, Œuvres
complètes*, ed. L. Aimé-Martin, 12 vols (Paris, 1818), xii. 145 ff.

[20] *Emile, Œuvres complètes*, iv. 825.

influence on the eighteenth-century literature of love. One authority has it that all Abélard's scholarly reading would have deflected him from the dangerous ways of love had Ovid's *Ars amatoria* not been a commonplace-book of his age;[21] but the story of his liaison with Héloïse reads more like Ovid's *Remedia amoris*. As a cautionary tale, Abélard's *Historia calamitatum* (the calamity being his castration by Fulbert's henchmen) has its poignancy. Far more moving, however, and the true source of all the eighteenth-century elaborations of the legend, is the true correspondence between the former lovers[22] written when Abélard had succeeded in persuading Héloïse to renounce the world and, after a period in the convent at Argenteuil, become abbess of his foundation the Paraclete, while he pursued a monkish life at St Gilda's. The theme of incomplete renunciation which emerges from Héloïse's letters is particularly prominent in Pope's influential poem *Eloisa to Abelard* (based only on Héloïse's first letter) and Rousseau's *La Nouvelle Héloïse*. Eloisa proclaims her continuing passion for Abélard even after he has been deprived of the part through which he sinned, and Rousseau's Julie similarly reveals that she cannot quell her longing for her former tutor and lover.

In both the original correspondence and Rousseau's eighteenth-century version there is a periodic reversal of roles in the master-pupil relationship, not least where matters of emotion are at issue. Julie recurrently lectures Saint-Preux, in the tradition of the *Remedia amoris*, on the nature of the senses and the desirability of transcending them to enjoy a disembodied communion of mind and heart with the object of one's love; and Saint-Preux, never one to argue with his lady, meekly accepts her wisdom and her reprimands on the subject of his sensuality.

Other adaptations of the Abélard–Héloïse story, however, emphasize the male's role as teacher in the art of love, if not necessarily in the incorporeal sense of which Julie would approve. Indeed, one such work, Mme de Beauharnais's novel *L'Abailard supposé* (1780), describes a man's successful effort to reclaim a woman for the delights of eros when she has apparently renounced them for intercourse of

---

[21] Jean-Pierre Letort-Trégaro, *Pierre Abélard* (Paris, 1981), p. 135.

[22] There has been, and remains, some debate as to whether or not Abélard wrote the entire correspondence. Etienne Gilson, *Héloïse et Abélard* (Paris, 1938), regards it as an authentic exchange between the two former lovers.

the heart and soul. The Comtesse d'Olnange, widowed at the age of eighteen, is as convinced as Crébillon's Hortense that male passion does not last, and is frightened of remarriage because of her earlier maltreatment at the hands of her husband. The young Marquis de Rosebelle undertakes to dispel her sexual fear by the roundabout method of feigning castration, allowing the story to be spread abroad that he was unmanned by a jealous husband after being discovered *in flagrante delicto* with the latter's wife. The natural compassion of the 'sensible' Comtesse is aroused by Rosebelle's supposed predicament, and she resolves to devote her life to consoling him. Like Laclos's Marquise de Merteuil, but with more sincerity, she dilates to her lover on the glories of the chivalric Middle Ages, 'ces temps glorieux où la beauté dictait sa loi à des héros, ne se donnait qu'à la vertu, et, faite pour l'inspirer, en était la récompense'.[23]

Rosebelle's virtue, meanwhile, is sorely tried as Mme d'Olnange adds insult to injury by requesting that a third party be in attendance when the couple meet on potentially compromising occasions.[24] He is persuaded to propose to her, is accepted (to the fury of the Comtesse's father, who assumes that he is to be denied grandchildren),[25] but postpones informing her of his true physical state until after the wedding, lest her old fear should reassert itself. Even after the ceremony the Marquis is bidden by his wife to refrain from entering the bedchamber until she can be certain that he is fully reconciled to his desexed condition, and that 'mon sentiment suffit à votre cœur'.[26] But she herself, ironically, comes to realize the unconquerable force of sensual love. She falls asleep, has an erotic dream, and wakens to find a manifestly uncastrated husband beside her. There can be only one outcome: 'tous les sentiments se confondent, une volupté céleste coule de veine en veine; l'âme y succombe, le cœur la respire, tous les sens s'y anéantissent.'[27] 'Le rêve', in short, 'est achevé.'

Mme de Beauharnais's story belongs to the familiar fairy-tale tradition in which a male (often, as in this work, a Prince Charming) awakens a dormant female, through trials and tribulations, to consciousness of and delight in her sexual nature. The vapid letter-novel by Restif, *Le Nouvel Abeilard, ou Lettres de deux amants qui*

---

[23] *L'Abailard supposé ou Le Sentiment à l'épreuve* (Paris, n.d.), p. 130.
[24] p. 152.
[25] In *ancien régime* France impotence was one of the few legitimate grounds for having a marriage annulled.
[26] p. 185.   [27] p. 195.

ne se sont jamais vus (1778), proceeds rather differently. The reader
is early warned of one important respect in which this book diverges
from the true story of Abélard and Héloïse, for a prefatory note
announces that its principal merit is to inspire reverence for the
institution of marriage.[28] Like Longus's Daphnis and Chloë, or
Saint-Pierre's Paul and Virginie, the two lovers of the title are clearly
'meant' for each other. Despite the evidence of this fact, however,
their parents believe that the children must be kept apart until the
proper time for marriage, and hit on the idea of having them make
love by correspondence. To this end the new Abeilard is despatched
to college at the age of seven, whence he sets about his intended's
education by letter. But the paternal role as teacher of sons has not
been forgotten: Abeilard's father has a fund of edifying stories to
relay to his offspring, and the boy communicates them in turn to
his Héloïse (who remarks justifiably that 'votre sexe aime assez à
prendre notre instruction sur lui').[29]

Abeilard's instruction about love is confined to regaling his
correspondent with his father's tales of conjugal happiness, drawn
from many sources. The most extraordinary for a priggish youth
to pass on to a virtuous girl (if less extraordinary to the reader familiar
with Restif's other novels) concerns the relationship between a
princess married to a Prince Charming—who has gallantly waived
his marital rights until he and his wife know each other's hearts—
and the daughter of her former wet-nurse, a story containing
explicitly lesbian scenes salaciously drawn out by Abeilard in the
telling. Another of these 'models', entitled 'L'Amour enfantin',
describes a nineteen-year-old man's selection of a child aged ten to
be 'formed' for eventual marriage to him—a scenario recalling both
that of Molière's *L'Ecole des femmes* (but without the same unhappy
consequences for the tutor) and the episode recounted in Rousseau's
Confessions of his adopting with a friend a Venetian girl, to be put
to sexual use by the two men when she had reached the appropriate
age.[30] (Neither of the friends, in the event, employed her for these
purposes.) Such tales, and others related in *Le Nouvel Abeilard*,
reveal a preoccupation with male sexual restraint shared by none of

[28] In her first letter to Abélard, Héloïse argues that the status of mistress is more
glorious than that of wife, for in the former woman gives freely a love and obedience
which in wedlock are enjoined on her.
[29] *Le Nouvel Abeilard, ou Lettres de deux amants qui ne se sont jamais vus*,
4 vols (Neufchâtel and Paris, 1778), i. 79.   [30] p. 323.

Restif's better-known novels, but which (albeit in a different sense) recalls the continence imposed on the historical Abélard by his castration. They also echo the mythic concern of Pygmalion with creating a beloved, a theme particularly close to Rousseau's heart: his account in the *Confessions* of inventing the 'belles âmes' of *La Nouvelle Héloïse*, his imaginative transformation of the real-life Sophie d'Houdetot into Julie d'Etanges, and his actual writing of a Pygmalion play[31] bear witness to this fascination.

The master–pupil relationship in two other novels by Restif, *Le Paysan perverti* and *La Paysanne pervertie*, is of a different order. Although both Edmond and Ursule refer to the Satanic Gaudet d'Arras as their Mentor, he has little in common with Telemachus's guide in the *Odyssey*, Télémaque's instructor in the novel by Fénelon, or Emile's tutor. Like Balzac's Vautrin, he is a man of parts, with respect to both his general influence on Edmond and Ursule and his sexual attitude towards them. Although the paths along which he tempts brother and sister ultimately lead to their destruction, he sometimes acts in what he judges to be their best interests. Since the two novels deal mainly with the sensual initiation of the principal characters and its aftermath, it is natural that Gaudet's educative influence should be closely related to their passions (or the deformation thereof). And it emerges that his friendship with Edmond is covertly or indirectly an erotic one (another anticipation of Vautrin). Although Gaudet is conscious of Ursule's charms, he is never so fond of his mistress Laure as when she has just left Edmond's bed.[32]

This element of self-interest in his relationship with his male pupil, as well as his corrupt intentions in forming the young man for society, most clearly distinguishes Gaudet from the mentors of Homer, Fénelon, and Rousseau. These three works show a teacher's effort to steer his charge away from the dangers of passion; Restif's two novels, the reverse process. In Gaudet's defence, however, it may be remarked that he rarely treats Edmond and Ursule as mere objects, tools enabling him to realize his private desires; and this perhaps makes his pedagogical activities less reprehensible than those of the *meneurs de jeu* in a novel Restif surely influenced, Laclos's *Les Liaisons dangereuses*.

---

[31] See Huntington Williams, *Rousseau and Romantic Autobiography* (Oxford, 1983), *passim*.

[32] See the introduction by Béatrice Didier to *La Paysanne pervertie, ou Les Dangers de la ville* (Paris, 1972), p. 18.

The Marquise de Merteuil and Valmont constantly refer to Cécile and Danceny as their pupils,[33] either in the art of love or in that of social intercourse and deception.[34] They value the young pair only in so far as they can serve their mentors' corrupt ends, enabling Mme de Merteuil to punish her former lover Gercourt by delivering his fiancée to the bridal chamber as damaged goods, and Valmont to wreak vengeance on Mme de Volanges for blackening his reputation in the eyes of the Présidente. As their evil activities gain momentum, it is true, the two libertines seem to lose sight of their immediate objectives in the sheer joy of imposing their will on society generally: hence the Marquise's pleasure in the *hors d'oeuvre* of humiliating the rake Prévan,[35] or Valmont's in procuring a night with the Vicomtesse de M. . . against the odds.[36] Mme de Merteuil's celebrated letter to Valmont about her self-education in deceptive arts of living and loving[37] is an apt reminder of her general, not specific, need to assert power over other people, a need rendered more acute in the female by virtue of the social convention that allows only men to proclaim their sexual conquests. But her and Valmont's employment of Cécile and Danceny as passive means to their own private ends is the most striking illustration of their abiding preoccupation with personal glory: even the Marquise's taking Danceny as her lover appears to be motivated by the desire to show Valmont her independence of him and contempt for his involvement with Mme de Tourvel. The 'education' of others is undertaken solely to serve that purpose.

As Letter LXXXI reveals, Mme de Merteuil had no master or mistress in the art of love. Although Valmont surely played a more active part in her sexual life than her teasing references to him as a 'preux chevalier' waiting on his lady's pleasure would suggest,[38] the burden of the letter is that she has nothing to learn from him or anyone else about the technique of conducting affairs. Responding in fury to Valmont's suggestion that she should refrain from dallying with the artful seducer Prévan, she underlines her superiority to the former in the strategy of love-making; hence her undisguised contempt for the slowness of his progress with the Présidente, and

---

[33] On the use of erotic pedagogical metaphors in the *libertin* novel, see Laurent Versini, *Laclos et la tradition* (Paris, 1968), p. 361.

[34] pp. 184, 220, 262, 298, 346.   [35] pp. 184 ff.   [36] pp. 140 ff.

[37] pp. 167 ff.   [38] pp. 13, 44, 243.

hence too her taunting him for his inferiority in matters intellectual, her telling him that he has not the 'génie de [son] état' and is incapable of inventing anything.[39]

Indeed, the most disquieting, but also the most characteristic, aspect of the libertine self-education she describes in Letter LXXXI is its cerebral nature. The Marquise's usual procedure, in sexual relationships as in other affairs of life, is based on principle, method, study, and technique. At the age of fifteen—Cécile's age when she is removed from the Ursuline convent to be married—she possessed, in stark contrast to her protégée, all the talents of a skilled politician. Never having attended a convent herself, or had a 'bonne amie' to enlighten her,[40] she had only vague notions of what love and its pleasures might be; natural processes had not yet provided their own lessons about sexuality when she decided to tackle the problem of its discovery intellectually. 'Ma tête seule fermentait; je ne désirais pas de jouir, je voulais savoir.'[41] The same approach governed her attitude to marriage: her wedding-night, from which she expected 'la certitude de savoir', was for her an opportunity, not for dropping her intellectual guard, but on the contrary for extending her knowledge, an 'occasion d'expérience' whose every occurrence, pleasurable or painful, she noted mentally and laid by for subsequent reflection. Her diverse sensations were 'des faits à recueillir et à méditer'. After widowhood had left her with scope for broadening her education, she decided to experiment with love, 'non pour le ressentir à la vérité, mais pour l'inspirer et le feindre', for which the talents of an author and an actress were the only ones required. The truism that love cannot be simulated had, she discovered, no basis in fact.[42] Her liaison with Valmont alone threatened her perfect intellectual control of love-affairs, and '[prit] un moment d'empire sur moi'.[43]

Her lack of formal education, then, seems scarcely to have inhibited Mme de Merteuil's intellectual development. Cécile's period in a convent, on the other hand, and friendship with a bosom companion there, evidently left her in almost complete ignorance of sexual matters. But it is perhaps too much to expect a girl of her background and day to have heard of contraception, or necessarily to know when she was pregnant (like Cécile) or had had a miscarriage. Valmont leaves her sexual enlightenment deliberately incomplete: he teaches her the technical words for sexual parts and acts, but not the social implications of using them, in the 'catéchisme de débauche'[44] he imagines the girl reciting to Gercourt on their wedding-night. The education with which convents provided women was often condemned in eighteenth-century France, but more frequently on account of its failure to inform them adequately about the wifely role they would be required to perform on leaving its walls than because it omitted all reference to the facts of life. It was for the former as well as the latter reason that Desmahis's *Encyclopédie* article 'Femme (morale)' criticized such institutions, emphasizing the deficiencies of those who taught there:

Des femmes qui ont renoncé au monde avant de le connaître sont chargées de donner des principes à celles qui doivent y vivre. C'est de là que souvent une fille est menée devant un autel pour s'imposer par serment des devoirs qu'elle ne connaît point, et s'unir pour toujours à un homme qu'elle n'a jamais vu.

The concern shared by men of the Enlightenment with seeing education fulfil a morally and socially instructive role justifies Desmahis's implied complaint, although it cannot be doubted that a part of his hostility derives as well from the *philosophes'* call for a general emancipation from Christian values, including such as would inevitably be professed in a religious institution.

Education through literature, as Rousseau's *Confections* make clear, was of course available even to those who lacked the (possibly dubious) advantage of instruction in cloister or college. We may leave aside for the moment the declared intention of a work like Restif's *L'Anti-Justine* to serve as a 'livre de chevet' passed on by wife to husband in the interests of conjugal edification,[45] or of *La Philosophie dans le boudoir* to offer needed information about matters sexual to young females ('La mère en prescrira [not 'proscrira'] la lecture à sa fille').[46] But Laclos, a more persuasive moralist, approvingly reports in the 'editor's' preface to *Les Liaisons dangereuses* a mother's opinion that she would do her daughter a signal service by presenting her with the novel on her wedding-day.

Can reading about affairs of the heart and senses develop our own awareness of the emotions, and literature thus become an

---

[44] p. 256.
[45] *L'Anti-Justine, Œuvres érotiques de Restif de la Bretonne* (Paris, 1985), p. 287.
[46] p. 15.

aid to moral growth? To judge by the tenor of prefaces to many eighteenth-century novels, there was a widely perceived need at least to claim that it could. Determined philosophers may reason, as Plato reasoned, that all art is an illusion and therefore of scant value in helping its consumers to understand real-life experiences; but the likelihood remains that most people will continue to find their imaginative grasp of reality enhanced by contact with the mimesis of representational art.

That this understanding involves an operation of the imagination was a fact only intermittently acknowledged by Rousseau. Making discoveries about conduct through literature must involve, to put it no higher, a process of transference and adaptation from the medium of book (or enacted play) to life. Identification, as the same Rousseau inconsistently put it, will never be more than partial. We know when we visit a theatre, he wrote in the *Lettre à d'Alembert sur les spectacles*, that that is precisely what we are doing, and that our reaction to whatever we are shown there is, morally speaking, a token. Although wicked ourselves, we may temporarily commiserate with the sufferings of virtue persecuted on stage, but our knowledge that drama is simulated life makes our response hypocritical: we make our routine obeisance to goodness, and then leave the playhouse to live as wickedly as before.[47] Rousseau, who in the process of artistic creation seemed susceptible of belief in the reality of his fictional characters, was not alone of his century in emphasizing the distance at which the beholder stands from art. The abbé Dubos had argued that we could not endure the artistic experience of baseness if we took it for real; but we know that poetry and painting are harmless counterfeits of reality. We find the skill involved in the counterfeiting pleasurable, yet are not deceived by it. Compassion, which tragedy characteristically arouses, demands only an imaginative dwelling on the situations presented, not a 'mistake' involving our construing them as circumstances as we have known ourselves, or are likely to know.[48] Great art, indeed, may have the property of taking us outside ourselves, freeing us from the prison of the ego and giving us a vivid realization of what it might mean to be another. It is perhaps in this sense that literature may be called educative.

---

[47] *Lettre à d'Alembert sur les spectacles*, ed. M. Fuchs (Lille and Geneva, 1948), pp. 33–4.
[48] See Lawrence Blum, 'Compassion', in *Explaining Emotions*, pp. 509–10.

Countless eighteenth-century novels tell us, in authorial prefaces, that literature may effectively promote the process of moral discovery. But this hackneyed repetition should not blind us to its validity, any more than disclaimers by writers like Rousseau should persuade us of its falseness. In any case, the argument Rousseau presents, especially in the second preface to *La Nouvelle Héloïse*, is seldom logical. There he declares in the same breath that a decent girl should not read books about love (the implication being that she will be corrupted by them) and that girls cannot understand a subject— standardly, the perverting influence of love on human nature—of which they have no direct experience.[49] It is odd, besides, that Rousseau should so fully discount the possibility of an imaginative transformation of the self through reading, when the *Confessions* makes clear how greatly he himself was affected by his early immersion in novels—although it may seem less curious when we reflect that Rousseau was a man much given to emphasizing how different he was from his fellow human beings. He himself offers one of the most celebrated examples of the interpenetration of literature and life: the idealizing of woman described in the *Confessions* as a result of his early reading, the 'creation' of Sophie d'Houdetot after the invented image of Julie, and so on.

Implicating the reader in a work of fiction (a process to which Diderot's imaginative writing perhaps gives the most complete eighteenth-century expression) has consequences for life. It is true that the nature of these responses varies from one reader to the next: where some will learn from Richardson's *Clarissa*, as Laclos supposed,[50] that the company of seducers like Lovelace should at all costs be shunned, others may draw useful conclusions about the best way of transforming a rake into a lover, and see the miraculous ability of love to turn life upside-down. Depending on the type of love-literature in question, the experience undergone by the reader may be disillusioning, edifying, or neutral, although his reaction will not necessarily be consonant with the ostensible 'message' of the book. Where literature teaches morally dubious lessons, it may be incumbent on the virtuous preceptor to draw attention to the fact. Early on in *La Nouvelle Héloïse* Julie praises Saint-Preux for scorning to use fiction as a means of seducing her: 'vouloir attendrir sa

---

[49] pp. 203–4.
[50] *De l'éducation des filles*, *Œuvres complètes*, p. 455.

maîtresse à l'aide des romans est avoir bien peu de ressource en soi-même.'[51]

Even without external assistance, however, readers may dissent from the values promoted by a writer in his imaginative presentation of love. Sade's determinedly anti-metaphysical portrayal of sexuality in *La Philosophie dans le boudoir* might cause some to conclude that there is a great deal more spirituality in erotic love than he implies. When Crébillon's Meilcour persuades himself that he is deeply in love with Hortense because novels have taught him that great adventures always start unexpectedly,[52] the fault is not really in the unreality of literature (as Rousseau would have it); rather, it is an imaginative failure, an inability to adjust and interpret, that is to blame. Meilcour unreflectively approaches life in terms of fiction, whereas the first important lesson he learns from Mme de Lursay is to think about the situations in which he finds himself, not blindly take them as paradigmatic with a literary model.

It is true, however, that the society he is entering sees human activity in terms of fixed orders. This way of organizing social existence has an obvious attraction to man, and it seems to mirror orders and hierarchies learnt in his earliest years. The rationality of emotions may be assured by the human ability to fit them into scenarios of a known kind and express them as a function thereof. Education partly consists in alerting the pupil to the existence of such models and teaching him the appropriate response to them. Meilcour's emotional immaturity resides in his unfamiliarity with the situations society commonly creates: his novel-reading has not provided him with material sufficiently like what he meets in the world to be of use. The task of Mme de Lursay, Mme de Senanges, and Versac will therefore be to alert him to the archetypal nature of the situations he encounters under their tutelage, so that he can in future react appropriately to related ones without needing their assistance. The rigidity of social etiquette in the world he is entering is both a hindrance and a help to the beginner. It is a hindrance in the very degree of its prescriptiveness, which makes deviation grounds for social ostracism or exclusion. (Mme de Lursay knows this, having had to make up for the flightiness of her younger days with a degree of propriety exceeding that demanded of other women.[53] Only the very bold, like the shameless Mme de Senanges, manage to breach

---

[51] p. 62.   [52] p. 36.   [53] pp. 180 ff.

the conventions with comparative impunity.) But the inflexibility of etiquette is a help in making the meaning of social moves readily apparent. The perfect teacher for Meilcour in this respect is Versac, who as pack-leader invents many of the rules governing conduct and imposes them on others, while all the time managing to persuade the world that every one of his activities is subordinate to the whim of women. He thus provides his pupil with two sources of insight, that of the female and his own supervening one.

According to this account of life in the world, a changed social milieu may lead to different emotional reactions. Des Grieux and Manon realize this when they leave France for Louisiana, put their risky Parisian ways behind them, and adapt to a less hedonistically inclined environment where hard work and neighbourliness are given a value not accorded to them by Paris high society. Experience, des Grieux tells his listener Renoncour, 'fit sur nous le même effet que les années'.[54] In other words, he and Manon have grown up; their emotional disposition has altered. But there is also the possibility, which Mme de Merteuil covertly exploits in *Les Liaisons dangereuses*, of altering the time-honoured scenarios in an attempt to eradicate certain ingrained social beliefs. We know in a general way that she invents new modes of ordering her world (and criticizes Valmont for being wedded to old schemes, and unable to invent); we know specifically that she intends to avenge her own sex against the domineering male by proving her superior rationality.[55] To do this effectively, though, she needs more support from other women than she gets. For a while she imagines that Cécile will be a useful ally, but is soon forced to admit that the girl is ill-equipped to assist her in her struggle, belonging to the great army of 'femmes à sentiment' who think nature has put their senses where their head is, and can only react as they have been conditioned to do. Ironically, in view of the Marquise's jealous contempt for her, the Présidente de Tourvel is the only female character in the novel who shows an ability comparable with Mme de Merteuil's own to adapt her code of behaviour so that it fits a new circumstance.

Certain dispositions of character which in a given culture are standardly called virtuous and unvirtuous may be susceptible of a different interpretation when determining factors change. Mme de Merteuil's words about the perceived morality of male versus female

[54] p. 190.   [55] p. 170.

sexual behaviour illustrate this point, as do Mme de Rosemonde's pronouncements on the irrelevance to females of the distinction between infidelity and inconstancy which is applied to males.[56] The patterns by which we learn to behave, in love as in other things, are not completely fixed and unchanging. Differences in social class and age, as well as in sex, may lead to socially acceptable variations.

If emotion appears irrational despite our best efforts to fit it within known schemas, this may be because of the very similarity between feelings which we are taught to regard as opposed to one another. Thus, as Valmont is forced to concede apropos of his sentiments towards Mme de Tourvel, 'que je hais et que j'aime avec une égale fureur',[57] there is a family resemblance between hatred and love. Mme de Lursay unconsciously teaches Meilcour something similar. After Versac has persuaded him that she is unworthy, for moral and other reasons, of his attentions, Meilcour ostentatiously signals his distaste for her company, but is subsequently chagrined to find that she has seemingly found easy consolation elsewhere for his abandonment of her. Even though he is aware that her stratagem is one of the oldest known to lovers, his indignation leads him to reflect on the bizarre proximity of two emotions which appear each other's antithesis. His emotional yielding to Mme de Lursay, he says, is unwitting, almost imperceptible; but vanity propels him from hatred to love, as cavalierly as on other occasions it pulls from love to hatred.[58] Yet in his still unenlightened state Meilcour ascribes this oscillation to caprice, of the kind men inconsistently reproach in women. He fails to realize that he is a victim of the fine gradation of sentiments, a topic on which Mme de Lursay has already given him instruction.

Crébillon's reader is aware that Meilcour's education in love must consist in his exposure to tests of this kind. It is an essential part of his growing up: if the novice cannot prove his mettle through acts of physical valour—and we remember that peacetime has given Meilcour his 'loisir dangereux'[59]—he may have to do so by successfully facing up to an internalized struggle. Meilcour does not invent situations of mock-heroism as a substitute for the real heroics denied to him by the times in which he lives; he is no Julien Sorel, who makes the clasping of a woman's hand a feat of Napoleonic

---

[56] See chapter 2, pp. 68ff. below.   [57] p. 226.
[58] p. 168.   [59] p. 13.

daring. But he needs to convince himself that winning a woman's favours, like a battle-campaign, is a difficult process demanding tactical skill, patience, and a sense of timing. There is an obvious resemblance here to Valmont's 'savantes manœuvres' which lead to victory over the Présidente de Tourvel's virtue, but an equally obvious difference: Valmont encounters real resistance, whereas the obstacles Meilcour meets are within himself rather than presented by the object of his affections.

Meilcour's 'belle inconnue', Hortense, provides the only exception to this general pattern, and the unfinished state of Crébillon's novel leaves the reader uncertain of his likely reaction to her aloofness once he has consummated his relationship with Mme de Lursay. Hortense expresses the same sort of fear concerning love and its threat to the individual's 'repos' as does Mme de Tourvel. Like Valmont, Meilcour feels that winning the object of his desires will bring him a kind of glory that obtaining the easier compliance of Mme de Lursay and Mme de Senanges will not. Realistically, however, he sees that paying his attentions to one of the latter two will be more likely to yield the experience he seeks, the more so since he believes Hortense's affections to be otherwise engaged.

The difficulty of gaining Mme de Senanges's favours, in his view, is insufficient for the initiation provided by her to be a source of self-congratulation. When he first meets her in society Meilcour is repelled by the obviousness of her advances and the simple victory she seems to offer.[60] It is for this reason that he is disconcerted by Versac's advice that he take her, rather than the other available women, as his teacher in love.[61] His unstated desire for a real test, a triumph achieved against the odds,[62] explains his indignation when Versac implies that Mme de Lursay is easy prey. (Valmont will later be infuriated by Mme de Merteuil's similar intimation that the Présidente is really as anxious to yield to him as were his other mistresses.) Versac's malicious observation that Mme de Lursay has 'le cœur fort tendre' (here a euphemism for something much less delicate), and his assertion that several men owe their education to her,[63] leave Meilcour with the firm conviction that her interest in him is opportunistic and base. He is eventually dissuaded from this view, but still has to be reassured that Mme de Lursay's giving herself to him has been a struggle—for her as well as for Meilcour himself. If earlier

[60] pp. 87 ff.   [61] pp. 150, 164.   [62] p. 32.   [63] p. 75.

on the difficulties seemed purely the result of the young man's own inexperience,[64] latter stages of his suit apparently meet with genuine obstacles, or what the experienced older woman is careful to present as such. Meilcour is 'enflammé . . . par sa résistance';[65] what he believes his victory to have cost makes it the more precious to him; and 'quoique je ne triomphasse, dans le fond, que des obstacles que je m'étais opposés, je n'en imaginai pas moins que la résistance de Mme de Lursay avait été extrême.'[66]

In fairy-tales and heroic romance, such obstacles to initiation and the satisfaction of love are externalized. Eighteenth-century fictions describing the education of emotion treat the old archetypes in terms of psychological rather than physical struggle. In Mme de Beauharnais's *L'Abailard supposé* the hero's task is to rouse the heroine from the sleep of sexual fear, not cut a path through a forest of thorns on the way to finding his 'belle au bois dormant' and waking her with a kiss from her physical slumber. Meilcour the 'ingénu', in search of his Holy Grail like the medieval 'niais' Perceval,[67] will encounter mental impediments rather than the material ones of his ancestor in romance, and (Crébillon's preface gives us to understand) will acquire ultimate wisdom through initiatory rites of the mind and heart as well as of the body. Meilcour's physical indoctrination is of far less interest to Crébillon, and is described in the most indirect terms.

Comparison of the ancient literature of initiation with eighteenth-century fiction is not fanciful, for just as there were links between romance and the world of faery (one of the earliest known forms of the 'Belle au bois dormant', or Sleeping Beauty, occurs in the fourteenth-century romance of *Perceforest*),[68] so elements of both traditions penetrated eighteenth-century literature and the works of *mondain* novelists like Crébillon. His story *L'Ecumoire, ou Tanzaï et Néardané* contains the archetypal themes of potency, healing, quest, and initiation found in Grail romance. In Chrétien's *Perceval* the health of the Fisher King is linked with the restoration of fertility to his land, the initiation is of the 'nice' ('niais') Perceval himself, but his quest fails at the Grail castle, leaving the king uncured and the land still barren. Tanzaï and Néardané are both robbed of their

---

[64] pp. 59, 62.   [65] p. 185.   [66] p. 186.
[67] The eponymous hero of the later romance *Perlesvaus* is also such an 'ingénu'.
[68] See P. Delarue and M.-L. Tenèze, *Le Conte populaire français*, vol. ii (Paris, 1964), p. 14.

sexual parts, Tanzaï's being replaced by the 'skimmer' of the title. They are healed after undertaking travel and performing tests, and are initiated into the mysteries of sexual love. From the fairy-tale tradition, which had entered *mondain* French literature towards the end of the seventeenth century, and in the early decades of the eighteenth became a favoured minor genre in aristocratic circles,[69] Crébillon borrowed the motif of a Beauty who has to kiss a Beast for the latter to take on human form, but with slight variations: Tanzaï is first required to plunge the handle of his skimmer (an obvious phallic symbol) into the mouth of an old hag, and then repeatedly make love to the hideous fairy Concombre, who turns out to be the old crone in a different form. Having done so, and thus proved his virility and right to enter the world of men, Tanzaï is frustrated in his desire to consummate his love for Néardané because her sexual parts have still not been restored (a parallel with the implied theme of *L'Abailard supposé*). When this last has been accomplished, Crébillon's story can progress to the traditional happy ending.

Other 'worldly' fairy-tales written in the first half of the eighteenth century similarly bridge the gap between the medieval initiatory romance and the fully-fledged novel of sentimental education. Such stories commonly show how the discovery of love gives knowledge to the initiate and equips him for adult life. A typical example is Duclos's *Acajou et Zirphile* (1744), whose moral—that love is a good teacher—is worked out through the device of showing two children, brought up in ignorance, achieving cognizance despite the wiles of a genie and a wicked fairy. Zirphile overcomes the curse of stupidity when she falls in love with Acajou, an instantaneous process mirroring that undergone by Arlequin and Silvia in Marivaux's short pastoral comedy *Arlequin poli par l'amour*. Like Arlequin and Silvia, Daphnis and Chloë, and Paul and Virginie, Acajou and Zirphile realize that they are made for one another. Like Daphnis and Chloë they begin lessons in love together, and like their forebears they achieve their goal of marriage and procreation—but only after overcoming the obstacles which fate, in the shape of the genie and the wicked fairy, places in their path.

The wit of the two lovers is not, Duclos carefully points out, the brittle and ostentatious wit of social butterflies. Acajou does succumb

---

[69] See M. E. Storer, *La Mode des contes de fées, 1685–1700* (Paris, 1928).

to this perverted mode under enchantment and becomes a 'fat', but eventually breaks the spell to regain wisdom. Zirphile illustrates the notion that 'ce qui donne l'esprit aux filles' is a deeper-rooted and more permanent thing than merely the physical initiation described in La Fontaine's *conte*,[70] being the product of romantic love rather than of sensual attraction. Prévost shows a similar process in *Manon Lescaut*. Although the naïve, emotionally unknowing des Grieux is granted a sudden, darting enlightenment at the first instant of his falling in love with Manon ('l'amour me rendait . . . éclairé depuis un moment qu'il était dans mon cœur'),[71] true and stable knowledge does not come until the couple's settling in the New World. From the state of false ingenuousness and calculation she displayed earlier in the novel, Manon too seems to advance in Louisiana to a wisdom that belies her years. In *Les Egarements*, to judge by his preface, Crébillon does not intend the debased 'esprit' of Versac to be the true goal of Meilcour's striving. What the latter must gain is the wisdom born of experience and knowledge, not the facile and insubstantial cleverness of his mentor. Valmont, finally, forgets his old principles of calculation in the brief idyll of his requited love for the Présidente de Tourvel. In all these cases, it is mature love that has been achieved, or will be achieved, through the process of testing and learning to which society, or the imagined world of fairy-tale and romance, exposes its initiates.

Education in love, although it may have the effect of turning the individual around, may equally confirm him in what he already knew, yet endowing him with certainty about its value. The latter experience can be as enlightening as the former. Much of the eighteenth century's love-literature is about a return, not the taking of a new direction. The end of Meilcour's sentimental education, according to Crébillon's preface, will bring him back to his beginning, restore him to himself: 'On le verra . . . rendu à lui-même, devoir toutes ses vertus à une femme estimable.'[72] Des Grieux, to take a more ambiguous example, will similarly be reminded of his origins, and seemingly confirmed in his allegiance to them. On the other hand, new emotion has given him new eyes: unlike Meilcour, who can only (and imperfectly) interpret love through literature,

---

[70] 'Comment l'esprit vient aux filles', in *Fables, contes et nouvelles*, ed. E. Pilon, R. Groos, and J. Schiffrin (Paris, 1948), pp. 547–50.

[71] p. 20.   [72] p. 11.

des Grieux becomes able to interpret literature through love. As he resumes his studies after Manon's first desertion of him,

Les lumières que je devais à l'amour me firent trouver de la clarté dans quantité d'endroits d'Horace et de Virgile, qui m'avaient paru obscurs auparavant. Je fis un commentaire amoureux sur le 4ᵉ livre de l'*Enéide*; je le destine à voir le jour, et je me flatte que le public en sera satisfait.[73]

The irony of the faithful, and much-betrayed, des Grieux's commenting on the story of an inconstant male lover, Aeneas, deserting a constant beloved, Dido, does not escape him.

None of the full-length novels I have been discussing ends with its hero or heroine having completed a successful education in the ways of love, although Crébillon's preface suggests that *Les Egarements* would have done so if completed. The original (1731) edition of *Manon Lescaut* described des Grieux's return to France, after Manon's death, in a state of divine grace and repentance, and his devoting himself to exercises of piety. The 1753 edition, however, shows the hero returning only to a mode of conduct consonant with his elevated birth, and determining to abide by his class's principles of honour.[74] Thus Prévost alters what appeared an unqualified renunciation of Manon's memory in favour of a more nuanced description of des Grieux's new condition, which makes his 'education' seem the less complete.

The educations in love we witness in the eighteenth-century French novel are, perhaps, educations primarily in the necessary loss of illusion. Love leads to a bitter knowledge; only fairy-tales show differently, which is why Mme de Genlis has the Baronne d'Almane write for Adèle and Théodore stories from which beautiful and persecuted princesses, princes handsome as the day, and ugly hags who try in vain to impede progress towards the inevitable happy conclusion, are resolutely excluded.[75] To be free of illusion, however, is to have arrived at a state where the rational perceptions on which lasting love can be built become possible. For des Grieux, disillusionment preceded fate's cruel removal of Manon in the American desert: long before that he had had the measure of her unreliable character,

---

[73] p. 38.
[74] Compare the two passages in the edition cited, p. 202.
[75] p. 85.

gauged with accuracy the potentially destructive consequences of her 'penchant au plaisir', taken what steps he could to ensure stability in their lives (notably by moving their household from Paris to Chaillot), but ultimately realized that he would have to live with her unpredictability. Love can coexist with disillusionment. That perception is perhaps a part of emotional growth: the 'complete' lover leaves the immature stage of idolatry behind to reach a willing acceptance of imperfection.

Manon is possibly mysterious to Prévost's reader in some respects, but her character is fundamentally understood by her lover; he knows that it imperils the stage of tranquil possession to which he aspires, but acknowledges that he must retain her with her imperfections. Unlike Tiberge, therefore, he will not decide to cut his ties with the humanly chancy and incomplete, and give himself over to the (distant) certainty of heavenly bliss. The lucidity with which he declares this intention to his friend at Saint-Lazare has a lasting poignancy. 'Je reconnais ma misère et ma faiblesse. Hélas! oui, c'est mon devoir d'agir comme je raisonne [i.e. having conceded that love of Manon cannot bring settled happiness]! mais l'action est-elle en mon pouvoir? De quels secours n'aurais-je pas besoin pour oublier les charmes de Manon?'[76] Taking the individual as an object of love, once an initial stage of uncritical adoration has passed (and even this stage may be absent from some love-relationships), means accepting that his or her individuality deserves to be honoured. The mature lover does not try to refashion the object of love in his own image, but takes it with respect, even if conscious that it has morally undesirable qualities.[77]

The love of a good woman, Crébillon tells us, will set Meilcour back on the path from which he strays in pursuit of worldly pleasures and knowledge. She will re-educate him, through emotion, to an awareness of the values which his mother taught him, and from which society's tempters and temptresses have seduced him with their offers of false knowledge. The Mephistophelean Versac, we may be sure, is the being principally designated in the prefatorial reference to the 'personnes intéressées à lui corrompre le cœur et l'esprit', under

---

[76] p. 93.
[77] Julie tells Saint-Preux, 'L'amour ne m'aveuglait pas sur vos défauts, mais il me les rendait chers, et telle était son illusion que je vous aurais moins aimé si vous aviez été plus parfait' (p. 346).

whose tutelage the once artless youth becomes 'plein de fausses idées, et pétri de ridicules'.[78] The malleability of the human character has its limits: the substance gradually hardens, the shape becomes fixed, and so the good work must be done when the material is still impressionable. If Versac is to be believed, Mme de Lursay took herself in hand too late: the 'égarements' of youth had left an indelible mark. Towards the end of the novel Meilcour becomes less sure that Versac's assessment of Mme de Lursay is correct; conversely, he resists his mentor's intimation that Mme de Senanges should be favoured with the task of educating him, finding her habit of corruption so settled as to be repugnant. The metaphoric value of the verb 'former' has rarely seemed so apt as it is in this novel, filled with images of modelling the soft substance of the social 'ingénu' into shape, and firing it in the oven of the 'beau monde'. And its euphemistic sense—of providing a sexual initiation—supports the image of physical moulding. Here, and still more in *Les Liaisons dangereuses*, it is with educating the body that high society is largely concerned. The moral education that interests Crébillon, and to whose principles his hero will eventually return, is another matter.

What Meilcour is to learn is not the debased 'philosophie' professed in Sade's boudoir, the knowledge of physical pleasure which is all that a liaison with Mme de Senanges seems to offer him. Such knowledge will be a necessary part of his education, for he will come to see for what it is the delusion that love can be 'Platonically' disembodied. But sexual desire, rather than being an absolute force, will come to seem a relative one, linked but not superior to the desire for wisdom. Both, he will see, are directed towards objects of the world—not Platonic ideas—and aim at holding and possessing them. The process of initiation into merely worldly pleasures will not bring stable enlightenment, for its values are contingent. Real insight will come with life-transforming love. Perhaps this will be a fulfilled love with Hortense (the unfinished state of the novel does not permit us to know), so that Meilcour's *coup de foudre* at first sight of her really will have represented illumination. Struck by a lightning-bolt, he will both resemble Prévost's des Grieux and differ from him: for des Grieux, Manon's brilliance is a source of instantaneous knowledge and at the same time an impulse with the power to wound, whereas for Meilcour the light of Hortense's beauty

---

[78] p. 11.

will bring, not immediate understanding, but the possibility of gradually approaching truth and permanence.

To borrow a distinction drawn by d'Alembert in his *Essai sur les éléments de philosophie ou sur les principes des connaissances humaines*, the knowledge which merely sensual lovers acquire is 'curieuse' rather than 'utile', being directed at bodily gratification instead of the needs of the community and the duties of its members.[79] D'Alembert's enlightened concern with sociability leads him to emphasize the need for all the members of society to work towards increasing the well-being of their fellows, and therefore to engage in activities that conduce to general, not particular, happiness. D'Alembert would not have denied the possibility that the arts, in teaching men as well as gratifying them, could be turned to ends as useful as those of the sciences; but he would have demanded that love, if it was an art, should concern itself with more than the sensation to which libertines reduce it. He was convinced as a *philosophe*, and not a philosopher of the boudoir, that his duty was to love humanity in general, and that an education in sensation was idle in so far as it failed to instil respect for communal values. Love (a concept for which, in its romantic sense, Sade's philosophers have as little time as Crébillon's worldlings) could contribute nothing to the advancement of mankind if it was unconcerned with moral worth. Learning more than was generally known about the technicalities of love might increase an individual's contentment, but it was a vain learning that remained captured in the prison of *amour-propre*.

However, the eighteenth century's concern with community values, often expressed in the educational theories it developed, should not be too closely linked with a theory of community love. The end of the century was to see an exaltation of the latter over individual love, sometimes formulated in terms of a devotion to the state rather than to persons, which was linked explicitly with the corporate ideals of the ancient Greeks and Romans. But it would be fanciful to argue that love for particular individuals was ever generally subordinated to love for the collectivity. Diotima's lesson in the *Symposium*, and its adaptation by eighteenth-century writers, is insufficiently related to average human desires to become part of popular wisdom or the

---

[79] *Essai sur les éléments de philosophie ou sur les principes des connaissances humaines*, *Œuvres philosophiques, historiques, et littéraires*, 18 vols (Paris, 1805), ii. 229; also Chavannes, op. cit. 33–4.

practical expression thereof. Plato's ladder is too difficult for most to climb, and the sacrifices which the Platonic theory of ascent demands are too great to be readily met.[80] If the ascent represents progress—and the theory of progress was one that signally pre-occupied thinkers in the eighteenth century—[81]then it is unclear that the progress is, humanly speaking, in the right direction. If love can be taught, perhaps the lesson should be learnt for the sake of real flesh-and-blood beings rather than those exceptions who can ignore the force of individual human ties.

This message, among others, emerges from *Manon Lescaut*. Tiberge argues that des Grieux's development of reason sufficient to combat his passions would be an advancement of his moral being. While admitting that his emotions threaten rationality, des Grieux refuses to concede that they should for that reason be suppressed (if that were within his power), for he sees eros as life-enhancing. If Tiberge can resist the appeal of the individual, his friend cannot, and would not want to if he could. The essence of the human, des Grieux argues, is to be non-godlike; while Tiberge worships his deity, des Grieux is irrevocably committed to the human, imperfect, partial Manon. He loves, not divine excellence or the immutable and perfect Platonic idea, but the changeable, individual, and mortal. If Manon is incomplete, that is a part of her individuality which des Grieux respects despite the suffering it engenders. The point about Manon is that she is like no one else; she cannot be seen as interchangeable with other human beauties. This is why her attempt to give des Grieux a substitute for herself, in the form of the young prostitute she sends to keep him company while she is pleasuring herself with the young G. . M. . ., is so hurtful.[82] It negates her particularity, which is the cause of des Grieux's feelings for her. Although he implicitly denies that particularity after Manon's first betrayal, affirming that 'je ne mettais plus de distinction entre les femmes, et qu'après le malheur qui venait de m'arriver je les détestais toutes également',[83] the mere presence of Manon in his room after his debating exercise at the

---

[80] See chapter 2.

[81] See, for example, Condorcet, *Esquisse d'un tableau historique des progrès de l'esprit humain* (Paris, 1822); J. P. Bury, *The Idea of Progress: An Enquiry into its Growth and Origin* (New York, 1960); John Passmore, *The Perfectibility of Man*, 2nd ed. (London, 1972).

[82] pp. 134–5.

[83] p. 37.

Sorbonne, which is a prelude to the resumption of their relationship, is enough to confute that assertion.

This fact, and des Grieux's reaction to her subsequent betrayals, would seem to belie the central role of experience in teaching humans wisdom, at least where love is concerned. Indeed, one of the truisms about love is that one never learns from it. This is contradicted, however, by another commonplace, that when people have been damaged by particular events they avoid laying themselves open to the damage a second time. The heroine of *L'Abailard supposé* illustrates caution of this kind.

But the cruellest example of the way the pain of misdirected love can hurt its victims occurs in *Les Liaisons dangereuses*, where the erstwhile innocents Cécile and the Présidente de Tourvel withdraw from the world as a direct result of Valmont's and Mme de Merteuil's manipulation of their emotions. Their loss of illusion—a perennial theme in the literature of love—is encapsulated in the Présidente's agonized realization when she receives Valmont's letter of rupture: 'Le voile est déchiré . . . sur lequel était peinte l'illusion de mon bonheur.'[84] Having vainly appealed to the experience of the elderly Mme de Rosemonde, who could offer only sympathy, she too finds that the wisdom of experience can be turned to no account: the lesson she has been taught by Valmont destroys her. What we learn from the human problems Laclos invites us to consider is unquantifiable, but it is a rare reader whose moral thought is not touched by this novel. If some truths about love seem only knowable by direct confrontation (as Crébillon suggests in describing Meilcour's worldly education), and if others seem obstinately to resist the lesson of experience, there remain areas of knowledge which literature can most valuably illuminate through its vivid creation of character, moment, and motive. An essential part of education in love will come through the imaginative identification of self with fiction, through the power of story-telling.

There is a positive side to the matter of sentimental education, and one which major novelists of the age were concerned to emphasize. Although the eighteenth-century French novel contains some memorably corrupt pedagogues (Crébillon's Versac, Restif's Gaudet d'Arras, and Laclos's Valmont and Merteuil), it also shows the beneficial influence exerted by well-disposed mentors. If the case of

[84] p. 330

Rousseau's Wolmar is ambivalent in this respect (for his goal is not self-evidently a straightforward good), others are clearly motivated by the selfless desire to steer their charges safely through the troubled waters of emotion. Prévost's Tiberge, who finally sees his efforts at educating his friend to responsible love bear fruit, is one such, even if his success seems the product less of a permanent change in his protégé's mental attitude than of a third party's removal from des Grieux's vicinity. Tiberge resembles the tutor of *Emile* in having his charge's best interests at heart, although the story of *Manon Lescaut* and the projected continuation to *Emile* suggest that the master's or friend's influence may prove insufficient in the face of emotional upheaval.

But what the reader learns from most of the works I have been discussing is the vulnerability entailed by emotion, the insecurity of happiness, and the link between grandeur and destruction. These have been the lessons of tragedy since antiquity. Being human involves undergoing risks that the gods do not run, but any attempt to overcome the risks may itself involve a significant sacrifice of humanness. If stability is to be bought only at the cost of ignoring human individuality, then ordinary humans may decide the cost is too heavy to bear.

Through love the Présidente and Valmont both learn the lesson, which Meilcour grapples with at the end of *Les Egarements*, but is not yet equipped to absorb, that this emotion permits the fusion, rather than the mutual destruction, of sentiment and sensation. Meilcour, as he is introduced to carnal delight by Mme de Lursay, agonizedly contrasts the 'ivresse' into which she has plunged him with the 'sentiments' he feels for the distant Hortense. His agony, he tells the reader, would not have been shared by a man better-educated in the ways of polite society; the sophistry of that world would have allowed such a person to plead innocence of intent to harm Hortense by invoking 'l'usage qui ne nous permet pas de résister à une femme à qui nous plaisons',[85] and so to salvage his heart from the disorder of his senses. Meilcour's 'formation', however, has not advanced far enough for this handy metaphysics (which he calls the quietism of the heart) to be available to him. He is left in a classic state of imbalance, oscillating between remorse and sensual pleasure, and lacking all certainty. The latter, according to Crébillon's preface,

---

[85] p. 187.

will be his only when the right tutor takes charge of him; and in a society where, as Versac has informed Meilcour, the female exercises an absolute power, it is appropriate that the tutor should be a woman.

Sentimental education, then, leads many astray in the eighteenth-century novel. It bids fair to make a 'fat' and a libertine out of Meilcour, crushes Cécile Volanges and Danceny, and turns Ursule and Edmond into sensualists who are barely redeemed by the force of familial and conjugal love. The stern wisdom imparted to Julie and Saint-Preux by the mentor-figure Wolmar, wholesome though it appears, proves to be cruelly out of touch with emotional reality; and Emile's privileged schooling is unable to save him from passional upheaval and marital conflict. For an age which aspired to revolutionary achievement in pedagogy,[86] the picture is a sorry one. If people learn anything about love, on this evidence, they do so primarily, and damagingly, through personal experience: coaching may help individuals to enjoy merely physical pleasure, but is manifestly inadequate where instruction about the intangible qualities of love is concerned. The lesson of literature, too, is well-nigh useless to the beginner, for whom the particular aspects of his passion are more impressive than its general properties. As Mme de Tourvel would surely concede, humans are fatally prone to take their own case as unique, and to spurn the counsel of those, like Mme de Rosemonde, who can see the overall pattern to which a lover's wooing will probably conform. While example may enrich the understanding of detached observers (like the readers Laclos imagines for his cautionary tale), it is usually heeded too late to save the suffering victims of love—Prévost's des Grieux, Rousseau's Emile, and Laclos's Présidente. And however determined humans may be to learn an art of loving that will preserve them from passion's catastrophic upsets, emotion often contrives, with tragic facility, to assert its dominance over the efforts of reason and reflection.

[86] See, for example, Mme de Brulart [Genlis], *Discours sur l'éducation publique du peuple* (Paris, 1791); Louis-René Caradeuc de La Chalotais, *Essai sur l'éducation nationale, ou plan d'études pour la jeunesse* (Paris, 1763); *Procès-verbaux du Comité d'Instruction publique de l'Assemblée législative*, ed. J. Guillaume (Paris, 1889); Harvey Chisick, *The Limits of Reform in the Enlightenment* (Princeton, 1981); Dominique Julia, *Les Trois Couleurs du tableau noir: La Révolution* (Paris, 1982).

2

Materialism and Metaphysics

DE *rerum natura*, the didactic poem by Lucretius which gives a full exposition of materialist philosophy, begins with an invocation to Venus, 'who alone governs the nature of things'. (The words were apparently written with feeling, for an ancient biography quoted by St Jerome relates that Lucretius wrote the poem in between bouts of insanity brought on by a love-philtre, whose effects eventually led him to commit suicide.)

Both the mechanistic philosophy of Lucretius and aspects of the Epicureanism he espoused—such as the doctrine that there is no such necessary conflict between reason and the passions as Plato's school and the Stoics had discerned—exerted a profound influence on French thought. The separation of science from metaphysics associated with the work of Gassendi, together with the latter's presentation of atomism as the system which best explained the phenomenal world, provided a basis for the mechanistic theories to be developed in the Enlightenment. Although some of the implications of materialist philosophy were to be resisted by thinkers like Diderot (notably its reduction of man's emotional life to a set of merely physical actions and reactions), the assumption that the universe and its components functioned according to unchanging laws of mechanical causation permeated the philosophy and literature of his age.

La Mettrie's *L'Homme-machine* (1747), to whose naturalistic account of man Diderot, d'Holbach, and Cabanis were indebted, extended to human beings Descartes's argument that animals were mere machines, and coloured contemporary and subsequent literary accounts of cognition, intention, purposive behaviour, and free will. Laclos's *Les Liaisons dangereuses* often invites interpretation in terms of the mechanistic and materialist speculation current at the time it was written, and for that reason will be considered at length in this chapter. It raises questions about responsibility and choice, particularly with respect to the emotions, which contemporary philosophy had made urgent matters for discussion.

Briefly, the materialists argued that human actions were necessitated by virtue of man's belonging to the closed system of the universe, whose states were determined according to mechanical laws (then seen as essentially those discovered by Newton). Whereas Descartes had tried to preserve human action from the realm of natural necessity by distinguishing mind from matter and claiming that natural laws applied to the latter alone, La Mettrie and his followers argued against metaphysical dualism on the grounds that mental processes were identical with their physiological causes. There is no distinction, they claimed, between conscious and voluntary processes on the one hand and the instinctual, involuntary ones of the human machine on the other. For Descartes, no machine or animal could have free will, the conscious intellectual deliberation peculiar to man. But materialist philosophers rejected the theory that the behaviour of animals differed from that of men in being merely reactive, without the consciousness that distinguishes most kinds of human conduct.

This deterministic view of the world also differed from that associated with Christian theology. Although some Christian apologists taught that the omniscience and omnipotence of God strictly precluded the notion that man's actions are freely chosen, most nevertheless argued that the human will was essentially free. But with the progressive discovery of natural laws in the seventeenth and eighteenth centuries it became possible to see the cosmos as self-regulating rather than directed by an immutable divine will. One consequence of this was the reductive assertion that, since man is a mere mechanism, he is subject to laws like all components and inhabitants of the natural world. Another, perhaps paradoxical, was that in a world free of transcendent governance it was open to man to try bending nature to the force of his will and intellect, irrespective of the possible futility of such an action in a materially programmed environment.

The choice between seeing man as a conscious, reflective being, a part of merely physical nature, or a creature conditioned by unnamed external influences, was explored in much imaginative literature of the eighteenth century, and frequently related to questions of love and sexual desire. Those who supported the notion that human beings are matter to be acted on by intangible outside forces often contended, like Prévost's des Grieux, that in affairs of sentiment will and determination are necessarily inoperative. People are driven to fall in and out of love by powers greater than themselves, and since they cannot choose their affections they may be blamed neither for the inception of passion nor for its end. Constancy and inconstancy in love, on this interpretation, are not properly subject to moral judgement, to approval or disapproval, for human will has no part to play in sentiment. Many writers in the 'sensible' eighteenth century, however, were unwilling to take such views to their logical conclusion. Although they were disposed to regard erotic love as a force of overwhelming urgency, a thunderbolt or a cataclysm which seemed to rob its victim of the ability to resist, sentimentalists of Rousseau's persuasion also liked to show *amour-passion* as under the dominion of intent. Only thus could they link their discussion of eros with that of morality, as Rousseau contrives to do in *La Nouvelle Héloïse*. But often, as that novel reveals, it was through sleight of hand alone that they managed to bring the realms of all-powerful passion and human voluntarism together.

The sentimental writer might share with the materialist his view of eros as an imperiously commanding power, but he could never countenance the materialist's reduction of love to a set of sensory reactions. For the man of feeling it was anathema to talk of his fellows as merely sentient beings, or human emotion as a compound of physiological causes and effects. Sometimes, it is true, the materialist view was espoused by novelists and their characters for the sake of belittling human capabilities rather than seriously proposing the equation of man and machine. Just as it was possible in the eighteenth century to be a hard-headed rationalist as well as a man of feeling, so it appears from certain novels of the period—*Les Liaisons dangereuses* and Diderot's earlier *Les Bijoux indiscrets* in particular—that a physicalist conception of love could coexist with a 'feeling' perception of its immaterial essence.

In *Les Liaisons dangereuses* it is often difficult to gauge the seriousness with which the vocabulary of materialism is used in connection with eros on the one hand and intellect on the other. The Marquise de Merteuil appears a woman of her philosophical times when she dismisses Cécile in a letter to Valmont as a 'machine à plaisir',[1] a female incapable of the intellectual activity required for intrigue, and good only for sexual commerce. This type of womanhood, for the Marquise, must share responsibility with the

---

[1] p. 244.

male sex for the latter's domination in society, a theme on which she elaborates in the eighty-first letter of the novel. Cécile belongs to the army of 'femmes à sentiment' and 'femmes sensibles' whose mindless relations with their lovers make their subjugation by the male inevitable. Like all women governed by feeling, she lacks the power of thought that has given Mme de Merteuil the dominion she boasts of in Letter LXXXI. Cécile is no more than a 'bel objet',[2] an example of La Mettrie's 'assemblage de ressorts';[3] and as the Marquise reminds Valmont, it takes little skill to learn the mechanical functioning of such a creature.[4] The prudent man therefore makes use of the machine as soon as he has discovered how it operates, and then destroys it. But although only thinking organisms can exert the force of thought and will on the merely mechanical, the quality of mind required to master such tractable matter is not high; hence the calculated tone of contempt in the Marquise's reference to Valmont's sexual activities. The working of the machine has, she implies, been almost automatic.

The means Valmont employs to procure his first night with Cécile underline the aptness of the mechanical metaphor, if metaphor it be. His insistent detailing of the care she must take in oiling the lock of her bedroom door preparatory to his insertion of the key is not merely lubricious, adding to the fund of *double entendres* that accumulates over the length of Laclos's novel, but emphasizes her status as a mechanism to be operated by her lover.[5] Yet the key, as Mme de Merteuil would not have been slow to point out, is itself only a 'part', a tool whose sole use is to make the mechanism work. It is not an operative possessed of intelligence, for none is required in the physical act which the combined operation of lock and key euphemistically describes. This is really the sense behind the Marquise's repeated taunting of Valmont. Although the apparent female desire for subordination ensures that the male is the 'key' to her efficient functioning, neither sex, in the kind of relationship Mme de Merteuil is criticizing, can rightly lay claim to a status above the mechanical. To that extent, both may be subject to manipulation by the rational being who stands outside and above the material realm.

---

[2] p. 15.
[3] La Mettrie, *L'Homme-machine*, ed. Aram Vartanian (Princeton, 1960), p. 186.
[4] p. 245.   [5] pp. 182–3.

For a later age, the eighteenth century's often demeaning assumptions about the limitations of machines, and particularly its view of animal automatism, perhaps require modification. The possibility that a machine might function purposively, even think and be creative, is now at least a matter for serious discussion. La Mettrie revised the Cartesian opinion that a machine lacks thought or feeling, taking account of the conscious and cognitive element in man's make-up, and his conclusions have been extended by later thinkers. It now seems less far-fetched than it would have appeared in the Enlightenment to argue that a machine—capable, for example, of solving theorems or playing chess—might be assigned thought and intelligence, even if a machine cannot itself be purposive (its purposes being those conceived by the engineer or programmer).[6]

Object-status, or the condition of being programmed to do what a higher, conscious power requests, is attributed in *Les Liaisons dangereuses* to humans of both sexes. This has important implications for the questions of moral responsibility and free will which Laclos's novel raises. On one occasion Valmont freely admits that he is an instrument rather than an instrument-maker or -user. He writes to Mme de Merteuil of a common female type, which he contrasts with the Présidente de Tourvel, that

pour beaucoup de femmes, le plaisir est toujours le plaisir, et n'est jamais que cela; et auprès de celles-là, de quelque titre qu'on nous décore, nous ne sommes jamais que des facteurs, de simples commissionnaires, dont l'activité fait tout le mérite.[7]

This is not very far from the situation the Marquise fancifully imagines as prevailing in a future relationship between herself and Valmont, where she will be a 'belle dame sans merci' and he her 'preux chevalier'. Then, the tradition of knightly gallantry and submissiveness will also ensure male obedience to the female will.

Neither type of relationship, Valmont implies, resembles that which holds between himself and the Présidente. If that is true, according to Mme de Merteuil, it is not because he exercises power over his beloved; he is wholly subjected to her, talk as he will of his scientific detachment from the process of pursuit. The Marquise

[6] See Margaret A. Boden, 'Human Values in a Mechanistic Universe', in *Human Values*, ed. Godfrey Vesey (Hassocks, Sussex, and Atlantic Highlands, N.J., 1978), pp. 135 ff.
[7] p. 309.

is not deceived by his claim to be merely an observer testing his experimental methods on an unfamiliar type of female, the 'femme délicate et sensible, qui fît son unique affaire de l'amour, et . . . dont l'émotion, loin de suivre la route ordinaire, partît toujours du cœur, pour arriver aux sens'.[8] (The Mangogul of *Les Bijoux indiscrets* is much closer to being such a 'scientific' observer of woman, as we shall see.) Mme de Merteuil realizes that Valmont, proud of his facility in bending women to his will, is now manipulated by the Présidente. And she recognizes that love is the one force capable of destroying human superiority to the man-machine.

Love procures a kind of involvement with the loved one that breaks down the barriers separating creature from creator. It ends the mental power a rational being may exert over a merely sentient one, because it brings with it a desire for reciprocation and equality that puts an end to the urge for domination. Mme de Merteuil senses as well that a lover's concern for the thoughts and feelings of the beloved renders her infinitely more valuable in his eyes than a pleasure-machine can be. The relegation of physical pleasure from its prime position forges a bond between lovers that is stronger than the relationship between master and machine. In direct opposition to the 'femme à plaisir' Valmont sets the woman who can 'sortir du plaisir tout éplorée, et le moment d'après retrouver la volupté dans un mot qui répond[e] à son âme'.[9] The effect is reciprocal, for 'Auprès [de cette femme] je n'ai pas besoin de jouir pour être heureux.'[10] Before the jealous Marquise destroys this woman as she might a machine, she is forced to acknowledge the power of a union that moves beyond carnal pleasure. She knows, after all, that the latter, 'qui est bien en effet l'unique mobile de la réunion des sexes, ne suffit pourtant pas pour former une liaison entre eux'. When desire is unalloyed with love, it is converted into repugnance. Mme de Merteuil declares this to be another of nature's laws which it is beyond any human agent to change, however powerful his intellect and will. And love, she owns, is not conceived at the individual's behest. Her conclusions may not command unqualified assent, for they ignore the elements of cognition and intention that may plausibly be regarded as essential parts of love. But their significance for Laclos's novel is that they drive her to impose control on the love-affair in the way she understands best, by forcibly ending it.

[8] p. 310.   [9] Ibid.   [10] p. 22.   [11] p. 305.

The irony of the novel is that this self-assertion leads to the destruction of the two principals who had delighted in their god-like superiority over other human beings. But before that final demonstration of mortal powerlessness, both Valmont and Mme de Merteuil have exulted in their ability to manipulate others. Their letters dwell with an often sadistic pleasure on the ease with which they make their victims means to their own perverted ends; hence the frequency of their references to systems, methods, inventions, projects, principles, designs, rules, prescriptions, 'ouvrages', and the like. Letter LXXXI is particularly rich in such formulations, intended by the Marquise to demonstrate her superiority to Valmont as well as to the objects of her manipulation. It is only his own inability to carry out her plans, she tells him, that makes him declare them impossible. He fails to consider the fact that his erotic victories are, by virtue of his sex, markedly inferior to hers, which are won despite social difficulties, obstacles, and prejudices which the male never encounters. She has achieved hegemony through the creation of means 'inconnus jusqu'à moi',[12] and contrasting completely with the ordinary ways of women. The war Valmont wages, principally directed at 'having' the Présidente 'pour me sauver du ridicule d'en être amoureux',[13] cannot stand comparison with Mme de Merteuil's campaign to lead her sex into total domination of the male.

Yet Valmont manifestly controls some of the lives into which he intrudes. With respect to Cécile and Danceny, the unknowing children, he operates as a supreme intelligence. Although he remarks at one point that 'entre la conduite de Danceny avec la petite Volanges, et la mienne avec la prude Mme de Tourvel, il n'y a que la différence du plus au moins',[14] Danceny's 'système' has little in common with Valmont's, being founded on the notion that a young girl's virtue must be respected. There is a Marivaudian ring to the older man's description of Danceny's transfixed emotional state faced with the enchantment of love:

En effet, si les premiers amours paraissent, en général, plus honnêtes, et comme on dit plus purs; s'ils sont au moins plus lents dans leur marche, ce n'est pas, comme on le pense, délicatesse ou timidité: c'est que le cœur, étonné par un sentiment inconnu, s'arrête, pour ainsi dire, à chaque pas, pour jouir du charme qu'il éprouve, et que ce charme est si puissant sur un cœur neuf, qu'il l'occupe au point de lui faire oublier tout autre plaisir . . .[15]

---

[12] p. 170.   [13] p. 18.   [14] p. 115.   [15] Ibid.

Valmont never forgets the strategic advantage he enjoys over the young man. Imitating the Marquise's Protean qualities, he is mentor, helpmeet, and apparent friend to Danceny—even his mistress, when he takes it upon himself to write Cécile's letters to her lover.[16] The Marquise is far from happy about Danceny's extreme inactivity, for her revised scheme of revenge depends on his seducing Cécile (given that Valmont has other sexual preoccupations). But whatever he does, Danceny will be a merely reactive instrument, seemingly incapable of willed self-assertion. His part in the game which his elders are playing will be a purely mechanical one, for he embodies the Newtonian principle of inertia. And according to Newtonian mechanics, an external force will be required to set him in motion (*omne quod movetur ab alio movetur*). Cécile proves too weak an impulse to bring about the requisite movement; hence the involvement of the two *agents provocateurs* in the affair.

Still more than Danceny, Cécile does what she has been pro-grammed to do. All the circumstances of her life with which we are familiar have the character of involuntariness—her proposed marriage to Gercourt, her infidelity to Danceny as Valmont's bedmate, even the conception and subsequent miscarriage of the latter's child. Her retreat into a convent at the end of the novel shocks by its unwonted decisiveness, for until then she has been a passive object: the marriageable offspring of a negligent mother, the Marquise's protégée, the pupil to whom Valmont teaches a 'catéchisme de débauche', and his own debauchee. Few of those who play a significant part in Cécile's life have her best interests at heart. All except Danceny ignore her feelings and thoughts, such as they are, and use her to realize their own purposes. Valmont's treatment of the girl is in this respect markedly different from his attitude towards Mme de Tourvel. For the latter he comes to show the human consideration he denies Cécile, acknowledging the Présidente's right to be treated as the moral being he refuses to see in the young girl. Machines, *qua* machines, make no appeal to our moral sensibility.

Cécile neither understands nor registers the process of change which external parties are imposing on her. Once Valmont and the Marquise have done their work she loses the grasp of moral distinctions which initially prompted her to remain faithful to Danceny and resist Valmont's advances. The web of relationships

---

[16] p. 270.

into which she has been drawn—engagement to Gercourt, sentimental attachment to Danceny, and sexual partnership with Valmont—is too complex for her undeveloped intelligence to grasp. She remains a child, unconscious that she herself has been got with child, unaware when she miscarries that her body has been bearing Valmont's offspring. And to the extent that she lacks cognition, she is a machine; she does not possess the capacities and capabilities of thinking organisms, and for this reason can scarcely be held responsible for her actions.

In so far as responsibility for Cécile's fate is admitted by any party, it is Mme de Volanges who concedes culpability in having chosen for her daughter a husband who is not the man she loves. But the measure of her admission is small indeed; Cécile's mother does not address the larger matter of her neglect of her daughter's upbringing, with its attendant consequences for the formation of the girl's moral and intellectual character. Yet this is a crucially important matter, not least because of its implications about the relation between adulthood and responsibility. Laclos makes Cécile's childishness into the essential aspect of her character. Her letter-writing style, to which both Mme de Merteuil and Valmont try to give greater maturity, is a trivial but telling indication of her undeveloped state.

It is natural to oppose the 'programmed' Cécile to the Présidente de Tourvel, in part because the latter's actions seem to have spiritual or mental causes rather than material ones. When she writes to Valmont that she cannot, as a respectably married woman, ever love him,[17] she is referring as much to moral impossibility as to the kind resulting from external constraint. She is thus conscious of an element of choice which Cécile does not perceive, an area in which free will and conscious decision must operate. Valmont represents external constraint, it is true, but of a less material kind than is suffered by Cécile: he withdraws from Mme de Tourvel's presence when she requests it, although he subsequently imposes a phantom presence by bombarding her with letters from Paris. And, significantly, it is by mental rather than mechanical means that he finally breaks down her resistance, resorting to the moral blackmail of reporting to her through a priest his repentance of his ways and renunciation of her. Inasmuch as her following actions are determined, it is hardly by circumstances outside her control. Although Mme de Rosemonde

_____

[17] p. 113.

(Valmont's aunt) exempts the young woman from blame for her adulterous affair, the Présidente acknowledges her own culpability. The old lady sees human frailties and shortcomings in a perspective that discourages stern reflections on desert and punishment.

Mme de Volanges finds nothing to indicate the workings of strict justice in the fates suffered by the various characters. Even had she been aware of the true events in which her own daughter and the Présidente were implicated, it seems unlikely that she would materially have altered her view that Valmont and the Marquise deserved punishment far more than those whose lives they tried to direct. In a just world, it may be felt, Cécile's ignorance of the relevant issues, together with the fact that the originating principle of action lay outside herself, should have rendered her liable to little or no punishment for the turn of events. In so far as Valmont and Mme de Merteuil set themselves up as godheads in their human world, and in so far as they can be seen as working 'against nature' in seeking, respectively, the corruption of the Présidente and the seduction of Cécile, they can be regarded as responsible (because self-willed) agents. This conclusion is reinforced by the evident extent to which their heads rule their hearts in their sexual dealings. But Valmont, no less evidently, becomes subject rather than ruler, sentiment rather than mind, in the course of his relationship with Mme de Tourvel, overwhelmed by the same 'charme inconnu' as fascinates Danceny. Even the Marquise seems at times accessible to involuntary feeling, if the emotion mingled with calculation in her caressing of her current lover can be so described: 'moitié réflexion, moitié sentiment, je passai mes bras autour de lui, et me laissai tomber à ses genoux.'[18] And towards the end of the novel, for reasons at which we can only guess, she reminds Valmont of the time 'où nous nous aimions, car je crois que c'était de l'amour', and when she had known a happiness which is not available to people whose actions are the product of 'ressource',[19] lacking the spontaneity of deep feeling.

In this compellingly analytical novel, the psychological complexity of the two main characters makes the just apportioning of blame with respect to thought and action far from simple. Our own age is readier than was the eighteenth century to see certain mental conditions as absolving an agent from responsibility for what he does.

---

[18] p. 30.   [19] pp. 306–7.

We cannot know whether social circumstances—such as the lack of a profession—or a disculpating degree of mental aberration drove Valmont and the Marquise to follow their cruelly destructive course. The pair may or may not be psychopaths. The difficulty in discussing psychological states is to know where to draw the line; and Laclos, so advanced for his literary day in dissecting the psyche, had a lesser weight of opinion with which to back up his speculations than is at the disposal of modern writers. It is perhaps for this reason that his novel, to the distress of some past and present readers, omits (rather than fails) to make the most apparently guilty parties suffer proportionately more than the less evidently guilty or the guiltless. The moral issues which he does not settle may be incapable of straightforward resolution.

Laclos does not broach the subject of anatomical and physiological determinism in his discussion of sexual activity. That had been done in the much earlier novel by Diderot, *Les Bijoux indiscrets* (1748), but in terms which made the author's true attitude to questions of responsibility and choice in erotic love difficult to determine. In one chapter of the book, 'Des voyageurs', Diderot parodistically presents a discourse on the crucial nature of an individual's bodily structure in matters sexual. The Sultan Mangogul is told how one race, a group of unnamed islanders, concludes marriages and ensures harmonious marital relations on the basis of physical compatibility. Male and female sex organs (the 'bijoux' of the title) are simply matched with one another according to shape. 'Un bijou féminin en écrou est prédestiné à un bijou mâle fait en vis.'[20] The grotesque spirit informing Diderot's account of sexual partnership can hardly be missed, here or elsewhere in the novel. He ridicules the anatomy and physiology of sex in terms which recall the fable told by Aristophanes in the *Symposium*, where the separation of the originally complete and self-sufficient being into two halves is described. Aristophanes emphasizes the oddity of the sliced-up creatures, with their exposed and dangling genital members seeming a punishment or a joke rather than the physical instruments of something profound called love. When reduced to the mechanical terms in which Diderot imagines it, desire is similarly robbed of all its dignity. The implication behind

---

[20] *Les Bijoux indiscrets, Œuvres romanesques*, ed. Henri Bénac (Paris, 1962), p. 53. This perhaps anticipates the story of the 'gaine' and the 'coutelet' in *Jacques le fataliste*.

both stories is that for intelligent beings to think the insertion of a projection into a hole a matter of the deepest concern smacks of absurdity—especially where, as in Diderot's novel, the sexual act seems commonly to have no utilitarian purpose. In the fable of Aristophanes, the splitting of originally perfect human forms was a punishment meted out by the gods for mortal arrogance; the new shape of the former androgynes is intended to be seen as a mutilation. Even if 'meant' for their opposite halves, the incomplete beings of the *Symposium* and *Les Bijoux indiscrets* are robbed of the grandeur that, in other circumstances, the notion of predestination might have conferred on them.

In Diderot's story, however, anatomical determinism is not in itself decisive. One form of incompatibility is allowed to outweigh the evidence of congruent bodily form, namely temperature. In a novel where experimental method is given hallowed status (Mangogul has a magic ring which enables him to establish with scientific accuracy whether a given woman is chaste or not), it is perhaps appropriate that thermometers should be a crucial diagnostic tool in assorting sexual partners. Even geometrical conformity, as with a male whose parallelipiped organ can be perfectly accommodated within a female's square one, may be deemed insufficient in the light of thermometrical evidence. Too great a dissimilarity of temperature leads to the proposed match being suspended: a 'cold' elderly man may not wed a warm-blooded young woman.[21] (Diderot does not suggest that certain bodily acts over which the individual conceivably has some control might affect the temperatures measured.) The question of an individual's will or choice entering into his sexual life is thus a carefully qualified one. Consideration of all the physical data necessarily precedes any final decision regarding sexual partnership.

This episode of *Les Bijoux indiscrets* is an exception, incidentally, in making males as well as females subject to experiment. A genie gives the magic ring to Mangogul so that he may amuse himself testing the virtue of court ladies by pointing it at their pudenda, which are provoked into indiscretion about their activities. It is not suggested that women are either more or less sexually promiscuous than men; Diderot simply accommodates himself to the tradition according to which a woman's lapse from virtue is more serious than a man's because liable to end in illegitimate childbearing (as well,

[21] p. 54.

perhaps, as to the Christian tradition of castigating woman's sexual rapacity). What he does not fully consider is the link between physical stimulus and response, and the supervening influence of will. In the course of the story a Brahmin delivers a tirade, worthy of Bossuet or Bourdaloue, on the guilt of incontinent women,[22] the conclusion to which is the assertion that the behaviour of the unchaste was the result of choice and is therefore punishable. But Diderot fails to distinguish between an (involuntary) consent of the body to sexual arousal and a consent of the will which alone, according to Christian theological doctrine, would represent culpable concession to the physical, and terminate in the sexual act. The author of *Manon Lescaut* had earlier introduced this notion into des Grieux's speech about the 'double concupiscence' of flesh and spirit,[23] and I shall return to it further on. Although Diderot examines the phenomenon of programmed bodily response in his dialogue *Le Rêve de d'Alembert*,[24] and to a limited extent in *Le Neveu de Rameau*, in his references to the 'fibres' which form the basic element of all live matter,[25] *Les Bijoux indiscrets* does not consider human determinism from this point of view.

It does, however, continue the implied theme of reducing humans to their purely physical selves in the chapter *Des âmes*, where the Sultan's favourite Mirzoza turns metaphysician and speculates about stripping the human body to its most-used organ. The soul would be deprived of all parts of its corporeal habitat superfluous to its life-giving activity, dancers being reduced to their feet, singers to their throat, and most women to their 'bijou'. That Mirzoza should suggest the latter reduction is surprising, to say the least. Intermittently throughout the novel, and emphatically at the end, she maintains the dignity of her sex against the evidence of Mangogul's experiments, implying that the proofs he has amassed are partial.[26] Her own final vindication as a virtuous woman goes some way towards disproving the materialist determinism she espouses in her metaphysical discussion.

---

[22] pp. 39–40.

[23] See A. H. T. Levi, *French Moralists and the Theory of the Passions* (Oxford, 1964), p. 209.

[24] *Le Rêve de d'Alembert*, *Œuvres philosophiques*, ed. Paul Vernière (Paris, 1964), pp. 320, 343.

[25] pp. 76, 89.

[26] See D. J. Adams, *Diderot, Dialogue and Debate* (Liverpool, 1986), pp. 90 ff.

In having Mme de Lursay describe love as a binding-together of souls rather than a conjunction of bodies, Crébillon may have intended to satirize Cartesian theories about the separation of mind from matter as well as neo-Platonist ones. Meilcour remarks on the inconsistency of such 'platoniciennes' as Mme de Lursay: they should scarcely proclaim the virtues of sexual abstinence, he observes, in view of their own proneness to fall victim to seduction despite their high principles.[27] Possibly, he muses, they have taken insufficient practical account of the dangerous pitfalls of love in their lofty theorizing. These are the reflections of the older man, however; the youthful Meilcour is unaware of the frailty of Mme de Lursay's 'system', and is unable to conceal his joy at her gradual yielding to his charms. He is the first, he remains convinced, to have won such a victory over philosophy, to have triumphed so completely over Plato.[28]

The requirement that love be cleansed of its fleshly impurities also bulks large in Christian doctrine, and often finds its way into the eighteenth-century novel in connection with the demands of monogamous marriage and the temptation of illicit love-affairs. This theme dominates the latter books of *La Nouvelle Héloïse*, although the call to abjure carnal pleasure is addressed by Julie to her lover Saint-Preux well before her marriage to Wolmar. She scolds him for his assumption that sexual continence comes more easily to the female than the male (also an implied theme of the historical Héloïse's correspondence with Abélard, for her letters reveal nothing more poignantly than the intensity of her longing for him even after his castration). Decency, Julie writes, is no less difficult and no less essential for one sex to preserve than the other. All so-called sexual needs have their source not in nature (that is, instinct) but in the willed depravity of the senses. Desires, once checked, learn never to recur; temptations multiply only through the individual's habit of yielding to them. She concludes that her lover is guilty of infringing the laws of decency, love, and reason.[29]

Julie's certainty that Saint-Preux's incontinence was governable, and may therefore be justly chastised, does not surprise her reader. *La Nouvelle Héloïse* as a whole is much concerned with the issue of choice—the choice of virtue and regularity in preference to vice and disorder, the choice of 'amitié amoureuse' against erotic love,

---

[27] p. 38.   [28] p. 59.   [29] pp. 301–2.

and ultimately the choice of death over life. All are interconnected, for Julie's *Liebestod* appears the willed renunciation of an earthly existence in which she is powerless to resist the force of her love for a man who is not her husband. This matter raises the central interpretative problem of Rousseau's novel. Are virtue and vice meaningful concepts, in the sense that man possesses the free will which enables him to choose between right and wrong action, or are they the empty words which some of Rousseau's contemporaries (among them, but intermittently, Diderot) declared them to be? The tension between Julie's professed belief in unconstrained choice, and the involuntary causation that seems to underlie her ineradicable passion for Saint-Preux, is never satisfactorily resolved.

Saint-Preux's adoration of Julie, at least during the periods when he is debarred from sexual intimacy with her, is that of a mere mortal for a goddess. If he worships her celestial charms, he tells her, it is 'pour l'empreinte de cette âme sans tache qui [les] anime, et dont tous tes traits portent la divine enseigne'.[30] Sometimes the idolized love-object seems scarcely worthy of such homage. For Prévost's des Grieux, schooled by Jesuits, it seems natural to address Manon in tones conventionally reserved for the godhead: he tells her that she is 'trop adorable pour une créature', and that all he has been taught about free will is an illusion.[31] He feels his heart carried away by a 'délectation victorieuse' which is the concupiscence of earthly desire, not the desire for divine grace of which his Jesuit mentors speak.[32] But it need hardly be said how ill the mantle of divinity sits on Manon's shoulders.

Prévost's novel is not the only one of its century to reveal the delusion underlying a lover's deification of his beloved. The obsessive nature of erotic devotion can often seem little different from the apparently more exalted homage paid to the real godhead.[33] At the end of *Les Liaisons dangereuses* the Présidente de Tourvel is well

---

[30] p. 41.   [31] p. 60.

[32] For men like Bossuet, earthly concupiscence, being of a body destined to become dust, was vain. Delectation awaited the lover of God, but it was the chaste delectation of eternal life. See P. Jacquinot, *Des Prédicateurs du XVIIᵉ siècle avant Bossuet* (Paris, 1863), p. 241. Adoration of the human seemed to Bossuet still more misplaced than it appears in Plato's *Symposium*, being a drug which dulled the spiritual faculties that should lead man towards God.

[33] See Jean H. Hagstrum, *Sex and Sensibility: Ideal and Erotic Love from Milton to Mozart* (Chicago, 1980), p. 131; Irving Singer, *The Nature of Love*, new ed., 2 vols (Chicago and London, 1984), i. 107, ii. 9.

punished for her mistake in transferring to Valmont the worship she had previously reserved for her heavenly maker; but Valmont had foreseen the 'dévote''s susceptibility to such an error early on, and determined to exploit it to his own advantage. His native arrogance assures him, before he has received any indication that Mme de Tourvel will succumb to his charms, that he will one day replace the old god of her universe; and so it comes to pass. The real-life lover of an unworthy man, Julie de Lespinasse, fully understood the narrowness of the divide between adoration of mortals and of immortals, and realized how easy it was for the pious to 'avoir la présence de Dieu sans distraction'.[34] The Présidente effects the transference with tragic facility, and is no less unreserved in her new devotion than she had been in her old. Happily, there are some literary examples of the felicity that may result from the substitution of earthly for divine love. The most straightforward is that provided by Marivaux's Mlle Habert, become Mme de La Vallée. Only one who has been for thirty years a 'dévote', reflects Jacob in *Le Paysan parvenu*, can immerse herself with such intense and religious zeal in the terrestrial bliss of marriage, and worship her spouse as though he were an icon. Other women profess their feelings lovingly; Mlle Habert does so worshipfully,

mais avec une dévotion délicieuse; vous eussiez cru que son cœur traitait amoureusement avec moi une affaire de conscience, et que cela signifiait: Dieu soit béni qui veut que je vous aime, et que sa sainte volonté soit faite; et tous les transports de son cœur était sur ce ton-là, et l'amour n'y perdait qu'un peu de son air et de son style, mais rien de ses sentiments; figurez-vous là-dessus de quel caractère il pouvait être.[35]

And although the result of her late marriage has been to increase her religious devotion, Jacob doubts whether that fact adds to her own merit in the eyes of God; for the soul of her redoubled piety is her possession of a twenty-year-old husband rather than pure adoration of her heavenly father.[36]

There is something of this charmed innocence in the Présidente de Tourvel too, as she confides in Mme de Rosemonde her 'consecration' of herself to the old woman's nephew. Unlike Mlle Habert, however, she is forced to acknowledge her delusion: Valmont is not deity but

---

[34] *Correspondance entre Julie de Lespinasse et le Comte de Guibert*, ed. Comte de Villeneuve-Guibert (Paris, 1906), p. 74.
[35] *Le Paysan parvenu* (Paris, 1965), pp. 163–4. [36] pp. 245–6.

devil. So, at least, it may appear to one such as Mme de Rosemonde (despite her fondness for her nephew) who has not followed the intricate tale of Valmont's rivalry with Mme de Merteuil, nor grasped the extent to which the latter bears responsibility for the Présidente's fate. If the Marquise and the Vicomte are a devilish pair, she is surely the archfiend. And yet, however culpable the machinations of these two, they have a terrible grandeur—the sublimity of evil, but sublimity none the less. They may be guilty of blasphemously confusing the sacred and the profane (although Valmont is as ready to call the Présidente his divinity as to insist that he will become hers), but they undeniably have a certain claim to our respect as begetters of grandiose schemes and architects of fate. The schemes come to nothing, and the hubris of the plotters is punished at least as emphatically as was the mortal arrogance of the creatures described by Aristophanes in the *Symposium*. But many readers of Laclos's novel will be left with the feeling that Plato's cautions are somewhat off the point. To warn against investing humans with the aura of divinity, whether with respect to love or in some other regard, is perhaps to ignore the real ways in which mortals can be godlike: not in omnipotence, omniscience, and therefore certainty, but in another kind of magnitude.

If the contrast between earthly and divine love, eros and agape, is a potent one in literary and philosophical traditions, so is another with which the eighteenth-century French novel is much concerned. This is the opposition of reason and feeling, and it is incarnate in the persons of Wolmar and Julie in *La Nouvelle Héloïse*. But are these qualities not, as the mature relationship between Manon and des Grieux might be argued to show, each other's complement rather than antithesis? In different ways, fiction perpetually examines the ways in which intellect may enter into love, and love into intellect. Marivaux is a subtle exponent of this theme in plays and novels, analysing both the cases where lovers are conscious of the deliberative element in their feelings, and those where they are not. Jacob, the peasant who has yet to 'parvenir', is drawn to Mlle Habert because she is physically appetizing (despite being thirty years his senior), but also because their relationship seems to bode well for him in material terms. Taking stock of his situation, he concludes that his feelings for a woman 'que je ne haïssais pas' come as close to love as makes no difference.[37] When Mlle Habert tenderly suggests that

---

[37] p. 85.

he may exaggerate his attachment to her, Jacob's disclaimers have
the effect of inducing in him the very feelings he is professing:

je tâchai d'avoir l'air et le ton touchant, le ton d'un homme qui pleure, et
. . . je voulus orner un peu la vérité; et ce qui est de singulier, c'est que
mon intention me gagna tout le premier. Je fis si bien que j'en fus la dupe
moi-même, et je n'eus plus qu'à me laisser aller sans m'embarrasser de rien
ajouter à ce que je sentais; c'était alors l'affaire du sentiment qui m'avait
pris, et qui en sait plus que tout l'art du monde.[38]

Thus Marivaux reverses La Rochefoucauld's maxim that 'l'esprit est
toujours la dupe du cœur.'

La Vie de Marianne gives further insight into the place of cognition
in emotional response. The first is similar to the case in Marivaux's
comedy *Le Jeu de l'amour et du hasard*: the orphaned Marianne,
who knows nothing certain about her parentage, falls in love with
the well-born Valville, and is persuaded by this fact and other
circumstances (such as the refinement of her beauty and her natural
grace) that her origins were as elevated as his own. The second
anticipates Wolmar's relationship with Julie in *La Nouvelle Héloïse*.
When Valville's desertion of Marianne becomes known, an older
man approaches her and presses his suit. Marianne is struck by
his sincerity in speaking to her (as well, typically, as by the fact
that he appears to be well-bred), and by his reassurance that allying
herself with him could have no such consequences as did her
relationship with Valville.[39] Her new admirer feels for her, not a
mad, inconstant passion, but the emotion awakened in a man of good
sense by someone possessing Marianne's soul. 'C'est ma raison qui
vous a donné mon cœur, je n'ai pas apporté ici d'autre passion.'[40]
However, in view of his age—nearly fifty, to Marianne's twenty—
he has no expectation of her loving him in return. But since she is
'raisonnable et généreuse', he has hopes of gaining her friendship and
even esteem.

If this settled affection resembles from its inception the kind of
feeling into which erotic passion characteristically resolves itself
in relationships of some duration, there seems no reason to deny it
the name of love. It comprises the sense of respect, admiration,
and particularity on which lasting contentment can be built, while

---

[38] p. 92.
[39] *La Vie de Marianne*, ed. Frédéric Deloffre (Paris, 1963), p. 421.
[40] p. 423.

attaching little or no importance to sensual appeal that may not have the capacity to last (although Marianne's admirer is careful to acknowledge her physical attractions). It is, in short, about as close to the intellectually and spiritually grounded love which Diotima recommends humans to cultivate as love for mortals can be expected to be. Above all else, it vindicates the opinion of those who maintain that love, far from being irrational and therefore a metaphysical mystery, is rooted in the will. Since it is subject to intention and judgement, it is a docile emotion whose *raison d'être* is always manifest. It permits man to live as a purposive being, not as one whose actions are necessitated in virtue of his subjection to higher forces. Knowing the rationality in which his feelings are anchored, such a lover need make no appeal to notions of cosmic causation (such as des Grieux intermittently voices) in accounting for his conduct under the influence of passion. For he is not the unresisting plaything of a *passio amoris*, but an active party who controls feeling rather than being controlled by it.

Julie's complacent rehearsing to Saint-Preux of the benefits attendant on such a 'mariage de raison' as hers to Wolmar makes clear its affinity with the union offered to Marianne by her suitor. Wedlock, Julie tells her former lover, should be close to a business arrangement, prompted less by sentiment than by rational criteria such as a woman's desire and ability to run a household, raise children, and perform the duties of civil life. If she and Wolmar appear different in character from one another, their very opposition in temperament serves to strengthen their union.[41] Although Wolmar does love his wife with passion (the only passion he has ever felt), too much love on his part, according to Julie, would have prompted him to pay importunate attention to her.

A similarly measured approach to matrimony is recommended in *Emile*, where the passion of love is nurtured, checked, and ultimately furthered by its alliance with wisdom, in the shape of Emile's tutor. Fénelon's *Télémaque* had earlier illustrated the same process, with the goddess Minerva assuming the form of Mentor to steer the young hero through such dangers as ungoverned emotion may occasion. The choice of a partner for life, Rousseau suggests in *Emile*, should be a rational one, and youth must remain ever conscious of what it is about in contracting such an alliance. This choice is as liable

[41] p. 372.

as others to be shaped by the good and bad habits learnt in a person's earliest years. Rousseau has no doubt that instinct plays its legitimate part in the process, but the tendential should be prudently allied with the cognitive. The attraction of the sexes, he writes, is a movement of nature, but time and knowledge are needed to make people capable of love.[42] Love follows on judgement as preference follows on comparison. Judgements may be made without one's noticing the fact, but are no less valid for that. The mental component in love ensures that, even if its impulses lead us astray, rationality intervenes to save the day. Rousseau argues that it is impossible to feel love without oneself possessing admirable qualities, and this is why people will always honour the true emotion. Our choice of beloved has traditionally been regarded as counter-rational, he observes, but is in truth a product of reason. 'On a fait l'amour aveugle, parce ce qu'il a de meilleurs yeux que nous, et qu'il voit des rapports que nous ne pouvons apercevoir.'[43] Rather than coming from nature, therefore, love is the curb and brake on our natural inclinations.

In so far as cognition is involved in emotional response, there is clearly a place within the domain of feeling for changing according to one's experience and knowledge. As we have seen, the growth to maturity in sentiment, as well as the failure to achieve such growth, is a main subject of the eighteenth-century French novel. In addition, contemporary fiction presents a detailed analysis of the way feeling responds to attention and attitude, a further respect in which it can be governed by the will. There are occasions and ages of man that are propitious to the development of emotion; regulating one's sentiment may be a function of the amount of attention at one's disposal to pay it. The Clarens estate in *La Nouvelle Héloïse* operates on the assumption that paying too much heed to love (erotic love, that is) has nefarious economic and social consequences. Given the singularly non-erotic nature of the Wolmars' own union, we should hardly have expected otherwise. The workers are rigorously segregated by sex during the day by being assigned different tasks and even pastimes.[44] Julie's conviction that man and woman are unsuited to constant proximity is at the root of the Wolmars' benign despotism: they permit married couples to consort with one another at night, but daybreak reimposes their strict separation.

[42] p. 493.   [43] p. 494.   [44] pp. 449–50.

The sentimentally luxuriating literature of the eighteenth century is eloquent testimony to the fact that, in circumstances encouraging them to do so, people will nurture emotion so that it flourishes virtually unchecked. The very distaste felt by later, more sober ages for much of that period's passional self-indulgence reminds us that the expression of feeling is to a high degree within our control; and the expression of feeling, as Marivaux's Jacob discovered, is often tantamount to feeling itself. *Les Liaisons dangereuses* suggests that even in the eighteenth century itself there were those who held back from the exhibitionism of their fellows in order to obtain influence over them, curbing as well as simulating emotion for their own calculating ends. Laclos's novel illustrates with unequalled force the power of choice and intention over feeling, even though it ends with the discomfiture of intellect and a kind of victory (albeit a pyrrhic one) of heart over head.

If the problem of the will's relation to the passions is put in sharp focus by the uncomfortable conclusion to this work, a different aspect of the matter is shown earlier in the book. Cécile's reaction to the fact of being in love with someone who is not her intended husband reveals not just her own emotional inexperience, but also the more general difficulty of bending the will to do one's bidding in affairs of the heart. It is much easier—particularly for those who lack imagination—to attend at will than to withdraw attention, because the target to be avoided announces itself to the mind precisely *as* an object of attention. Ridding oneself of unwanted emotion is infinitely more difficult than engendering a state of desired emotion within oneself. Cécile's attempted solution, as Mme de Merteuil maliciously reports to Valmont, is to pray to God constantly for the power to resist thinking about Danceny.[45] Thereby she neatly, although unconsciously, procures for herself the very state of perpetual preoccupation with her beloved which she is ostensibly trying to escape from, and simultaneously squares her conscience with heaven. The Marquise, experienced in all the petty deceits of love, remarks that this is 'une de ces ressources qui ne manquent jamais à l'amour'. It is perhaps with similar, and similarly unconscious, motives that Rousseau's Julie takes refuge with God when the world and her marital vows to Wolmar prohibit her from dwelling on her passion for Saint-Preux: she makes confession at the supreme tribunal

---

[45] p. 105

not just because only her maker can absolve her from sin, but also because such confession allows her to continue thinking and talking of her forbidden love.[46]

There is doubtless a rational element in all this: the tyranny over the heart described, in their different ways, by both *La Nouvelle Héloïse* and *Les Liaisons dangereuses* adequately explains why such recourse should be had to conscious or unconscious forms of deceit and self-deception as a way of handling emotion. That knowledge about the object of one's love has its limitations in teaching one how to live with it is a commonplace, and perhaps illustrates the kind of dominion of heart over head that is most familiar to ordinary people. In *Les Liaisons dangereuses*, it is rarely a simple matter to determine the extent of an individual's self-delusion with regard to emotion. Is there a strategic point to the Marquise's assertion in her letter to Valmont that the two of them were once lovers (in the romantic rather than merely physical sense of the word)? Is she jealously intent on denying the special force of his 'new' feelings for the Présidente by intimating that her own relationship with him had once partaken of the same quality? Is this the reason why, in her letters to Valmont, she is so contemptuous of the figure the Présidente cuts in the world?[47] Is the Présidente deceived by appearances when she forms the belief that the Valmont she knows is a new man, a regenerate who has abjured the errors of his old libertine ways? Or does the fatal letter of rupture he sends her reveal only that he is unprepared to admit to the Marquise, the woman who has been his rival in amorous calculation, the truth of his attachment? Laclos does not answer these questions. But it would seem imprudent to read the Présidente's tormented avowal to Mme de Rosemonde about the rending of the veil on which the illusion of her happiness was written as a balanced assessment of the circumstances.

Julie prides herself on the fact that no illusion clouds her and Wolmar's view of one another; the feeling that unites them is not the blind transport of passionate hearts, but the constant attachment of two decent and sensible beings who are happy at the prospect of spending the rest of their lives together.[48] As the end of the novel reveals, there is an immense irony in this self-deception. But what is

---

[46] See also Chateaubriand, *Le Génie du christianisme*, *Œuvres complètes*, 22 vols (Paris, 1833–5), xi. 345.

[47] p. 18.   [48] p. 373.

Rousseau about in concluding the book as he does? There can be little doubt that he meant to present in Julie an ideal woman; but there is scant likelihood that he can have been unaware of the subversion of moral values implicit in Julie's *Liebestod*. The theme of illusion is sufficiently present in the novel to suggest that Rousseau intended to make a more deliberate revelation of its embodiment in the central figure than criticism has generally allowed.

His stated belief in *Emile* that true love presupposes illusion and delusion deserves attention in this regard: 'Qu'est-ce que le véritable amour lui-même si ce n'est chimère, mensonge, illusion? On aime bien plus l'image qu'on se fait que l'objet auquel on l'applique.'[49] The matter of the lover's delusion as to the real object of his worship is further examined in Rousseau's short play *Narcisse*, which describes how the Narcissus-figure, Valère, falls in love with his own image when he sees a portrait of himself touched up so that it appears to show a woman.[50] In *La Nouvelle Héloïse*, however, the rapturously imaginative Saint-Preux proves unable to conjure illusion in his picturing of Julie's absent cousin Claire. The reasons, evidently, are psychological ones, and they well illustrate the power of (unconscious) will over imagination and hence emotion. Saint-Preux finds that Claire appears more beautiful in the flesh than he visualizes her in his mind's eye; her charms are more dangerous at close quarters than from afar. Yet the fact that this process is reversed in Julie's case, as he writes to her, provides an explanation. Julie's effect on him is that of 'enthousiasme', the divine madness of the ancients, with all its attendant exaggerations and irrationalities. For Claire he can muster only respect and calm admiration, which offer little scope for the play of the imagination.[51]

Based on 'enthusiasm' or no, Saint-Preux's love for Julie can scarcely have been regarded by his creator or himself as the fruit of a central delusion. Her lover is as clear-sighted about her modest beauty as was Rousseau about the plain looks of Sophie d'Houdetot. In neither case does emotion seem to have induced that 'mistake' described by Stendhal in *De l'amour*, the crystallizing effect which endows an ordinary object with foreign brilliance, as a branch thrown into a salt-mine is rendered dazzlingly unrecognizable by the

[49] p. 656.
[50] See Williams, op. cit. 46 ff.; also Levi, op. cit. 225–6.
[51] pp. 675 ff.

encrustation of crystals.[52] On the other hand, such 'crystallization' may well have given a deceptive glitter to other experiences of love which Rousseau describes in the *Confessions*, most notably that between himself and Mme de Warens, and the relationship between Rousseau and his 'maman's' lover Claude Anet.

The limitations of knowledge where love is concerned are nowhere better illustrated than in *Manon Lescaut*. Although mental awareness is a far lesser component in des Grieux's initial response to Manon at the Amiens posting-inn than in hers to him, he quickly enough comes to a lucid understanding of her character and its defects. At Amiens des Grieux experiences a *coup de foudre* that, while affording him a new enlightenment about the nature of love,[53] leaves him incapable of responding with intellectual clarity to his predicament. It is Manon who has the presence of mind—a quality in her that almost disconcerts des Grieux on this occasion—to fabricate the story which frees her from the tiresome attention of the 'argus' who is accompanying her. Furthermore, it is suggested, she sees the expediency of behaving as though she were as stricken with love for des Grieux as he is for her: des Grieux 'cru[t] [s'] apercevoir qu'elle n'était pas moins émue que [lui]'.[54] As time goes by, however, des Grieux loses his illusions about the character of this girl, whose leanings towards pleasure cause her to commit multiple infidelities. About his own folly in pursuing the affair with Manon he is perfectly clear: he is, to borrow Locke's comparison, like a drunkard who knows that his use of liquor is bad for him, but whom mere knowledge cannot dissuade from gratifying his appetite.

Being prepared both to see imperfections in the beloved and to chastise them when they are encountered is a way of showing the importance of will in emotional relationships. The desire to see improvement in the loved one is a token of one's assumption that love should last; and this assumption itself rests on the belief that love is something more than a feeling.[55] Feeling can come and go, or change in aspect—vacillation is often a feature of youthful emotions. But the commitment to constancy, of the kind des Grieux offers Manon, presupposes a degree of self-determination, stemming from will, in the person who makes that commitment. Against such

[52] Stendhal, *De l'amour* (Paris, 1965), pp. 34–5.
[53] p. 20.   [54] p. 22.
[55] See Erich Fromm, *The Art of Loving* (London, 1975), p. 51.

an intention, eighteenth-century *libertins* elaborated a creed of hedonism, which rested on the belief that what counted in erotic relationships was the moment, alone capable of affording intensity. Long-term commitment, in their view, diluted pleasure as all kinds of duration must do. Marivaux's Cupidon gave expression to this belief in the need to seize and enjoy the passing instant: 'Je blesse, aïe, vite au remède . . . Allons, dit-on, je vous aime, voyez ce que vous pouvez faire pour moi, car le temps est trop cher, il faut expédier les honneurs . . .'[56] In a conversation with Hortense, Crébillon's Meilcour combats her implied view (later to be repeated by Laclos's Mme de Rosemonde) that the male is by temperament more given to inconstancy than the female.[57] But Hortense, whose pessimism about the outcome of love has been made sharply apparent, and whose conviction it is that the essential happiness of her life depends on the resolution to this issue, will not be persuaded.

In the determinist's universe, talk of changing the personality through an effort of will is meaningless. Male or female fickleness, on this interpretation, is a character trait like others, and can no more be altered than involuntary muscular activity can be governed. But if inconstancy is a constant theme of the eighteenth-century novel, it is rare for authors seriously to suggest that it is an inevitable feature of human love. *La Nouvelle Héloïse*, for example, contrasts the French race with the Swiss in this regard. Exiled in Paris, Saint-Preux writes to his beloved in Switzerland about the disunity which prevails in the Parisian *beau monde*, the complete absence of a focal point in social life there. This lack of focus leads to dissipation rather than concentration, 'papillotage' rather than a gathering-together of social forces. The very banishment of the word 'chaîne' (along with 'amour' and 'amant') from the vocabulary of love is symptomatic of the revulsion against cohesiveness and stability. If '[le] cœur n'y forme aucune chaîne',[58] the reason is that the over-indulged constantly seek new relationships to stimulate their jaded palates, and must therefore remain perpetually 'disponibles'. There has been little change, it seems from the days described in *Les Egarements du cœur et de l'esprit*, when love-affairs were settled at first glance, and rarely lasted more than a day.[59] Although the code of propriety imposes its own structure on the social world, the fundamental order of

---

[56] *La Réunion des amours*, scene 1.
[57] p. 116.   [58] p. 270.   [59] p. 15.

morality is absent from it; according to Saint-Preux, 'tout l'ordre des sentiments naturels est ici renversé',[60] so that fundamental values are dissolved. The only concern of this world is freedom, with which the fixity of commitment to a single partner is clearly incompatible.[61] In *Emile et Sophie* Rousseau describes the spurious union of the couple Emile and his wife initially spend most of their time with. Although ostensibly closer to one another than are Emile and Sophie, who agree out of a residual mutual respect to go their separate ways, they are in fact further apart. Their show of conjugal harmony is merely a front erected to hide their complete indifference to each other.[62]

The subjection of social life to a system, in short, is far from investing it with the sort of stability that the inhabitants of Clarens seek in their country solitude. The education which Paris provides, being exclusively of the senses, can teach nothing about durability. Lovers meet, Saint-Preux says, 'pour le besoin du moment'; a 'liaison de galanterie' lasts scarcely longer than a visit, and physical desire is gratified at the very moment of its inception.[63] First come, first served is the order of the day; a man is always a man, and all are essentially as serviceable as each other. Perhaps this view is a cynical parody of the argument developed in the *Symposium*, where Diotima traces the individual's progression in his search for stability and permanence from loving the beauty of particular things to seeing their family resemblance with other beauties, and subsequently regarding these related beauties as one and the same. 'Et puis,' Saint-Preux continues,

à certain âge tous les hommes sont à peu près le même homme, toutes les femmes la même femme; toutes ces poupées sortent de chez la même marchande de modes, et il n'y a guère d'autre choix à faire que ce qui tombe le plus commodément sous la main.[64]

This denial of individuality is a denial of love in the conventional sense (as contrasted with Diotima's ideal sense). It effectively reduces the loved one to the non-particular level of a functional object—a physical machine which is merely required to perform, and which is qualitatively interchangeable with other machines. Malfunction

---

[60] p. 270. [61] p. 271. [62] pp. 887–8.
[63] p. 271. [64] p. 272.

results in the machine's being cast aside, if not (as Laclos's Mme de Merteuil would wish it) destroyed altogether.

In *Les Egarements*, Mme de Lursay describes resistance to a fixed, long-lasting relationship as particularly characteristic of youth, the age when wild oats are sown. If she and Meilcour became lovers, she claims, she would merely be an amusement to him, whereas 'vous me fixeriez.'[65] Earlier in the book Meilcour had elaborated the view that older women, conscious of their fading charms, try to capture a permanent lover against the solitude of impending age: love stops being a momentary preoccupation for them, and becomes instead their 'unique ressource'.[66] In similar spirit Mme de Lursay now argues that Meilcour would soon grow tired of her attentions, and would come to see in her less an 'amante sensible' than a 'personne insupportable'.[67] Later on, Versac declares that a grand passion is to be avoided because it leads a couple only to 's'ennuyer longtemps l'un avec l'autre'.[68] The heart, he proclaims, should never be strait-jacketed, and he himself is always inclined to roam when he becomes aware that a woman means to capture him.

For one libertine, however, instantaneous and evanescent pleasure is not the ultimate goal of seduction. At an early stage of *Les Liaisons dangereuses*, Mme de Merteuil observes contemptuously to Valmont that he has been pursuing the Présidente for a fortnight without tangible result, and her scorn has greatly intensified by the time he has triumphed over his victim's virtue—a process which has taken him over three months. The true explanation for behaviour so untypical of the libertine, she repeatedly declares, is that Valmont is in love. The evidence against him is too damning to be overthrown by his single denial;[69] for he had admitted in his first letter of the book that 'J'ai bien besoin d'avoir cette femme, pour me sauver du ridicule d'en être amoureux.'[70]

Time is not given to Valmont to prove or disprove Mme de Rosemonde's contention (which she herself admits not to be universally applicable) that man is inherently more flighty than woman.[71] According to the old lady, as we have seen, proof lies in the fact that a distinction between infidelity and inconstancy is drawn only in the case of men. In eighteenth-century parlance, the lover whose ardour for his mistress had cooled, but who had not yet attached

---

[65] p. 66. [66] p. 29. [67] p. 67. [68] p. 78.
[69] p. 320. [70] p. 18. [71] p. 304.

himself to another woman, was an 'inconstant', whereas the man who took a new mistress was an 'infidèle'. Perhaps surprisingly, however, the 'galant' tradition was more tolerant of 'infidélité' than of 'inconstance', for the latter derived from the very nature of the lover, and was therefore an irremediable weakness, while 'infidélité' was simply a *de facto* betrayal that did not exclude faithfulness of the heart.

Had tradition extended the application of this distinction to women, it would perhaps have decreed that Prévost's Manon was guilty only of the lesser crime in her relations with men other than des Grieux. Her tenderness towards him appears genuine, and its physical expression is interrupted only by the call of material pleasure which he is unable to provide. She herself regards faithfulness of the heart as supremely important, and faithfulness of the body as relatively insignificant. The view that Manon's unreliability is dispositional, an ingrained trait rather than a chance accretion, and thus a reprehensible example of inconstancy rather than infidelity, might be countered with the evidence of her reformed behaviour when the couple settle in New Orleans, and past social temptations are absent. On the latter interpretation, Manon possesses a fundamental stability of character.

Mme de Rosemonde's doctrine of male weakness is undeniably a discouraging one. She will allow no female to seek refuge in a distinction which she clearly regards as sophistical, save 'ces femmes dépravées qui . . . font la honte [de notre sexe], et à qui tout moyen paraît bon, qu'elles espèrent pouvoir les sauver du sentiment pénible de leur bassesse'.[72] She attempts to disabuse Mme de Tourvel of her chimerical belief in a perfect happiness attainable through love. That this belief is chimerical is not necessarily borne out by Valmont's desertion of the Présidente shortly after the writing of her cautionary letter; for his inconstancy appears the product less of his nature than of contingency, namely the pique he feels at the Marquise's taunts about his romantic fixation on the Présidente. It is far from evident that his devotion to Mme de Tourvel has lessened.

If this account is accepted, the disintegration of the little world Laclos depicts appears less a product of the destabilizing male influence over human emotion than of the application of falsifying arts, emanating from the female as well as the male, to a province

---

[72] Ibid.

of life where perverted artistry can ultimately exercise no power. The 'science' developed by the libertine, which has been the undoing of his victims, cannot save him from the pain of conflict—with his artful peers—or the disorder of passional weakness. All Laclos's characters are, in their different ways, vulnerable. The 'gens du bon ton' clad themselves in a kind of armour in an effort to lessen such vulnerability—the brittle armour of social etiquette, which prohibits deep attachment to another person and prevents impulsive movement. It is not proof against insidious assault, as Versac is to teach the apprentice Meilcour, but it affords protection against straightforward frontal attack.

The goal which true lovers seek is that of reliable stability, safe from the depredations of fortune or the unsettledness contingent on immaturity of character. Until Nemesis proves otherwise, des Grieux thinks that he has found refuge from the chance-ridden society which he and Manon left behind in France, where fire destroyed belongings, servants stole money, lovers tempted Manon away, and nothing was fixedly possessed. Tanzaï and Néardané, and Acajou and Zirphile, are happier: they eventually acquire freedom from the ungovernable and unpredictable forces of enchantment. As we have seen, 'libertin' philosophy excludes the idea of peaceful duration in erotic relationships as running counter to the principle of intensity, and rejects repose because stability cannot provide excitement. 'Egarements' signify deviation from what is settled, a departure either from the moral rule-book or from the 'route du bonheur'. That they may also connote the loss of one's true self in the 'branloire pérenne' of worldly intercourse is suggested in Crébillon's preface to his novel of the same name.

At the beginning of the eighth book of *La Vie de Marianne* Marivaux, accused by all of Paris for having made Valville unfaithful, contrasts the 'aventures d'imagination' which occur in novels with the true 'instabilité des choses humaines' which the memoir-novel must faithfully record.[73] To have shown a Valville who was unvarying in his devotion to Marianne would have been to enter the world of fiction; faithfulness is chimerical not because of an ingrained tendency in the male character (as Mme de Rosemonde would have it), but because the human condition is to be unstable. Valville's fault, if fault it be, is extreme susceptibility (most often seen as a feature

[73] p. 376.

of youthful temperaments) and impressionability to the passing instant. He is like a fruit that rots quickly because it has ripened too early. Patience, or the ability to wait for things to mature naturally, is not in general a virtue of youth, and the person hungry for new experience, like Valville or Meilcour, is unlikely to possess it. Their personalities will more probably be marked by 'inquiétude', or a kind of spiritual fidgeting. They will find time long when it is short, and tire of engagements that do not progress rapidly. The conscientious instructor will therefore be called upon to teach tranquillity, the only climate in which settled emotion can grow. In *Le Cabinet du philosophe* (1733) Marivaux contrasted the balanced man whose sentiments rarely vary, and who is constant in love, with the 'cœurs ardents et sensibles' which 'ne se donnent pas le temps de faire un fonds' and 'dissipent presque tout leur amour à mesure qu'il vient'.[74] Inconstant hearts, he concludes, are in reality merely childish ones: the implication (which remains only an implication in the unfinished *Vie de Marianne*) is that Valville may return to Marianne when he has grown to emotional maturity.

The word 'tranquille' has three senses in *Les Egarements du cœur et de l'esprit*. The first describes the state of one who has not been roused into love, and Mme de Lursay uses it thus of herself when contrasting her placidity and 'repos' with Meilcour's imagined condition of lovesickness.[75] Meilcour notes of the same occasion that he was 'surprised' but not 'touched' by the situation in which he found himself, being preoccupied by his new feelings for Hortense. Had Mme de Lursay been aware of them, he reflects, she would have stirred herself more in order to press home her advantage. Then, faced with her availability against Hortense's distant reserve, he would doubtless have opted for the 'amusement tranquille' the older woman promised, declining the laborious task of inspiring love in a person who, at least initially, would certainly have rebuffed his advances. According to the account Meilcour gives here, one can be tranquil when one's emotions are engaged (though surely not in the passionate state which Meilcour believes to be excluded in his relations with Mme de Lursay), but also seek to remain disengaged for the sake of continuing to be 'tranquille'. Violent emotion, at all events, is deemed incompatible with tranquillity. It

[74] *Journaux et œuvres diverses*, ed. Frédéric Deloffre and Michel Gilot (Paris, 1969), pp. 342–3. [75] p. 44.

will almost certainly entail a sense of 'inquiétude', of physical and moral disequilibrium, and for those who regard the attaining of inner peace as the true goal of human striving it is to be shunned. People who have known both states, however, may ultimately choose intensity rather than settledness: thus Julie de Wolmar, who confesses that 'le bonheur m'ennuie.'[76]

The third sense of 'tranquille' in Crébillon's novel is one of opposition to the 'égaré': to enjoy tranquillity is to remain constant, not deviating from one's prime duty of fidelity. It has more strongly moral connotations than the other two senses, and is applied by Meilcour on the last page of his narrative to the state of untroubled stability and purpose which fidelity to the idea of Hortense—at present it is no more than an idea—would have brought with it. As such it is the antithesis of the guilt-ridden condition into which he has plunged by starting a physical relationship with Mme de Lursay. Meilcour does not point out as he mentions the 'criminal' implications (with respect to his moral conscience) of his possessing Mme de Lursay that it has produced in him the reverse of the tranquillity he earlier believed it would bring. There has been a complete about-turn: Hortense, the 'princesse lointaine', now represents calm and balance, Mme de Lursay 'égarement'. But he is at least aware of the contradiction that has entered his life:

Dérobé aux plaisirs par les remords, arraché aux remords par les plaisirs, je ne pouvais pas être sûr un moment de moi-même . . . quelquefois je me justifiais mon procédé, et je ne concevais point comment j'avais pu manquer à Hortense, puisqu'elle ne m'aimait pas . . . Je persuadais assez facilement à mon esprit que ce raisonnement était juste; mais je ne pouvais pas de même tromper mon cœur. Accablé des reproches secrets qu'il me faisait, et ne pouvant en triompher, j'essayai de m'en distraire, et de perdre dans de nouveaux égarements un souvenir importun qui m'occupait malgré moi. Ce fut en vain que je le tentai, et chaque instant me rendait plus criminel sans que je m'en trouvasse plus tranquille.[77]

In a state of contradiction there can be no peace. At the point at which we leave Meilcour, he has clearly achieved no such enlightenment as his sentimental education promised to bring.

That 'inquiétude' is often a concomitant of worldliness is a lesson which emerges from other novels of the period. In *Emile et Sophie*

---

[76] p. 682.   [77] p. 188.

it is the characteristic state of one who, like Emile, abandons fidelity to the old ideals inculcated by his moral mentor and dissipates himself in 'galanterie'.[78] Paradoxically, and surely with ironic intent, Restif has Gaudet d'Arras propose to Edmond in *Le Paysan perverti* that to devote himself to Mme Parangon, the paragon of virtue, would entail the 'agitation violente' characteristic of 'amour-passion', to which repose of the heart is preferable.[79] Yet as we later see, it is Mme Parangon who represents stability in Edmond's world, and the relationships in which he engages under Gaudet's tutelage that signal his moral breakdown. The latter are the 'égarements' which ultimately destroy Ursule and Edmond himself. At the end of their lives, however, brother and sister do attain a state of repose. Ursule reforms her ways and is welcomed back to her father's house as the prodigal daughter, before Edmond, under a misapprehension about her activities, murders her on her return from a charitable mission. Edmond, now one-armed and one-eyed, makes benefactions to his parish church and marries Mme de Parangon, so seeming to settle his moral being.

In des Grieux's case, as ultimately in that of Mme de Tourvel, emotional tranquillity is subordinated to another kind of imperative. For des Grieux, existence is ruled by a necessity that stands outside himself. The necessity is wholly compelling; it may lead to destruction, but the power of its attraction means that it is obeyed without regret. The same is true for the Présidente de Tourvel, once she has realized that she is necessary to Valmont's happiness. That value becomes an absolute against which her old values are measured and found to be wanting. By giving herself to Valmont she renders herself vulnerable to chance, and is ultimately its victim; but the call to completion—she comes to believe that she and Valmont are each other's complement, the severed parts of the androgyne described by Aristophanes in the *Symposium*—cannot be resisted.[80]

Love and merit are not necessarily matched to one another. A person may be deserving but unloved, or undeserving but loved with passion. In the Christian tradition there is a kind of love that is indifferent to merit and value. *Les Liaisons dangereuses* depicts the devout

---

[78] p. 886.
[79] *Le Paysan perverti*, ed. François Jost, 2 vols (Lausanne, 1977), ii. 134.
[80] p. 308.

Présidente de Tourvel's struggle to accommodate her awareness of biblical teaching on this subject to her own sense that she has deserved better of God than his treatment of her would suggest. As part of his strategy for seducing her, Valmont has let it be known that he repents of his libertine ways and is returning to the path of godliness. The Présidente's entirely human feeling of pique at the ease with which he has apparently abandoned her mingles with unchristian indignation at the fact that Valmont has been granted a grace denied her—morally the more deserving party—in her efforts to forget him. In vain she reminds herself of God's charity towards sinners, enshrined in the parable of the Prodigal Son:

Je sais qu'il ne m'appartient pas de sonder les décrets de Dieu: mais tandis que je lui demande sans cesse, et toujours vainement, la force de vaincre mon malheureux amour, il la prodigue à celui qui ne la lui demandait pas, et me laisse, sans secours, entièrement livrée à ma faiblesse.[81]

Agape is bestowed where it pleases the giver to bestow it; God's love is uncaused and indiscriminate, and may as freely be extended to the sinful as to the good. Perhaps this is essentially the same as the tolerant love of a des Grieux for his Manon, with the distinction that erotic love does demand specific elements that are present in some people—the beloved—and not in others. It therefore lacks the complete randomness characteristic of agape.

What these elements consist of is the implied subject of many an eighteenth-century novel. The interest of the age in the concepts of sympathy, attraction, and the elusive 'je ne sais quoi' testifies to its abiding preoccupation with those aspects of love which cannot readily be explained in empirical terms. Antiquity had evolved a theory of cosmic sympathy which, building on an insight into the natural interrelationship of things, deduced the doctrine of their mutual interaction; and eighteenth-century fiction continues to investigate the phenomenon of metaphysical attraction. In *La Nouvelle Héloïse* Julie develops the notion that two souls may have an immediate sympathy for each other independently of body or senses,[82] to which Rousseau appends a footnote mocking Richardson for denying people's instantaneous attraction on the basis of 'conformités indéfinissables'. It need not be added that this type of union differs fundamentally from the anatomical one imagined

[81] p. 284.   [82] p. 330.

by Diderot in the chapter 'Des voyageurs' of *Les Bijoux indiscrets*. It is a conjunction of spirits which closely resembles that described by Marivaux in *La Vie de Marianne*, when the heroine meets Mme de Miran, Valville's mother, in church.[83] The contrast between the encounters detailed by Rousseau and Marivaux respectively is seemingly that of love or 'amitié amoureuse' on the one hand and friendship on the other. However, Marianne's subsequent declaration to Mme de Miran that the latter means more to her than does Valville[84] somewhat blurs the distinction.

The elusive 'je ne sais quoi' is related to this quasi-mystical sympathy. Despite being frequently mentioned in eighteenth-century literature, it was far from being an invention of that period. Aristotle's 'tí esti' expresses the indefinable essence which the French phrase attempts to capture, as do the Augustinian 'nescio quid', the 'aliquid in anima' of the scholastics, and the mystics' 'scintilla animae'.[85] Nor does the concept necessarily refer to matters involving emotion: Diderot, to take just one example, uses it in the *Salons* to help explain the magical, transforming quality of Chardin's still-life paintings, which manage to invest common objects with an inexpressible heightened significance. But most usually the 'je ne sais quoi' is linked with the ability to arouse feeling, often of an erotic kind. Given the favour enjoyed by the 'je ne sais quoi' in French literature from the Renaissance to Romanticism, it may seem surprising that Prévost fails to use the term to describe the ineffable quality of Manon's charms, and which, if we are to believe her lover, des Grieux is powerless to resist. Yet the story of the couple's life together fully bears out the observation of Pascal's *Pensées* that

Qui voudra connaître à plein la vanité de l'homme n'a qu'à considérer les causes et les effets de l'amour. La cause en est un 'je ne sais quoi' (Corneille), et les effets en sont effroyables. Ce 'je ne sais quoi', si peu de chose qu'on ne saurait le reconnaître, remue toute la terre, les princes, les armes, le monde entier.[86]

In Bouhours's *Entretiens d'Ariste* (1671) the 'je ne sais quoi' appears as a force that causes sympathy, love, and all human

---

[83] p. 147. [84] p. 343.

[85] Erich Köhler, '*Je ne sais quoi*: Ein Kapitel aus der Begriffsgeschichte des Unbegreiflichen', *Romanistisches Jahrbuch*, 6 (1954).

[86] *Pensées*, ed. Louis Lafuma, 3 vols (Paris, 1952), i. 232 (number 413). The reference is to Corneille's *Rodogune*, I. v.

relationships. Ariste defines it as a secret inclination which makes one individual feel for another what he feels for no one else. His interlocutor, Eugène, attempts another definition which both links and contrasts this essence with the quality des Grieux is to perceive in Manon: for Eugène the 'je ne sais quoi' has strong religious connotations, being an impression of the spirit of God, a 'plaisir victorieux', a 'sainte concupiscence'.[87] For des Grieux, however, Manon's charms are of a pagan nature, calculated to bring the whole world back to idolatry. There could be no greater opposition between Eugène's Christian concupiscence and the 'victorious' or supervening state of (earthly) delectation experienced by des Grieux at Saint-Sulpice, when Manon comes to visit him and their affair resumes its course. In the later eighteenth century, with discoveries about magnetism and the popular enthusiasm for Mesmer,[88] such attraction was sometimes described in more prosaic physical terms. There is little reflection of this shift in literature, although Louvet's *Six Semaines de la vie de Faublas* provides one non-serious example: Faublas falls into the hands of Mesmerists (whose business proves to be run by his father's former mistress Coralie) and finds himself the object of their erotic experiments in 'animal magnetism'.[89]

However different in other respect, the ineffable qualities with which both Bouhours and Prévost are concerned have a shared psychic component at their heart. Despite some indications to the contrary, it is not perverse to speak of a 'metaphysics' of emotion in connection with *Manon Lescaut*, for emphasis on love of the heart is at least as strong as emphasis on love of the flesh. Manon herself, it will be recalled, tells her lover that the fidelity she wants from him is 'celle du cœur'. When she says to him that 'dans l'état où nous sommes réduits, c'est une sotte vertu que la fidélité',[90] she means that bodily constancy is a luxury that must be dispensed with when their pressing need for money demands a resumption of her prostitution to wealthy men. Though the ease with which Manon separates carnal from sentimental love is a source of pain and incomprehension to des Grieux, there are many indications that she does feel deep devotion to her lover. After their reunion at

---

[87] [Bouhours], *Les Entretiens d'Ariste et d'Eugène* (Amsterdam, 1671), p. 276.

[88] See Robert Darnton, *Mesmerism and the End of the Enlightenment in France* (Cambridge, Mass., 1968).

[89] *Les Amours du Chevalier de Faublas*, in *Romanciers du XVIIIᵉ siècle*, ii. 777 ff.

[90] p. 69.

Saint-Sulpice she informs him with apparent sincerity that even amidst the material pleasures procured for her by M. de B . . . she missed the true love he alone could provide her with. Her need for heartfelt sentiment and union, it would appear, is at least equal to the desire for physical satisfaction which so early mars their relationship. On the evidence Prévost provides us with, it is the psychic phenomenon of love that causes des Grieux to see it as the supreme good. Emphasizing the spiritual element in his union with Manon also allows des Grieux to enlarge the significance of that union. Physical impulses are aggrandized by being attributed to overwhelming abstract forces such as the 'ascendant de [la] destinée', the 'volonté du Ciel',[91] and the indescribable, spellbinding 'charme' (of Manon's being) which des Grieux so readily invokes.

The question whether he has in any sense chosen his emotion with respect to Manon is clearly a critical one in a novel that purports to tell an exemplary tale. To escape the imputation of culpable responsibility for events (and also to magnify his narrative), he repeatedly intimates that an outside force prevented his free action; but when he suggests to Tiberge during the latter's visit to him at Saint-Lazare that he lacked divine grace to act virtuously, his Jesuit friend, shocked, rejects his claim as Jansenist special pleading.[92] The idea of intentionality features explicitly in the novel at one other point, when des Grieux shoots and kills a warder during his escape from prison.[93] He is not guilty of any crime, he declares, because he did not know that his gun was loaded—a fact which, if true, would prompt the impartial observer to ask why he then thought it worth pulling the trigger. Yet the overriding issue of intentionality— that is, des Grieux's freedom or otherwise to love or not to love Manon—should perhaps be reformulated. What really matters, it may be felt, is not the question whether his passion was freely chosen, but the fact of his consent to it. As we have seen, he denies that his will was involved in the matter ('Tout ce qu'on dit de la liberté à Saint-Sulpice est une chimère'),[94] so denying the theological principle that the will cannot be forced to consent to fleshly pleasure because the gift of grace enables man to follow the way of the spirit.[95] But more important, surely, is the manner of his reaction to the state of being in love. General moral rather than theological

---

[91] p. 20.　　[92] p. 93.　　[93] p. 97.
[94] p. 46.　　[95] See Levi, op. cit. 209.

attitudes are probably of greatest moment to Prévost; in comparison with them, the hair-splitting of dogma appears little more than a metaphysical obfuscation. If des Grieux 'refuse d'être heureux', as the *avis* of the Homme de Qualité has it, his refusal can most valuably be discussed in terms of those areas where free will undoubtedly has space to assert itself—in the patching-together of life after such shattering blows as the *coup de foudre* and more material assaults, rather than in those murky regions whence the thunderbolts of love are often thought to issue. When judged by these more prosaic standards, the des Grieux who slips into fraud and homicide for the sake of keeping an unreliable mistress by him may well be found wanting.

It is not hard to find literary support both for and against des Grieux's view of passion (or, as he would probably prefer it, *his* passion, which he regards as different from most other people's in virtue of his refined perceptions, the natural delicacy which he believes to stem from his advantages of birth and upbringing).[96] The eponymous hero of Diderot's *Jacques le fataliste*, to take just one illustration, asks his master the question 'Est-ce qu'on est maître de devenir ou de ne pas devenir amoureux? Et quand on l'est, est-on maître d'agir comme si on ne l'était pas?',[97] to which he evidently expects the answer 'no'. But the determinist theory Jacques espouses was contradicted by the philosopher who exercised considerable influence on Diderot at the time he wrote the novel, Spinoza. Spinoza distinguished between 'actions' and 'passions', the former demonstrating man's freedom and the latter his passivity, his being driven by impulses of which he was not aware. Love, according to Spinoza, was an action, the practice of a human power which could be effected only in freedom and not under compulsion.

The belief that humans elect to love is unambiguously confirmed in Crébillon's *Les Egarements du coeur et de l'esprit*, whose hero declares in tones which Constant's Adolphe will later echo that 'A force de me persuader que j'étais l'homme du monde le plus amoureux, je sentais tous les mouvements d'une passion avec autant de violence que si je les eusse éprouvés.'[98] His involvement—a vacillating one—with Mme de Lursay bears all the marks of this act of volition, and there are strong suggestions that Meilcour has made

---

[96] p. 81.   [97] *Jacques le fataliste, Œuvres romanesques*, p. 498.
[98] p. 42

a 'wrong' choice, from the point of view of temperament if not of morals. But his later *coup de foudre* for Hortense seems unplanned, and the lack of forethought which attends it perhaps gives it a more positive ethical value than his manufactured feelings for Mme de Lursay.

Laclos's Valmont, at least to some readers, seems a clear instance of one who initially wills an involvement, only subsequently to find himself involuntarily gripped by it. Does the latter state make him more, or less, responsible for the act by which he effectively kills the Présidente de Tourvel? To what extent is he forced into action by a higher power, in the shape of Mme de Merteuil (and what value has the word 'forced' here)? Is the Présidente freed from some of the guilt that attaches to adultery in a Christian society because she puts up such a long fight against Valmont? Does the intensity of his campaign exonerate her?

Les Liaisons dangereuses obliges its reader to consider the different kinds of causation that lie behind actions. In this it resembles *Manon Lescaut*, whose author seems finally unprepared to apportion responsibility for events to any one impulse, whether cosmic or earthly, material or immaterial. Literature has often turned such uncertainty to account. We do not know whether Béroul's Tristan and Iseut continue to love each other after three years because of the potion they drank or not, or whether the 'Vénus tout entière à sa proie attachée' signifies Phèdre's subjection to the immutable will of the gods or just the intense force of merely physical passion within her. Is the goddess of love metaphor or reality, spirit or substance? It may be that such unclarity is somehow connected with our view of these works as examples of great literature. Were we able to explain the motivation of their protagonists solely in terms of Kant's abhorred 'empirical self', interpreting them as products of a biological constitution or a neurological structure, an important element in our appreciation of those characters and thus of the works in which they figure might be removed. The notion that physical explanations are the only valid explanations of emotional conduct, because only they deal in unquestionable reality, is reductive: it goes against the imaginative truth which many readers wish to ascribe to literature and its creations. As far as feeling is concerned, physicalism seems to represent a capacity for bodily pleasure or sensation rather than for the less tangible emotion of love, and in this sense to remove humanity from the persons of whom it is predicated. The search for ideal values that marks des Grieux, Julie and Saint-Preux, the Présidente de Tourvel and the seemingly regenerate Valmont, is lessened if it is redirected at the merely empirical.

The view that love is dehumanized when stripped to its material being lies behind Diderot's statement in a letter to Mme de Meaux that their relationship is founded on more than the stimulus–response mechanism of physical organisms.[99] Although in the *Lettre sur les sourds et muets* Diderot dismisses metaphysics as unreliable and introspection as impossible,[100] and although he held that biology alone could lead to a better understanding of man, the metaphysical 'chimère' of union with Sophie Volland which he describes to her gives a more accurate measure of Diderot's thoughts concerning love.[101] There may be 'un peu de testicule au fond de nos sentiments les plus sublimes et de notre tendresse la plus épurée', as he wrote to Damilaville,[102] but Diderot was aware that a crucial sense of human individuality is removed when the universe and its inhabitants are explained away in purely material terms. In his *Etudes de la nature* Bernardin de Saint-Pierre was later to observe that in the new world imagined by philosophers a lover trying to inspire affection in his beloved would have to discourse on magnetic attraction, fermentation, electricity, and other determining physical causes.[103]

The dignity of man, which is bound up with the notion that he possesses at least an element of free will and an attendant concept of moral responsibility, makes it possible for us to feel sympathetically with him, and thus to take the imagined lives of literary creations as having human meaning. Individuality, what is intrinsic to a specific human nature and makes one person uniquely himself or herself, is central to the phenomenon of love. The other person is loved as a particular, the 'other half' described by Aristophanes in the *Symposium*. It is perhaps possible to regard physical bodies as qualitatively interchangeable, given the fairly common experience of sexual promiscuity or sexual desire that is not directed at a specific person; but this view excludes love. Failure to respect an individual's interests, which is the crime of the protagonists in *Les Liaisons*

---

[99] *Correspondance*, ix. 154.

[100] *Lettre sur les sourds et muets*, ed. Paul Hugo Meyer, *Diderot Studies*, 7 (1965), p. 95.

[101] *Correspondance*, ii. 283.   [102] Ibid. iii. 216.

[103] *Etudes de la nature*, *Œuvres complètes*, ed. L. Aimé-Martin, 18 vols (Paris, 1820), vi. 147.

dangereuses, is to treat him mechanistically; and machines can only in a trivial sense be found lovable. Seeing the individual as the sovereign good, on the other hand, entails a care for him that excludes the possibility of subordinating him to one's own will, using him as one would a mechanical contraption. The limitations of the materialist interpretation of love are summed up in that formulation.

In his *Essais sur la peinture* Diderot makes the metaphysical attraction between two beings, to which he and his contemporaries gave the name of sympathy, into an ethical phenomenon:

> qu'est-ce que la sympathie? j'entends cette impulsion prompte, subtile, irréfléchie, qui presse et colle deux êtres l'un à l'autre, à la première vue, au premier coup, à la première rencontre; car la sympathie, même en ce sens, n'est point une chimère. C'est l'attrait momentané et réciproque de quelque vertu. De la beauté naît l'admiration; de l'admiration, l'estime, le désir de posséder, et l'amour.[104]

This recalls a letter to Sophie Volland in which Diderot stresses the need for him and his beloved to strive ever harder to deserve each other through developing their moral characters.[105] But in so doing it points to the link between a metaphysics of love and the notion of love as necessarily involving will. Virtue stems from the human act of choice; it is not a supernatural gift, which is why Rousseau and the lovers of *La Nouvelle Héloïse* found it hard to practise. If it plays a central role in the act of loving, as Diderot argues (even in the pornographic *Bijoux indiscrets*), then that act transcends the routine physiological process described in his first novel and becomes a union which presupposes more than the coupling of material bodies.

But in so far as such love involves the will (as did the idealistic *amour courtois* of the Middle Ages), it is also elevated above the merely passional. Sentimental love may be divorced from volition, being a flow of emotion before which the lover is passive; but the intervention of reason gives completeness to the emotion. Thus Lucretius and the Epicureans argued, against Plato and the Stoics, that passion and reason belonged together. The introduction of reason to des Grieux and Manon's love (albeit too late to save Manon) is what lends stature to the emotion presiding over their

---

[104] *Essais sur la peinture*, *Œuvres esthétiques*, ed. Paul Vernière (Paris, 1968), p. 700.
[105] *Correspondance*, iii. 99–100.

new life in Louisiana. Des Grieux's description of his earlier state of love, for all his efforts to dignify it by imputing it to cosmic forces, suggested that its occurrence was despite himself, in a way that split his very personality. In New Orleans the rift is healed. It emerges from the last pages of Prévost's novel that there is no necessary conflict between what is dictated by passion and what by the mind. The same resolution, reached from the opposite direction, marks the culmination of Valmont and the Présidente's love in *Les Liaisons dangereuses*. In both cases the consequence is an ethical enhancement of love which stems from a recognition of man's dual nature as a creature of both biological and psychic impulse, whose actions are not merely the consequence of natural necessity.

3

Medicine, 'Mediocritas', and the Pathology of Passion

IN THE last pages of *La Fin des amours de Faublas*, Louvet describes
the mental collapse of his swashbuckling hero. Blaming himself for
the death of the Marquise de B***, his former mistress, and that
of her successor the Comtesse de Lignolle, Faublas goes mad with
grief and is committed to a lunatic asylum. When his father arrives to
remove him to the family home, he finds his son chained up, bruised,
bloody-faced, and in a state of delirium. Willis, an English doctor
famous for treating the insane, is summoned, and Faublas is locked
inside a padded room in the parental house under the constant
surveillance of six strong men. On Willis's recommendation the
Baron de Faublas takes a house outside Paris with a vast landscaped
garden in the English style, hoping that the tranquil setting and fresh
air will have a beneficial effect on his son. The treatment gradually
works: Faublas slowly progresses from frenzy to melancholy,
recovers his memory, and finds solace in the tender forgiveness of
his wife Sophie and the writing of his life history. The family
emigrates to Poland, where Faublas settles into a state of comparative
domestic felicity. The novel concludes on a homiletic note, advising
its reader that 'pour les hommes ardents et sensibles, abandonnés
dans leur première jeunesse aux orages des passions, il n'y a plus
jamais de parfait bonheur sur la terre.'[1]

Among the examples of lovesickness furnished by eighteenth-
century literature, that of Faublas is extreme. He has lived life to the
hilt, and has suffered in corresponding measure. He becomes, like
Prévost's des Grieux, 'un exemple terrible de la force des passions'.
Another such is the lesbian Mother Superior of *La Religieuse*, whose
physical sickness and mental 'égarement' are described by Diderot
as the result of frustrated love—a very different case from that of
the philandering Faublas. Few other novels of the age approach these

[1] p. 1222.

two in the melodramatic intensity with which they depict the disease
of passion and its consequences. But it is not uncommon to find in
fiction a consideration of the manner in which an individual's physical
health might depend on his moral being, and the other way about.
The *Liebestod* features prominently in the literature of sensibility,
and is often shown as consequent on the kind of illness that would
nowadays be called psychosomatic. The eighteenth century's efforts
to found a science of man, as well as its consciousness of ushering
in a new age in the history of medicine, led to a more profound
discussion in imaginative literature of the clinical symptoms associated
with states of emotion than had been possible before. Love might
be regarded as threatening the individual's state of physical health
or, conversely, as a universal panacea; as a disease or as a force which
enhanced the body's well-being. Health, equally, might be seen as
issuing from habits of erotic moderation or as promoted by the
cultivation of intense emotion; passion could appear either as
destructive of mental and physical equilibrium (particularly if the
lover was a woman) or as a force which enabled its subject to rise
above the dull routine of the everyday.

Writers of the period could draw on an ancient tradition of
portraying social states and, by extension, states of moral being in
pathological terms. The familiar notion that certain forms of human
society are 'sick' and others 'healthy', implicit in Rousseau's diagnosis
of the civilized world, can be found in the philosophy and literature
of antiquity: the extended comparison drawn in Plato's *Republic*
between the health of the soul and the health of the political constitution is perhaps the best known of such analogies. In *La Nouvelle
Héloïse* Rousseau shows how, with a mixture of goodwill and
deliberate self-restraint, men can create a society that functions as
efficiently as a human body which has not been abused by the
immoderate desires most humans are prey to. The miniature state
of Clarens, with its willed simplicity, temperance, and moral hygiene,
mirrors Rousseau's own ideal of self-denying happiness, even though
the pure contentment it procures for its members ultimately proves
too insipid for the passionate Julie.

In *La Religieuse* we are shown the denaturing effect of certain
institutions of the Catholic Church on its adherents, driven by
its constraints into a state of sickness. According to Diderot, the
convent induces aberrance in both the beliefs and the affective lives
of its inhabitants by removing from them the essential freedom of

self-determination. The deprivation entailed by its structure, he argues, leads to physical irregularities (not merely sexual ones, although Diderot's focus on lesbianism inevitably acquires great prominence in the novel) and moral turpitude.

Another environment described by the eighteenth-century novel as fostering a sickness which called for extirpation was that of *mondanité*. Laclos deals with it in *Les Liaisons dangereuses*. The notions of health, moderation, and destructive excess permeate the book, with the author blaming the hyper-civilized and idle world of the aristocracy for the wanton perversion of moral values and human lives depicted in his novel. Sex and love appear as threats to most of those seen indulging in them, representing either a form of social weakness to be exploited by individuals whose intellects control their own and others' senses, or a force inimical to the moral order of existence implied by the traditional Christian code of marital fidelity. They are threats precisely because they cause the people they afflict to go beyond the bounds set by a prudent concern for self-restraint and moral repose. If they momentarily appear to offer a heightened state of existence, their power ultimately works against itself and turns into a canker.

In another sense, too, Laclos's novel invites interpretation in the medico-ethical terms not just of its own age, but of earlier and subsequent ones as well. Without deciding for the moment whether it is the immoral tract that Church and State condemned or the 'livre de moraliste' Baudelaire declared it to be,[2] one may observe in the feelings it arouses something close to the effect which ancient theory had declared as desirable in tragic drama—catharsis, or the clarification of passions by their arousal, to which Aristotle gave a part-medical, part-aesthetic value in the *Poetics*.[3] Whether the outcome of *Les Liaisons dangereuses* evokes a predominant sense of tragic pity in the reader on account of the worthwhile lives it shows being destroyed, or of fascinated solidarity with the wicked but compelling protagonists Valmont and Mme de Merteuil, it stirs the emotions in a way many other eighteenth-century novels no longer can.

---

[2] 'Notes sur *Les Liaisons dangereuses*', *Œuvres complètes*, ed. Claude Pichois, 2 vols (Paris, 1975–6), ii. 68.

[3] On the idea that *kátharsis* means 'clarification' rather than 'purgation', see Martha Nussbaum, *The Fragility of Goodness: Luck and Ethics in Greek Tragedy and Philosophy* (Cambridge, 1986), pp. 389 ff.

The notion that love is a disease of the soul with leisure, which Laclos's book may prompt in the reader's mind, is also confirmed by *La Nouvelle Héloïse*. In this novel idleness among the workers on the Clarens estate is expressly associated with the arousal of immoderate passions, and its danger averted by the Wolmars' determination to fill every moment of their time usefully. That idleness promoted indulgence in erotic love was firmly stated by writers on medicine in seventeenth- and eighteenth-century France as well as by non-professional commentators. In his *Histoire de la médecine* of 1696, Daniel le Clerc drew attention, like many of his predecessors, to the fact that only the leisured classes could afford the luxury of concentrating on their bodily states, including those involved in the enjoyment of erotic love; sickness has never been an interest, although it has necessarily been a preoccupation, of those who have to work hard for their living.[4] With remarks like le Clerc's in mind, thinkers of the Enlightenment who aspired to develop a science of man therefore concentrated their attention on the individual's formative background as well as on the more or less unchanging functions of his bodily machine.

The idea that the healthiest state for humans is one of freedom from passion, including the passion of love, extends back in time at least to the Stoic philosophers. Seeing passion as destructive in the way that disease is destructive has long been a literary commonplace. In the French tradition it is impossible not to think of seventeenth-century illustrations—the many sufferers in Mme de Lafayette's fiction, for instance, or Racine's tragic heroes and heroines, contrasting with the strong-willed characters of Corneille who successfully maintain their hold on themselves despite the promptings of emotion. But eighteenth-century literature is different from seventeenth in the attention it pays to the detail of passionate arousal and the latter's bearing on the ill health or well-being of the individual.

Ancient philosophers, again, had examined this matter closely. In the *Symposium* the doctor Eryximachus declares that the human body contains two loves, different from one another and having contrasting desires. The good and healthy elements should be indulged, but the bad ones—including immoderate desire—discouraged. This is the job of the physician; for medicine amounts to a knowledge of the body's proclivities, and the doctor's task to the separation of fair

---

[4] *Histoire de la médecine* (Geneva, 1696), pp. 89–90.

from foul, or the latter's conversion into the former.[5] Skilful practitioners know how to eradicate or implant love as required, and make opposing elements 'fall in love' with one another. Medicine is thus under the direction of eros.

Later ages considered that medicine was an art in which the Greeks had come as close as possible to complete knowledge; it had enabled them to understand with some exactness the processes of nature which later preoccupied the eighteenth century. The kind of scrutiny to which the historian Thucydides subjected plague victims was later directed by Diderot and like-minded writers at the victims of claustration. Laclos effects something similar in his description of the Présidente de Tourvel's delirium before her death, and Louvet in the detailing of Faublas's madness.

It was commonly assumed in the eighteenth century that while a body in an unhealthy state called for medical attention, and an unhealthy soul for the attention of a moralist, both states influenced one another. Doctors were therefore encouraged to study the passions in order more effectively to treat the physical ailments of their patients.[6] Some of them also saw it as their duty to issue warnings against the bookishness whose early effects on himself Rousseau describes in the *Confessions*, but less because of the consequences he mentions—a premature acquaintance with the passions, gleaned from the reading of novels—than on account of the adverse physical results of prolonged bodily inactivity.[7]

The physician Roussel voices particular concern about protracted bouts of study by females, partly for the reasons invoked by Rousseau in *Emile* (that a woman knows enough when she has learnt how to look after a husband and family), but partly too because Roussel, in common with his age, holds that a woman's constitution is less robust than a man's, so that even sedentary and intellectual activity may have an undesirable physical outcome.[8] It is no surprise to discover that Mme de Merteuil, a critic of female debility, scorns

---

[5] See Anthony Kenny, 'Mental Health in Plato's *Republic*', *The Anatomy of the Soul* (Oxford, 1973), p. 3.

[6] See, for example, C. J. Tissot, *De l'influence des passions de l'âme dans les maladies, et des moyens d'en corriger les mauvais effets* (Paris, 1798).

[7] See e.g. Robert Burton, *The Anatomy of Melancholy*, ed. Holbrook Jackson (London, 1972), p. 242.

[8] *De la femme considérée au physique et au moral* (Paris, 1788), pp. 129 ff., 144 ff.

the notion that woman is naturally vulnerable to the male by virtue of her physical constitution. Women have made themselves subject to man's will out of laziness, she writes, gladly playing on their supposed organic susceptibility in order to procure for themselves the pleasure of luxuriating in sentiment. If woman has a less robust constitution than man, then she should, like the Marquise, compensate by cultivating the intellect that permits her still to make the male her plaything. According to Mme de Merteuil's own beliefs, it is nonsense to claim that either sex is superior in strength to the other; one type of force should simply be traded against another, and it is convention alone that has made the alleged bodily weakness of the female an argument for her general subordination to the male. The really determined woman can both have her cake and eat it. Like the Marquise in the Prévan episode, she can turn what is assumed to be physical vulnerability to tactical advantage, exploiting the conventional sensibility of her sex to ensure complete moral victory over the brute male aggressor. The fact that Prévan is a man of proven intellect as well as the possessor of undoubted physical attractions makes Mme de Merteuil's triumph over him all the sweeter.

Quite apart from believing a woman's 'machine' to be a frail construction, eighteenth-century doctors held the time-honoured view that females were more excitable and less stable than males, possibly because they were endowed with a womb. According to many commentators, the womb was the seat of sensibility, liable to react spasmodically and unsettlingly to states of tension and fatigue. Diderot, who contends in the *Encyclopédie* article 'Génie'[9] that uncontrolled sensibility debars humans from achieving great things in life, remarks in his *Paradoxe sur le comédien* that the average woman, a more 'sensible' creature than the male, is condemned for that reason to inferiority in the art of acting:

Voyez les femmes; elles nous surpassent certainement, et de fort loin, en sensibilité: quelle comparaison d'elles à nous dans les instants de la passion! Mais autant nous le leur cédons quand elles agissent, autant elles restent au-dessous de nous quand elles imitent. La sensibilité n'est jamais sans faiblesse d'organisation. La larme qui s'échappe de l'homme vraiment homme nous touche plus que tous les pleurs d'une femme.[10]

[9] In fact written by Saint-Lambert, but undoubtedly influenced by Diderot.
[10] *Paradoxe sur le comédien*, *Œuvres esthétiques*, p. 311.

Almost anything, it seemed, might over-stimulate the delicate female organism. When the women servants gather together at Clarens for a festive Saturday collation, we read in *La Nouvelle Héloïse*, they are forbidden anything which might over-excite them, particularly wine and the presence of the opposite sex. The dairy products and sweetmeats of which the meal in this 'gynaseum' consists are sufficient to their desires, and symbolize the innocence and sweetness of their kind.[11]

The classic illness of excitable females, of course, was hysteria. In his *Histoire naturelle de l'homme* Buffon describes hysteria as the gravest consequence of the impossibility, for the chaste of either sex, of expelling seminal fluid from the organism. Allowing the liquid to stagnate, he writes, can result in severe illness, or in irritations so violent that human reason and religious principle will barely suffice to contain them. He reports that girls of twelve are not too young to escape the effects of such seizures, although in their case the treatment should consist not in marrying them off, but in ensuring that they live only among members of their own sex.[12] But as we shall see, such measures seem not to have been successful in at least one of the convent communities depicted in *La Religieuse*.

According to some authorities, hysteria was a pathological state that could be cured by completely excluding the experience of solitude from the sufferer's day-to-day life. To be brought back to oneself, to regain self-possession, it was sufficient to find oneself a mate. Perhaps there is a relic of the androgyne myth in this form of therapy: the patient's sense of incompleteness causes a febrile 'inquiétude' which can only be cured by her finding her other half and being restored to her original completeness. 'Uterine fury' was regarded as a seizure in which the victim had slipped the moorings that held her body and soul together, a turmoil of the vital principles or—in the language of eighteenth-century medicine—of 'fibres' gone mad.[13]

In its section on madness as the indirect cause of illness Zimmermann's *Von der Erfahrung* describes uterine fury as often accompanying the

[11] pp. 451–2.

[12] *Histoire naturelle de l'homme*, *Œuvres complètes*, 25 vols (Paris, 1774–8), iv. 258–9; see also Ilza Veith, *Hysteria: The History of a Disease* (Chicago and London, 1965).

[13] See Paul Hoffmann, 'Le Discours médical sur les passions de l'amour de Boissier de Sauvages à Pinel', in *Aimer en France 1760–1860 (Actes du Colloque international de Clermont-Ferrand)*, ed. Paul Viallaneix and Jean Ehrard, 2 vols (Clermont-Ferrand, 1980), i. 350.

ecstasy of mystical love or religious passion in women.[14] The 'orgasm' (a word then used for convulsive seizures of an unspecific kind) experienced by women in this state, he writes, is continuous and long-lasting, and ends by draining the body of all its strength. (Significantly, Zimmermann also uses the word 'enthusiasm' for this condition.) [15] Although he principally blames constitutional weakness for inducing it in women, he mentions as well the part played by constricting clothing in bringing on hysteria, deep melancholy, and even apoplexy,[16] and thus joins the army of eighteenth-century writers who advocated loose garments on account of the medical benefits they could bring. But among the women most evidently subject to physical and mental frenzy in eighteenth-century France, nuns at least suffered no such sartorial constraint. Nor, presumably, were they in much danger from another popularly accepted cause of uterine fury, the reading of licentious novels.

The origin of the lesbian Mother Superior's seizures in *La Religieuse* apparently lies in the excessive inwardness and ostensible celibacy of life in the convent (a life which Guilleragues's Mariana, in *Lettres portugaises*, regarded as uniquely suited to the fostering of earthly as well as heavenly devotion in its subjects). That Diderot intended the convulsive states the Mother Superior experiences to be interpreted as orgasms in the modern sense of the word there can be no doubt. What is interesting in his detailing of these states is its similarity with descriptions of hysterical attacks in contemporary works on medicine. Perhaps Diderot was prudently mitigating the shock of direct portrayal by closely relating it to that of a common medical condition.

Such descriptions may be compared with the section on hysteria in the translation by Diderot, Eidous, and Toussaint of Robert James's *Medicinal Dictionary*, whose original version was published in 1743–5. In the French translation we read that the passion of hysteria has as its symptoms a tightening in the throat, a feeling of suffocation often followed by fainting, a loss of the power of speech, and a general sensation of 'assoupissement'. The sufferer feels in her stomach a kind of mass rolling about and rising up: some women, we are told, wrongly imagine this to be the uterus itself. As the attack

---

[14] I have used the French translation, *Traité de l'expérience en général, et en particulier dans l'art de guérir*, 3 vols (Paris, 1774), iii. 316–17.

[15] p. 323.　　　　　　　　　　　　　　　　　　　　[16] pp. 342–3.

approaches its climax the subject feels a violent pain in her forehead, temples, and eyes, and is liable to burst into tears. Her sight becomes clouded, and she has a sensation of oppression, terror, and mental confusion. These symptoms are succeeded by trembling, shivering, and palpitations of the heart. The face becomes pale, the head and limbs move convulsively, and the sufferer sometimes clenches her fists.

In *La Religieuse* Diderot does not, of course, allow the innocent Suzanne to describe what she observes in the Mother Superior with any degree of comprehension. But to the less naïve reader it is clear enough what will probably follow on her preliminary fondling of the girl.[17] In the words of Suzanne herself, the woman first kisses and caresses her and dandles her on her knees. This is followed by her asking Suzanne whether she loves her, and a tightening of her grip. She presses Suzanne's face to her own, sighs, trembles, and sheds tears. As the Superior's excitement mounts, she places Suzanne's hand on her bosom, falls silent, and appears to feel intense pleasure, which increases as Suzanne kisses her again. Her hand wanders all over the girl's body, from the tip of her toes to her waist, and she presses Suzanne in different places; her voice becomes deep in tone, and she implores Suzanne to continue her caresses.[18] Eventually the climax comes (Suzanne is uncertain whether the Superior is experiencing pain or pleasure), and the woman turns deathly pale, closes her eyes, and stiffens; there is a tightening of her mouth, which foams slightly, and her lips part. Heaving a profound sigh, she appears to die. Thinking her to be ill, Suzanne gets up quickly to call for help; but the Superior opens her eyes, reassures her 'innocent' as to her state of health, then loses the power of speech and motions the girl to sit on her lap. On both this and a later occasion, Diderot hints at Suzanne's own sexual arousal: she trembles, has palpitations, breathes with difficulty, feels confused, constricted, and agitated; physical strength deserts her, and she fears she is about to faint. Yet she cannot claim that it is pain she is feeling. The two of them remain in a state of torpor, the Superior as though dead, and Suzanne as though she is about to die. When she tries to play the harpsichord shortly afterwards, she finds that she is trembling too much to read the music.

There are three further explicit descriptions of similar arousal in the Mother Superior. On the first occasion Suzanne, alerted by what

---

[17] *La Religieuse, Œuvres romanesques*, p. 343.   [18] p. 344.

happened previously, fears a renewed onset of the woman's 'malady', observing the symptoms of distracted gaze, confused speech, wandering hands, and convulsive embrace; and again the girl herself feels fear, trembles, and believes she is about to faint, which all confirm her suspicion that the illness is a contagious one.[19]

The second time the Superior finds her way to Suzanne's cell and gets into her bed (which Suzanne has been unwilling to let her share), weeps, holds her tightly, speaks haltingly, trembles, and complains of cold. She tells Suzanne that the latter's warmth will 'cure' her; but once in her bed she is seized by the same convulsions as before, loses the power of speech altogether, and is unable to move.[20] Later in the novel, faced with further advances, Suzanne details the unsettling symptoms she herself is afflicted by at such times, and which make her unwilling to do as the Superior asks her: physical agitation, an inability to meditate or pray, a sense of 'ennui' to which she had never previously been subject, and an uncharacteristic desire to sleep during the day.[21] She tells the woman that she had herself believed these symptoms to be merely the effects of a contagious disease, but that her confessor had interpreted them very differently.[22]

Madness is a leitmotiv in this novel. There is the transformation of healthy women into 'vierges folles' by convent life, the accusation of mental derangement levelled at Suzanne by the Mother Superior of the Longchamp convent when the girl tries to obtain her freedom, the repeated references to the 'folie' of Sœur Thérèse at Arpajon when she has been supplanted by Suzanne as the Superior's favourite, the story of the crazed novice at Sainte-Marie, when she escapes from her cell to appear before Suzanne in a hideous state, half-naked, in iron chains, wild-eyed, tearing her hair, mutilating herself, screaming, and threatening to jump to her death from a window; and, most terribly, the description of the lesbian Mother Superior's decline into delirium and death as a result of her frustrated passion for Suzanne. All that had previously been written about the hyper-sensibility of women, the dangers of love, the perils of solitude, and the extravagances fostered by religion is encapsulated in this woman's condition. Passion robs her of her self-possession, makes her lose her wits, carries her 'outside herself'—alienates her, as the French has it. Her mental equilibrium is upset by a physical obsession as,

[19] p. 348.   [20] p. 355.   [21] p. 371.   [22] p. 372.

during her sleepless nights, she pictures the calvary of her loved one at the Longchamp convent, feet bleeding from the glass other nuns have strewn on the ground, a torch in her fist, and a rope round her neck.[23] The Superior's temper changes from gaiety to melancholy; piety yields to delirium, conviviality to utter solitude.[24] Dom Morel diagnoses her condition for Suzanne as being that of any creature who enters a way of life for which nature did not intend him or her.[25] Constraint turns the victim from healthy to unhealthy appetites, he tells her; already deranged, the woman will descend ever further into madness. Her mania finally expresses itself in an appearance mirroring that of the crazed novice at Sainte-Marie: she walks barefoot with her hair dishevelled, foaming at the mouth, running about her cell with hands pressed to her ears, and eyes closed.[26] After she has spent several months in this state, she is finally released by death.

Some of these symptoms of frenzy are also caused in Claire after Julie's death.[27] She is told by Wolmar to take a grip on herself; but when her fit of maddened grief has seemed to run its course, she still displays the marks of affliction which the loss of a loved one can bring: her unsound mind, although less weakened than that of the lovesick Mother Superior, is the product of a shattering upset. As in *La Religieuse*, the notion of mental health has a more than metaphorical value. Like Diderot, Rousseau suggests that the desires of an unwell soul—Claire's longing for her own oblivion, or the Mother Superior's illicit longing for union with Suzanne—should be restrained, just as the immoderate appetites of a diseased body should be curbed. Balance and harmony must be allowed to prevail. According to the ancient theory of medicine, such an equilibrium was itself a state of love. This is the meaning of Eryximachus's remarks in the *Symposium*. Medicine was concerned with reconciling hostile, disorderly forces so that the patient's body could be restored to a state of temperance and well-being.

Although the eighteenth-century novel suggests that physical weakness and sensibility are more commonly to be found in women than men, it occasionally depicts them in the latter too. But as often as not such frailty is nothing but a pretext for arousing the sympathy of

---

[23] p. 354.   [24] p. 375.   [25] p. 381.
[26] p. 385.   [27] pp. 738–9.

a particular female. This is the service it performs for Valmont in *Les Liaisons dangereuses*. That this arch-seducer should himself allege physical indisposition as a part of his campaign to win the Présidente de Tourvel is unsurprising in a novel where ill health is so frequently invoked by females as a reason for not seeing people they wish to avoid; both Mme de Tourvel and Mme de Merteuil, for instance, use attacks of migraine as a pretext for ridding themselves of a man's attentions. The indisposition Valmont adduces as the cause of his keeping to his room is more surprising: not the headache he feigns elsewhere in order to have the opportunity of writing a letter to the Présidente,[28] nor the 'violentes agitations' he claims to be suffering as a result of the Présidente's rebutting his advances[29] (for he wants to appear a lover no less 'sensible' than Rousseau's Saint-Preux), but an attack of vapours. Although Mme de Rosemonde praises Valmont's knowledge of things medical as she encourages him on one occasion to take Mme de Tourvel's pulse[30] after the young woman has alleged a bout of feverishness in order to evade Valmont's pursuit, he himself cannot have been unaware that vapours was almost exclusively a female condition, much commented on in the medical literature of the day. Restif de la Bretonne considered it to be born of woman's idleness, and wrote that it served as a pretext 'pour écarter les maris incommodes'.[31]

Much fiction of the period implies that women valued the settled repose that a well-regulated life procures more than men, and were quite willing to see it as most readily obtainable through marriage. Their search for stability might derive from precisely that constitutional proneness to passional upset which the (male) medical establishment detected in them, or from some other cause. Sometimes the embracing of absolute values, or what they perceived as such, gave the female characters of the eighteenth-century novel the anchorage in peace which their restless temperaments craved, and sometimes it did not.

The case of Rousseau's Julie is instructive in this respect. Her life contains two absolutes, to each of which she returns and pledges a new allegiance after a period of absence. The first is that of virtue—the virtue of her unsullied maidenhood, compromised by the irregularity of her liaison with Saint-Preux, but restored with her

---

[28] p. 53.   [29] p. 87.   [30] p. 55.
[31] *Le Ménage parisien*, 2 vols (The Hague, 1773), ii. 65.

marriage to Wolmar and renunciation of her lover. The second, however, is that which the lover represents—the passionate feelings she abjured in agreeing to become Wolmar's wife. In the long letter to Saint-Preux in which she explains the emotional revolution she experienced on getting married, Julie tells her former lover how susceptible she was to the fevers of love:

Vous le savez, mon ami, ma santé, si robuste contre la fatigue et les injures de l'air, ne peut résister aux intempéries des passions, et c'est dans mon trop sensible corps qu'est la source de tous les maux et de mon corps et de mon âme.[32]

Like the Présidente de Tourvel after her, she considers the ardour of passion to be a sign that it cannot last, that it will burn itself out and risk destroying its victims in the process. But also like the Présidente, she is forced to retract, and to pledge her whole being to its irresistible force. That she can manage none the less to end her life in a state of comparative grace, and the novel in an odour of comparative sanctity, is a tribute to the legerdemain of which Rousseau was so skilled a practitioner.

When she dies she has achieved both clarity and purity, a state of catharsis in which she is enlightened about her emotions, freed of muddy misconception. Catharsis and learning are linked: the former signifies an unimpeded rational state of the soul when it is cleared of the troubling influences of sense and passion.[33] Julie will no longer be concerned about her emotions, because they are about to be legitimized; she can give up the struggle against them because her death will break her moral and legal bond with her husband and free her to love Saint-Preux again. Rousseau, a man of his age, meant this to be edifying. To the modern reader it may not seem so, for it smacks of the sophistry that appears in some of the author's other writings; but the modern reader is probably less likely than was his eighteenth-century counterpart to want his reading of imaginative literature to be edifying. On the other hand he may, to continue the Aristotelian line of argument, be interested in developing his self-understanding through attending to the values that govern the lives of fictional characters. To conclude that Julie is a fake, as he may well do, will not damage this process. Indeed, the very blatancy of her duplicity may assist the clarity of his vision. However, it is

---

[32] p. 351.   [33] See Nussbaum, op. cit. 388 ff.

perhaps over-harsh to put all the blame for Julie's artifice on Julie herself, and even to talk of duplicity. For on the one hand she is a victim of Wolmar's method, based on the misguided theory that passion—which he, a Stoic, sees as aberration and threat—can be uprooted if the right pedagogical approach is adopted, and the sufferer himself cleared of the infection. And on the other she is a victim of emotion, which, while it can give us access to new knowledge (as des Grieux discovered), can just as easily lead to distorted judgement.

The Présidente de Tourvel mistrusts all extremes in life, partly because she sees them as threatening her state of marital contentment, and partly because she regards any form of excess as a danger to the equilibrium of her moral constitution—a balance which, as we discover in the novel, finds physical expression. It is unsurprising that in a work concerned in large part with sensual and sentimental love the notion of *mediocritas* (the mean) should occupy the prominent position it does.[34] Writers on conjugal and other forms of love had long stressed the medically beneficial effects of self-limitation. Nicolas Venette's influential *Tableau de l'amour conjugal* (1687), which by the end of the eighteenth century had run through at least thirty-one editions in France, advises partners in marriage to indulge in strictly limited pleasures.[35] Moderation, according to the humoral principles of medicine which Venette professes in the book, will have positive consequences for the constitution: it will alleviate weariness, help clear the moistness accumulating in the brain, cure bad dreams, make sight sharper by drying out the eyes, improve digestive processes, and give women a fresher complexion. Venette also dilates on an aspect of love-making which much preoccupied earlier writers, namely the 'expenditure' of seed[36] this activity entailed in men and (according to beliefs then current) in women, and the possibility that being overly self-indulgent or spendthrift might seriously weaken either partner.

In Mirabeau's *Le Rideau levé* (1788) there is a moral tale concerning the fate of a female with an unbridled sexual appetite, Laure's

---

[34] See Robert Mauzi, *L'Idée du bonheur au XVIIIᵉ siècle* (Paris, 1960), pp. 115 ff.

[35] *De la génération de l'homme, ou tableau de l'amour conjugal* (Cologne, 1726), pp. 340 ff.; see also Roy Porter, 'Spreading Carnal Knowledge or Selling Dirt Cheap? Nicolas Venette's *Tableau de l'amour conjugal* in Eighteenth-Century England', *JES*, 14 (1984).

[36] See Stephen Heath, *The Sexual Fix* (London, 1982), p. 14: the common Victorian term for the climax of sexual pleasure (today's 'come') was 'spend'.

erstwhile companion Rose. This young libertine falls victim to her excessive indulgence in sexual pleasure, and suffers a catalogue of woes. She never starts to menstruate, falling into a languorous state which is succeeded by a violent fit of vapours. This has the unexpected effect of weakening her sight; and as time goes on other forms of debility announce themselves. She becomes a shadow of her former self and eventually dies.[37] Her story contrasts with that of Lucette, whose appetites are responsibly channelled, and for whom as Laure's father informs her, the release afforded by moderate sexual activity is wholly salutary. She is, Laure is told, of an age when the retention of her seed would have an adverse effect on her constitution: flowing into her blood, it would either cause ravages there or stagnate and become fetid, upsetting the blood's circulation. As a consequence she would be overtaken by the vapours, fits of vertigo, and finally insanity—a phenomenon which may be observed, her lover adds, in certain convents whose inmates find no release for their pent-up sexual appetites.[38]

This cautionary account is designed to alert Laure to the need for proper moderation in sexual matters. Mirabeau shares the general view of his age that female sexual maturity comes earlier than male (an opinion endorsed by Saint-Pierre in *Paul et Virginie*), but issues a warning against premature sexual experience. Nature's operations, Laure is told, must not be hurried by human agency. The female's sexual being is not fully developed until she reaches the age of seventeen or eighteen, and any hastening of its progress towards completion by untimely sexual play leads to the damaging ejaculation of substances which should have aided her general growth.[39] Females who stray into such practices ahead of time—unlike Laure, who wears a chastity belt for several years before her menstrual cycle begins—either die, remain stunted, become enfeebled and languishing, or decline into marasmus. In the latter case they become so thin that their chests are weakened, their periods are disturbed, and they become prey to the vapours, nervous tension, vertigo, and weakened eyesight.

Although Laure's father adds that youths who yield to their sensual urges prematurely are afflicted likewise and often die young, and although the story of Rose is paralleled by a similar cautionary tale

---

[37] *Le Rideau levé*, Œuvres érotiques de Mirabeau (Paris, 1984).
[38] p. 339.   [39] p. 337.

about her brother Vernol, Mirabeau's emphasis is primarily on the
danger to women of yielding to immoderate sexual appetites. This
fits well into the tradition of female libidinousness which had for
centuries been fuelled by religious and secular writings, and which
the image of womanly purity presented in several eighteenth-century
novels could not entirely dispel.[40] Some of the best-known writers
of the age, men not otherwise known for their unsympathetic view
of women, contributed to the body of misogynistic works which
attacked the sexual greed of the female. In *Les Egarements du cœur
et de l'esprit* Crébillon portrays Mme de Senanges as a woman who
is unconcerned with exciting love in the male provided that she can
arouse his desire,[41] and contrasts her with the more circumspect and
sentimental Mme de Lursay and the sexually unresponsive Hortense.
The same author's *Tanzaï et Néardané* depicts, in the transposed
form of the fairy-tale, the classic type of the devouring female feared
by the male, on whom she inflicts the punishment of impotence.
Following in the same centuries-old tradition were those works of
medicine which described woman as a debased product of intercourse,
and advised on the best methods for avoiding conception of a female.
Venette's *Tableau de l'amour conjugal* announced that debilitating
activities like repeated intercourse at the start of a marriage (probably
demanded by the insatiable female) were more likely than not to
produce a girl, presumed to be both less desirable than a boy and
inevitably, given the circumstances of her conception, a weaker
vessel. Moderation, conversely, was conducive to the begetting
of males.

   None of these cautions, one feels, is appropriate to the case of
Mme de Tourvel, a childless married woman whose every instinct,
to judge by her discouraging letters to Valmont, draws her towards
the ideal of *mediocritas*.[42] For her, the repose she enjoys with her
husband is by definition preferable to the excitement Valmont offers,
both because it is a known quantity (and the agitation of passion
unquantifiable), and because the very constraint of marriage is
something which a devout woman may not call into question.

[40] See K. L. M. Rogers, *The Troublesome Helpmate: A History of Misogyny in
Literature* (Seattle, 1966).
[41] p. 100.
[42] On the influence of this notion in antiquity, see Theodore James Tracy, S. J.,
Physiological theory and the Doctrine of the Mean in Plato and Aristotle (Chicago,
1969).

Immediately after assuring Valmont in Letter LVI that she cannot answer his feelings, she mentions the torment that love for him would inflict upon her.[43] She then appeals to his own moral conscience as an arbiter in the case: his esteem for her as a woman of principle and virtue should reveal to him, if further proof were needed, the necessity for her to repel his advances. Her assertion that tranquillity is essential to her being is no answer to the declaration which Valmont has previously made to her, that her coldness has driven his soul into torment and distraction, plunged him into the fever of delirium.[44] He follows this description of his medical state with a letter to Mme de Merteuil which continues the metaphor: 'Voici le bulletin d'hier,' he writes, reporting on the current state of play in his campaign against the Présidente's virtue.[45] For Valmont is no respecter of tranquillity, and it is in the very placidness of her relationship with her husband that he professes to find a reason for her to escape into the more exciting existence he offers. The fact of having a husband, albeit an absent one, gives the tempted female a powerful reason for begging her tempter to desist; but the stronger motive still appears to be fear of a passion that might destroy her state of calm.

But as Rousseau's Julie comes to realize, along with other heroines of the eighteenth-century novel, the avoidance of passionate states on account of their agitating effect on the organism may diminish the individual's sense of existence. As we have seen, the crisis of her married life comes with her acknowledgement that 'le bonheur m'ennuie.' To seek contentment of the soul by forgoing violent emotion, it is implied, may be to pay too high a price. Such a suggestion, from a heroine whose regularity and self-control Rousseau has encouraged the reader to regard as exemplary, is not far short of astonishing; but it should prepare us for what is recounted in the final pages of the novel, with Julie's death and the admission contained in her posthumous letter to Saint-Preux.

For Julie's Wolmar, without a doubt, regularity and freedom from disturbance are the *summum bonum*. He almost converts Julie herself to this view, but the ultimate failure of his effort raises the question whether the therapy he practises on her as well as Saint-Preux was either reasonable or at all likely to succeed. Physicians should observe caution in experimenting on their patients. Was not Wolmar, a man

---

[43] pp. 113–14.   [44] p. 53.   [45] p. 55.

professedly without passions, unjustified in attempting as he did to 'tranquillize' the febrile pair in the deliberately constructed 'ménage à trois' at Clarens? In an age which understood many of the relationships between emotion and physical health, it might reasonably have been hoped that the doctor himself should possess an understanding of deep feeling. But this the cold Wolmar is apparently without, despite the lessons he could have learnt from his perusal of the lovers' correspondence.

From an early stage in the novel Rousseau establishes the medical metaphor as one that will adumbrate the development and curtailment of Julie and Saint-Preux's love. Julie is aware of the 'poison' that courses in her veins as a result of her fatal encounter with her lover.[46] But Saint-Preux is dissatisfied with her constitutional robustness.[47] He declares that his frustrated love for her will drive him to an early grave,[48] and finds Julie's manifest health a source of irritation, believing that she should suffer as he does. Soon the pair exchange the 'baiser mortel' that seals their fate.[49] On his beloved's lips Saint-Preux has drunk the poison that brings ferment to his blood, makes it boil, and bids fair to destroy him. Julie's cousin Claire later reports to him that his mistress has finally succumbed to the disease he exposed her to: her health has declined as a result of the efforts she made to distance herself from Saint-Preux.[50] She aggravated her condition by talking of him to her father, and finally received the *coup de grâce* from one of Saint-Preux's ill-timed missives. Shocks, both greater and lesser, continue to assail them. Saint-Preux experiences palpitations of the heart after receiving a particularly virtuous letter from Julie;[51] her choleric father, meanwhile, has hit her in a rage, causing her to fall downstairs and suffer a miscarriage of Saint-Preux's child;[52] and subsequently Julie's mother dies after the shock of discovering her daughter's clandestine correspondence with her tutor.[53] (Claire also blames the Baron d'Etanges's bad temper for having worn down his wife's resistance over the years.) Later on Julie, like her predecessor in epistolary fiction, Richardson's Pamela, contracts smallpox; like her forebear (and unlike Laclos's Mme de Merteuil), she suffers no serious impairment to her beauty from it. In a lover-like gesture Saint-Preux tries to catch the disease from her,[54] but

[46] p. 39.   [47] p. 47.   [48] p. 37.
[49] p. 63.   [50] pp. 93–4.   [51] p. 227.
[52] p. 178.   [53] p. 307.   [54] p. 333.

merely succeeds in inoculating himself—the 'inoculation d'amour' which, according to Claire, simply makes his looks more rugged and virile.[55] Before being despatched on a world voyage to cure him of his passion, Saint-Preux toys with the idea of suicide, but comes back to his friends in Switzerland proclaiming that he has been restored to full emotional health.

Subsequently he too appears to incline to the ideal of calm and passionlessness upheld by the two Wolmars, enjoying the peace which Julie's Elysean garden at Clarens infuses into his soul,[56] and contrasting the concealed order of its layout with the 'ordre social et factice' that had once made him so unhappy:[57] for the solitary Elyseum offers to the beholder only the sweet aspect of uninhibited nature. It is no surprise to the reader to learn that Wolmar has a natural taste for order, his guiding principle in life.[58] He reveals two-thirds of the way through the novel, after Saint-Preux's apparent 'guérison' seems to have justified his attempt to govern the lovers' emotions, that he has a liking for reading into the hearts of other beings. Through the feverish crises of body and soul that Julie and Saint-Preux undergo, Wolmar himself remains obstinately healthy, sound in spirit and in physique (Julie admits that his age is 'un peu avancé' but says that he looks far younger than his years).[59] With his settled disposition, he is free to perform his unsettling experiments on the young pair. Julie never seems to hold against him his presumption in daring to mould their moral characters, in part, presumably, because she reveres Wolmar as a wise teacher rather than a mate. Comparing her passion for Saint-Preux with her profound affection for her husband, she tells the younger man that Wolmar represents constancy rather than the instability and changeability she fears in passion. What they feel for each other is a temperate attachment born of respect and admiration.

Throughout the letter in which Julie proffers this information runs the tacit assumption that, while Saint-Preux represented the climate of fever, with its swift changes of temperature, Wolmar stands for the equilibrium of health. According to Julie he exerts only a salutary influence on her, precisely because passionate love is absent from their relationship. She graphically describes the febrile state of 'inquiétude' which she now, as Wolmar's ready pupil, regards as

[55] p. 427.   [56] p. 487.
[57] pp. 471–2, 486–7.   [58] p. 490.   [59] pp. 369, 374.

characteristic of passion, a state giving rise to destructive emotions like jealousy.[60] (Julie shares this perception with the Présidente de Tourvel.) The ardour of passionate love does not cleanse the individual cathartically, but burns him out.[61] He is then free from illusion, but in a state of terminal decline. With Wolmar, by contrast, she shares 'l'immuable et constant attachement de deux personnes honnêtes et raisonnables'.

Because Wolmar is a being governed by temperance and reason, his solution to the problem of 'curing' Julie and Saint-Preux of their illness must be less radical than the one which suggests itself to Saint-Preux, namely the total severance effected by death. Although, ironically, it is Julie who will finally remove herself from life, Saint-Preux has earlier to be dissuaded from the view that to cure oneself of unhappy love by committing suicide is exactly similar to curing oneself of gout by taking the appropriate medical remedies.[62] The régime which Wolmar invents for the couple involves some of the traditional anaphrodisiacs, in particular physical separation and abstinence. But other aspects of his curative system are peculiar to himself, and bear the mark of his severely rational character. His plan for purging Julie and Saint-Preux involves, as well as Saint-Preux's geographical removal from his beloved, their reintroduction to the places most associated with their guilty passion. In particular, he leads his patients back to 'ce même bosquet où commencèrent tous les malheurs de [leur] vie'.[63] Once in the fatal grove where the pair sealed their love for each other, Wolmar sets about sanitizing it by declaring his belief that the three of them will live happily together, and their mutual friendship be the comfort of his approaching old age.

He also reveals his intention that the former relationship between his wife and Saint-Preux should be regularized rather than destroyed— a fatal miscalculation, as it turns out, since the propinquity he encourages and the healthier climate of Clarens both become dangers for the couple, the first by perpetuating a close relationship, and the second by encouraging a calm 'bonheur' which Julie eventually finds frustrating. Wolmar's physicianly perceptions inform him that forcing the patient to fight his disease can irritate the symptoms,[64] and that violent effort, while it exercises the soul, may inflict on it torments

---

60 p. 372.   61 Ibid.
62 p. 383.   63 p. 489.   64 p. 495.

capable of destroying the organism. His diagnosis of the lovers' condition consequently leads him to choose the homeopathic method (similar to Saint-Preux's 'inoculation de l'amour') of administering carefully reduced doses of the very substance that caused their fever and thereby rendering it harmless. By moderately 'feeding' their illness, he tells them, he kept alive its beneficial qualities while removing its life-endangering ones.

As a later letter of his to Claire reveals, however, Wolmar is himself guilty of a certain intemperate hubris in his medical practice. He rightly sees that his 'jeunes gens' are more in love than ever, but is wrong in asserting that they are nevertheless completely cured.[65] He admits that his diagnosis and prognosis are more readily applicable to Saint-Preux than to Julie, for he sees that his wife's heart is so thoroughly veiled in virtue and decency that the human eye can no longer penetrate it. Just how well Julie has covered over her real feelings is revealed in her posthumous letter to Saint-Preux; but it is ironic that Wolmar, at a comparatively early stage, can deduce a part of her true state from the observation that she is practising a moderate form of deception: 'elle ne cesse de chercher en elle-même ce qu'elle ferait si elle était tout à fait guérie, et le fait avec tant d'exactitude que si elle était réellement guérie, elle ne le ferait pas si bien.'[66]

It is hard to believe that Rousseau himself did not regret the eradication of fervent love that Wolmar tries to engineer. The latter's calm nature, however clinically it enables him to set about curing amorous obsession, always appears less appealingly human than the 'sensible' intensity manifested by Julie and Saint-Preux. Wolmar clearly establishes the link, to which professional writers on medicine in the eighteenth century drew attention, between passion, health, and morality. He believes that the lovers' feelings for one another, which always included an affection for virtue, 'tenai[ent] à tant de choses louables qu'il fallait plutôt [les] régler que [les] anéantir'.[67] But for Rousseau's novel to have made an unambiguously moral point, it would have been necessary wholly to extirpate the disease of the lovers' passion. The affective purpose of *La Nouvelle Héloïse* precluded such a solution; and the positive assertion of passionate involvement, combined with and finally

[65] p. 508. [66] p. 509. [67] p. 495.

supervening over virtuous sentiment, was probably what gave the book its phenomenal contemporary success.

Other eighteenth-century novels describing the psychological treatment of passional upset unsurprisingly suggest that it is more reliably entrusted to professional hands than amateurs. In *Faublas* Willis successfully employs a form of therapy widely practised in the eighteenth century, organizing a 'réalisation théâtrale' of his patient's condition which mingles real images with those of his imagination so as to bring on the salutary moment of crisis. (By giving their patient tangible, externalized counterparts to the substance of his diseased fantasies, doctors attempted to transmute their threatening inward quality into something apparently more real, and therefore reassuring.)[68] Willis assembles an array of familiar objects and people in an effort to bring Faublas back from hallucination to a grasp of reality either comfortingly known or obscurely feared. The garden provided for the young man is the tranquillizing scene into which the human and inanimate 'players' are led to work on the patient's delirious mind. The tombstones erected to the memory of the Comtesse de Lignolle and the Marquise de B*** remind him of the deaths he believes he caused, but become less fearsome and more majestic as he gradually comes to venerate in them the memory of two loved companions. The river that runs through the park similarly ceases to be a terrible warning of the perilous element in which the Comtesse lost her life, and loses its remaining terrors as Faublas plunges into it to swim. Finally, the abandoned but never-forgotten Sophie dramatically intervenes to console her peccant husband in tones of love and forgiveness. All these actors and props, magisterially deployed by Willis, help to bring Faublas's delirium to its climax and so free him from it.

Roughly comparable scenes occur in other works describing states of high emotion. Like *Manon Lescaut* on the memorable occasion of Manon's visit to des Grieux at Saint-Sulpice, *Les Liaisons dangereuses* contains tableaux which highlight states of bodily collapse in the face of extreme emotional pressure. Although Valmont refers to the 'violentes agitations' that assail him as a result of Mme de Tourvel's inhumane treatment,[69] it is usually the Présidente

[68] See Michel Foucault, *Histoire de la folie à l'âge classique* (Paris, 1972), pp. 350 ff.   [69] p. 87.

herself who falls victim to such pathological seizures. Valmont details their physical symptoms in various letters to Mme de Merteuil. In Letter XCIX he clinically notes the pathognomic and physiological accompaniments to her intense emotional state in his presence—the dulled look in her eyes, her faltering voice, and her collapse into his arms, followed by a convulsive movement as she struggles out of them, with eyes now wild and hands raised to heaven.[70] She sinks to her knees (a posture much beloved of Richardson's Pamela), appears to be suffocating, weeps, embraces Valmont's own knees, and all but immobilized him in a powerful grasp reflex, as uninhibited as the clutching of a child.

Looks often appear as giveaways in this novel—if not Mme de Merteuil's, then those of Cécile or the Présidente. In Letter XCIX the latter's lowered gaze, together with the faltering of her voice, are to Valmont the unmistakable symptoms of a crisis he can turn to his own advantage. But when he holds her in his arms as she loses the power of speech, becomes rigid, and has a fit of violent convulsions, he is bound to admit that he himself is 'vivement ému'. On a later occasion he is better able to preserve self-control, while yet offering the Présidente signs of emotion to rival her own. In Letter CXXV he describes how he judges from her quavering voice that his victim will shortly suffer an emotional collapse from which he can profit: he sinks dramatically to his knees (but cannot produce the tears on which he had counted) and achieves the desired result by altering his vocal inflection, staying on his knees and boldly meeting her eyes.[71] The effect on Mme de Tourvel is electric: she promptly rises and wriggles out of his embrace, whereupon Valmont himself adopts an upright posture, fixes wild but attentive eyes on her, and observes her promising condition. Her uncertain attitude and rapid breathing, the contraction of her muscles, and the trembling of her half-raised arms, sufficiently demonstrate to him that he is producing the hoped-for effect. Then, subtly modifying his tone of voice, Valmont almost imperceptibly draws closer to the Présidente, takes her by surprise through producing a collection of her own letters to him, threatens suicide, and captures her in his arms before she can recover from the shock of this announcement.[72] Valmont now introduces into his voice the accents of 'enthousiasme' suited to the state of religious intensity he wishes to convey to his victim, bids

[70] p. 223.   [71] p. 291.   [72] p. 292.

her an eternal farewell, and notes how her heart duly palpitates, her facial expression changes, and tears nearly suffocate her.[73] As he feigns departure, she grasps him almost involuntarily and falls into his arms in a dead faint. When she regains consciousness, Valmont has won physical possession of her.

The letter immediately following this one of Valmont's, written by his aunt to the Présidente, reveals a similar attention to the detail of pathological passionate states; but its writer employs the vocabulary of medicine in order to encourage Mme de Tourvel to resist the 'disease' of love which Valmont threatens to communicate to her. Mme de Rosemonde knows nothing of her nephew's stated desire to observe the death-agony of his quarry's virtue;[74] but she has witnessed the young woman's precipitate departure from the château, caused by the dangerous proximity of Valmont, and has divined the state of her emotions.[75] The best she feels able to do for her, Mme de Rosemonde writes to the Présidente, is to act as support and consolation:[76] she cannot relieve her pain, but by sympathetically sharing it she can mitigate its effect and soothe the Présidente's inner turmoil. Ironically, Mme de Merteuil hypocritically plays the same sort of role *vis-à-vis* Cécile, faced with the girl's distraught condition after her sexual induction by Valmont.[77] But Cécile has obstinately failed to react as planned to the 'treatment' Valmont has administered to her:[78] although the loss of her virginity should have dispelled her 'sotte ingénuité', Cécile continues as naïve as ever. Henceforth, the Marquise informs Valmont, he must increase the doses of his medicine; if he fails to do so, the plan of humiliating Gercourt and Mme de Volanges will have no chance of success. And yet, Mme de Merteuil continues, Cécile's apparently incorrigible childishness may indicate that she suffers from a fatal weakness of character, and is therefore of no use to the plotters.

In the letter Mme de Tourvel receives immediately after yielding to Valmont, but before Mme de Rosemonde has been apprised of this state of affairs, the old woman counsels her protégée to bear with fortitude the pain she is experiencing at Valmont's (feigned) renunciation of her.[79] (Laclos's choice of vocabulary may remind the reader of Mme de Merteuil's words of consolation to Cécile after the latter's first night with Valmont: she tells the girl that the

---

[73] p. 293.   [74] p. 220.   [75] p. 233.   [76] p. 234.
[77] pp. 214 ff., 239 ff.   [78] p. 244.   [79] p. 296.

'peines de l'amour'—which Cécile will connect with her unhappiness at having been unfaithful to Danceny, but which the Marquise is also relating to the virgin's physical pain at the moment of penetration— are like the shame which love induces. Both are experienced only once, although they can be fabricated subsequently. Here Mme de Merteuil presumably has in mind the use of such lotions and astringents as are also revealed to the young Faublas in Louvet's novel.) Mme de Rosemonde writes that the Présidente has been saved by Providence, as a skilled doctor saves a patient from the depredations of disease. The discomfort the young woman now feels at having been abandoned by Valmont will soon lessen; it will respond to spiritual consolation, like an illness being overcome by the physician's remedies.

Mme de Rosemonde's medical turn of phrase becomes still more pronounced as she asks the Présidente's indulgence for her temerity in broaching the subject of the danger which her friend has just evaded. While it is cruel to be honest with the desperately ill, who are beyond any help bar that of consolation, it is wise to enlighten convalescents as to the perils they have left behind them so that they are impressed with the need still to observe caution. 'Puisque vous me choisissez comme votre médecin,' she goes on,

c'est comme tel que je vous parle, et que je vous dis que les petites incommodités que vous ressentez à présent, et qui peut-être exigent quelques remèdes, ne sont pourtant rien en comparaison de la maladie effrayante dont voilà la guérison assurée.[80]

She informs the Présidente that her dear nephew is an evident threat to her friend's well-being, despite his many charms and attractions. Buy by influencing Valmont for the good, Mme de Tourvel has contributed to his healing, his salutary return to God. For this reason, the old lady concludes, she can join to the repose she has earned from an easy conscience the tranquillity that comes from victory over life-threatening disease.

Mme de Rosemonde's tone must evidently change when she discovers what has since happened between her nephew and the Présidente. But in her next communication to Mme de Tourvel her tone is still that of the kindly physician, one who must now offer consolation in a near-hopeless case. The pain underlying the

[80] p. 297.

Présidente's admission, she writes, was much less an affliction to the old woman than to herself.[81] To Mme de Rosemonde her young friend is in the dangerously self-deluded state of many invalids in being erroneously persuaded that she may enjoy perfect happiness (with Valmont). In these circumstances his aunt can do little but try to lessen the Présidente's pain;[82] for in untreatable illnesses advice must be confined to specifying the regimen which the patient is to follow. But the old lady ends by imploring Mme de Tourvel to remember that feeling pity for the sick is a very different thing from blaming them; there is little justification for any of God's imperfect creatures condemning one another. And she reminds her correspondent that not even the state of love, which the Présidente rightly sees as resulting in a loss of selfhood, obliges others to forfeit their claim to partial possession of the 'lost' lover. To be 'outside' onself is not to be without relation save to the loved one.

The illness Mme de Tourvel diagnoses in Valmont is that of ungodliness, which leads him wilfully to destroy the reputations, and sometimes the lives, of the women he seduces, instead of treating them in a Christian spirit of fraternity and charity. Too unworldly to know the peril of liaisons—which the well-meaning Mme de Volanges tries in vain to alert her to—the Présidente sets about attempting to cure the sinner and convert the infidel. But there is a danger, which she does not perceive until it is too late, in attempting to heal the sick: the would-be healer may catch the very fever he is attempting to treat. Rousseau shows the frustration of similar medical ambitions in Julie and Wolmar, the former suffering a relapse into the same passional condition as overtakes the Présidente. Unaware that the disease of her love for Saint-Preux is merely dormant, and that it may resurface in conditions favourable to it, Julie sets out to change her former lover into a brother (or a cousin, when she has conceived the plan of marrying him to Claire). She also has ambitions to purge her husband of his atheism.[83]

Wolmar, in the meantime, turns his therapist's skills to the relationship still evident between his wife and Saint-Preux, and tries to render it benign. When Julie's death has revealed the failure of this project, Wolmar finds himself called upon to act as doctor to Claire, whose passionate grief knows no bounds; and in turn he appeals to the absent Saint-Preux to join the shattered community

---

[81] p. 303.   [82] p. 304.   [83] p. 699.

at Clarens: 'Venez partager et guérir mes ennuis.'[84] Saint-Preux had earlier written to Julie apropos of his determined absence from Clarens that his 'weakness' for her persisted, despite the best efforts of Wolmar, an eternal impression which neither time nor attention could erase. A wound heals, but the scar remains, a seal which protects the heart against further attack.[85] After Julie's death Claire writes to Saint-Preux, like Wolmar, expressing the hope that he is sufficiently recovered to face a journey to Clarens to be with his grief-stricken friends, and suggesting that the sight of common affliction may bring relief to his own.[86]

The final lesson of *La Nouvelle Héloïse* is that for some kinds of love there is no effective remedy. Furthermore, the ending carries with it the ironic reminder that would-be physicians should attend to healing themselves before trying to heal others. Julie may be an effective 'prêcheuse' in small things: the book is full of illustrations to that effect, such as the occasion on which she criticizes Saint-Preux's involuntary drunkenness after he mistook wine for water in a Paris brothel. But her inability to purge either herself or Saint-Preux of their great passion demonstrates the limits of her healing powers. The symptoms become less apparent only because the disease has lodged itself deeper in the organism. And the aspirations of the Présidente (repeatedly called a 'prêcheuse' by Valmont) to cure the Vicomte of his rakishness become reduced to the ambition simply to administer salutary 'bonheur' to her lover, once he has convinced her that his lovesickness will otherwise kill him.

The dangers of love could not have been more graphically depicted for the eighteenth-century reader than by the story of the historical Héloïse and Abélard, recurrently adapted by French writers of the age; nor could the drastic remedies available for suffering lovers have been more clearly portrayed than by the *Historia calamitatum*. The tendency of the passions to ravage the human 'machine' was too insistently stated by writers on medicine in the eighteenth century to be ignored. The old Stoic deprecation of the passions, continued in the pessimistic works of Racine and Mme de Lafayette, was too well rooted to be easily overcome by the eighteenth century's belief in the positive good that might also issue from strong feeling. Nor could dispassionate observers set much faith by some of the remedies or pseudo-remedies fashionable in the age of Enlightenment.

---

[84] p. 740.   [85] p. 675.   [86] pp. 743–4.

Wolmar, the cold atheist, speaks dismissively of one favoured treatment for psychotic malady when he describes religious devotion as an opium for the soul.[87] Taken in moderation it cheers, animates and sustains the sufferer; but its excessive use, as by the ecstatic mystics whose fervour Julie had formerly blamed, brings with it the threat of unconsciousness, frenzy, or death.

According to an ancient theory, however, the kind of frenzy exhibited by the unbalanced soul might itself be a positive good: not a debilitating and corrupting state of disharmony, but a way of acquiring near-divine perception. This is the theory of *manía* (related to the 'enthusiasm' of Valmont), and its most famous exponent is Plato. In the *Phaedrus*, the main subject of which is love, the dialoguists discuss the belief of the rhetor Lysias that erotic *páthos*— being carried away by love, pulled outside oneself by an uncontrollable force—is a condition to be rejected in favour of enjoying mere sensual gratification.[88] Lysias holds that the former emotion is inimical to order, reason, and health, or *sōphrosýnē* (the state of the soul in which intellect rules over other elements). He argues not for abstention, the potentially arid impassiveness of the Stoic, but for conscious desire, desire without illusion—pleasurable indulgence without the dangers inherent in erotic self-forgetfulness. But Socrates proposes instead that man's greatest good is procured by the gift of *manía*, when he is infused by divine madness (*enthousiázō*). Not all erotic emotion has this divine essence; but *theía manía* permits the man who yields up his self-control to achieve ultimate wisdom.

In the *Phaedrus*, then, Plato revises the stern warnings of the *Symposium* and the *Phaedo* against sexual feelings and emotions, as against the artistic creations of poets, and adopts the view that *manía*, or the ferment of the personality, leads to the highest kind of insight, and that intellect alone cannot guide man to the greatest good. Far from being mere urges for sensual gratification, the passions are seen as furthering his understanding of what is truest and best. The person in love, according to this argument, is not sick, but in a condition which it would be madness to want to cure. This state is infinitely preferable to that of a person enjoying a well-controlled sensual friendship of the kind Lysias advocates. Contrary

---

[87] p. 697.

[88] See Josef Pieper, *Love and Inspiration: A Study of Plato's 'Phaedrus'* (London, 1965).

to Diotima's teaching in the *Symposium*—that the individual should cultivate self-control and develop the clarity of his intellect—Socrates proposes that having one's wits about one can lead to narrowness of vision and a crippling of the personality. Love is not a crude instrument for obtaining physical pleasure, but a part-spiritual means of passage to the best life; and as such it deserves to be treated with respect and awe.

In *La Nouvelle Héloïse* Julie dies knowing that the greatest blessings (to which she looks forward in the world beyond) accrue from *manía*, and that to suppress it means settling for a state of irremediable ordinariness. For true health resides not in her husband's god, self-possession, but in surrender to the divine power of enthusiasm. In yielding to emotion Julie is carried away, transported by the benevolent power of *theía manía*. The most conscientious efforts of the 'sound' mind, represented by Wolmar, are incapable of lifting man to the godly heights that make accessible fullest comprehension of the true and good.

Self-control and rational insight are worshipped even more fervently by Valmont and the Marquise de Merteuil than by Wolmar, because they know that their power depends on having no romantic illusions or physical weakness. Both have been able to boast of their self-possession and autarchy. But in conceiving passionate love for the Présidente Valmont loses his balance, and becomes subject to the gravely unsettling forces of emotion. And when she realizes that this has happened, Mme de Merteuil allows her deep contempt (or perhaps simply her jealousy) to show.

The Marquise reveals her conviction that people in love are 'sick' by her dismissive account of the romantic feelings shared by Cécile and Danceny and also by her mocking catalogue of the 'symptoms' Valmont is displaying. According to Mme de Merteuil, the Vicomte has reverted to the childhood state in which Cécile and Danceny still exist;[89] his heart is duping his mind, and he is obsessed by the unknown charms of fresh emotion. 'Ou ce sont là, Vicomte, des symptômes assurés de l'amour, ou il faut renoncer à en trouver aucun.'[90] He is ill with love, rather than enjoying her own condition of *sōphrosýnē*; and she takes bitter pleasure in detailing his loss of clarity, his self-deception, and the mindless urges that now rule his life. Valmont's submission to the non-intellectual elements of passion

[89] p. 312.   [90] p. 313.

means, for the Marquise, that he is in a bad way. Her advice, like that proffered by Diotima to Socrates, is that he should re-cultivate the state of self-possession, which alone conduces to exercising the intellect with clarity. Valmont is to purify himself by purging all traces of passion from his being.

When he proves unwilling to swallow her bitter medicine, the Marquise retorts by sending him the 'petit modèle épistolaire' which he forwards to Mme de Tourvel. The letter which the Marquise wrote as a 'défi', he transcribes and sends to the Présidente as a 'défi'—but less a challenge to her, it appears, than one to Mme de Merteuil. He knows that he must demonstrate to the latter that he is not addicted to Mme de Tourvel, that he is without the craving for such joys as only she can provide. Whether we are to suppose him cured of his love Laclos does not choose to make clear. Mme de Tourvel, at any rate, feels that with the end of his love she no longer has any reason to exist. The ravings that prefigure her death, the loss of her wits, stand in tragic contrast to the beneficial *manía* of which Socrates speaks in the *Phaedrus*. But before her terrible end the Présidente has come to conceive of love as the divine revelation praised in Plato's dialogue. This, for her, legitimizes the tones of worship with which she speaks of Valmont. Like him, she had once had a different ambition from enjoying the adoringly worshipful activity of fulfilled love; she had aspired to be his spiritual healer, to effect the catharsis of his impure passions. In the darkness of her soul after his betrayal she perhaps sees fit punishment for her hubris.

The best of human experiences, as Diotima suggested to Socrates, is shifting and uncertain. By investing all she has in an exclusive attachment to another being, inherently changeable and additionally weak, the Présidente suffers the grief brought first by alteration and then by death. Valmont's desertion upsets their almost preternatural harmony, and plunges Mme de Tourvel into a state of mental turmoil. She emerges from delirium only to hear that Valmont has been killed in a duel, after which news she swiftly dies herself. It is ironic that after losing herself to Valmont, as the 'best' lover of the *Phaedrus* yielded himself up to all-consuming eros, she cannot regain self-possession; and she dies 'outside herself', divested of her reason.

Valmont perhaps dies in something approaching a state of grace. He effects a reconciliation with Danceny after the latter has mortally wounded him in the duel, and attempts to bring the Marquise to justice by releasing their entire correspondence. This state may be contrasted with the ignominious end reserved for Mme de Merteuil, who is expelled from the society in which she flourished, loses her court case, and is horribly disfigured by smallpox. She flees the country bankrupt, cast out in the way that Plato deems appropriate for the sick in the *Republic*: for they are a drain on the state's resources, contribute nothing to it, and should not be permitted to continue living amongst ordinarily healthy citizens. The heavy-handedness of her treatment by Laclos, a writer of refined perceptions, may awaken an unease which in the modern reader is perhaps compounded by a mistrust of retributivist justice. Such justice as is meted out in the conclusion to *Les Liaisons dangereuses*, besides, is manifestly partial. In the words of Mme de Volanges: 'Je vois bien dans tout cela les méchants punis; mais je n'y trouve nulle consolation pour leurs malheureuses victimes.'

Mme de Tourvel has seemingly compromised some of her conventional 'goodness' in contracting an extra-marital liaison with Valmont. But it is possible to argue that she is spiritually justified in having done so. The view, which some will deem sentimental or romantic, that Laclos wished his reader to judge Valmont's relationship with the Présidente in terms of moral salvation, however provisional and short-lived, is not self-evidently a mistaken one. That the process of salvation should be so abruptly terminated is perhaps just a tragic consequence of the Marquise's, and to a lesser extent the Vicomte's, psychopathic disposition, a disorder of their minds which leads to conduct of a horribly abnormal and aberrant kind. The eighteenth century knew little of the medical treatment of psychopathic states which is an accepted part of modern therapy. But even if it had, it would surely not have wanted to assimilate Valmont's and Merteuil's mental condition simply to those properly treated by the physician. Baudelaire's judgement that Laclos's novel is a 'livre de moraliste' must be allowed to stand. Despite the proliferation of medical metaphors in the book, the eighteenth-century reader would probably have been as unwilling as the modern one is to allow medical to replace moral diagnosis. The two 'meneurs de jeu' are not just sick, but vicious too.

The introduction to modern law of the concept of psychopathic disorder was greeted with the observation that it was a phenomenon characteristic of the age, one much inclined to allow behaviour formerly consigned to the moral realm to be absorbed by the scientific or technical.[91] The discomfort which may be felt at this development would surely be a salutary response to any attempt similarly to reduce morals to medicine in *Les Liaisons dangereuses*. Laclos does not appear to be arguing that the mental, and hence behavioural, aberrations of the two main characters could be effectively treated by a dose of the right physic. That is why he destroys them, despite allowing Valmont to enjoy a spell of 'treatment' that appears to promise success.

As the Présidente de Tourvel discovers, there are no drugs which can replace the loss of happiness. Nor, as Mme de Rosemonde perceives, do doctors provide a weapon against 'crèvecoeur': they can counsel fortitude, but only the patient's inner being can supply it. Cécile, who becomes pregnant but suffers a miscarriage, might one day have discovered that technology offers a substitute for chastity (the 'precautions' which, Valmont boasts, are the only things he has not taught her about sexual activity). But Laclos seems not to want moral qualities to be supplanted by medical nostrums and aids. His novel is a sharp, brilliant, and lastingly impressive commentary on the interplay between medicine, states of the mind, and states of the heart that fascinated his age.

[91] See Kenny, op. cit. 25.

4

Nature and Love

AN influential essay on gardens which Addison published in the *Spectator* of 6 September 1712 has some lines from an ode by Horace as its epigraph:

> . . . An me ludit amabilis
> Insania? Audire et videor pios
> Errare per lucos amoenae
> Quos et aquae subeunt et aurae. (II. iv. 5–8)[1]

[Does the madness of love deceive me? I am seen listening and wandering through the haunted groves, into which stray pleasant streams and breezes.]

Addison's essay is largely devoted to describing what the eighteenth century knew as the 'English' style of garden, one characterized by a cultivation of irregularity which, Addison tells us, a foreigner might initially mistake for wilderness. Its 'beau désordre' is antithetically opposed to the geometrical regularity of the French type, exemplified by the gardens Le Nôtre created at Vaux-le-Vicomte and Versailles. In the course of the eighteenth century such orderly constructions came to be seen as contrary to the spirit of nature which the modern gardener's art must attempt to capture: they suggested, not the abundance of the Garden of Eden, but man's presumptuous effort to transform the harmony of the natural world into a pattern of his own devising.[2] In failing to respect the 'genius loci',[3] contemporary theorists declared, they prevented that union between man and his surroundings to which the age of sensibility should aspire (and to which Rousseau would later give celebrated expression). Addison's own garden, he writes, would puzzle an experienced gardener: for how should it be designated, and was it a garden at all? The varied topography and flora seemed to owe little to the horticulturalist's

---

[1] Joseph Addison, Richard Steele and Others, *The Spectator*, 4 vols (London, 1970–3), iv. 11.

[2] See Michèle Plaisant, 'Poésie et jardins anglais (1700–1740)', in *Jardins et paysages: le style anglais*, ed. A. Paireaux and M. Plaisant, 2 vols (Lille, 1977), i. 112.

[3] See Ronald Paulson, *Emblem and Expression* (London, 1975), p. 20.

conventional skills, and yet its non-uniformity was the product of an art that concealed itself. Nature[4] had been disposed by man, but at the same time respected; and Addison implies that this respect permits a bonding of humans and the natural world which the formality of the old style precluded.

His essay suggests that a garden fills the human mind with calm, putting turbulent passions at rest.[5] It conveys a sense of stability, and so contrasts with the swift and unsettling revolutions of emotion. If there is changeability in the vegetable world, its inhabitants are unaware of it because they lack the painful endowment of memory, or have it in short measure. Dr Bordeu translates this idea into more prosaic terms in Diderot's *Le Rêve de d'Alembert*: short-lived organisms believe in the immutability of things.[6] Trees and flowers, which know little or nothing of pain, console the sufferer for the torments of the world. Lovers can take refuge in the consecrated 'pleasant place', the 'locus amoenus' which from the time of Horace and Virgil had furnished a principal theme of nature-description in literature.[7] The amenity of such a retreat is a balm to the body and soul of those made unhappy by passion.

Nature wears many aspects in the imaginative literature of eighteenth-century France. Although the vegetable world may offer consolation to humans in a state of passional flux, the savageness of an environment which man has not tamed or cannot tame may pose a threat to him in his search for stability and tranquillity. The countryside may seem reassuringly free from the denaturing pressures of the town, and to offer a form of honesty which civilization and sophistication are necessarily without; but as more than one heroine of the period's fiction is brought to realize, it may harbour threats at least the equal of those contained by the city. Moreover, its very freedom may depend on forms of constraint which the unschooled and innocent know nothing of. To follow the course of 'nature' unheedingly, the eighteenth-century novel sometimes implies, may be to give oneself over to forms of irregularity which cannot be

---

[4] On the multifarious senses of the word 'nature' in the eighteenth century, see Arthur Lovejoy, ' "Nature" as Aesthetic Norm', *Essays in the History of Ideas* (New York, 1955). I use it here in the sense of 'the natural world'.

[5] p. 14.   [6] p. 304.

[7] E. R. Curtius, *European Literature and the Latin Middle Ages*, trans. Willard Trask (London, 1953), pp. 192 ff. The verb 'amare' and the adjective 'amoenus' are etymologically related.

tolerated in a world governed by humane values. Works like *Paul et Virginie* and *Faublas* suggest that if to follow one's instincts is to be 'natural', then that type of spontaneity is best checked until such time as emotional maturity has been reached. For men and women bear only a limited resemblance to the animals and plants, the fauna and the flora, of the natural world. A supervening order is proper for rational creatures to observe, and is ethically superior to the impulses of instinct and ungoverned desire. Such is the message not just of 'intellectual' novels like *Les Egarements du coeur et de l'esprit*, but also of 'feeling' ones like *Manon Lescaut, La Nouvelle Héloïse*, and *Faublas*.

The 'hortus inclusus' of *La Nouvelle Héloïse*, Julie's Elysée, has been created as an antidote to the threat of disorder represented by her relationship with Saint-Preux. Unlike Horace's, this retreat specifically excludes the 'lucus', or 'bosquet'; as Wolmar reminds Saint-Preux, it was in just such a place that Julie and her lover exchanged their fatal first kiss, which is why the walls of the garden do not enclose the adjacent groves.[8] Although the Elysée is a pleasance, it was not designed solely for pleasure; it also serves a useful purpose as a 'remedium amoris' which shows how the products of nature (vegetation on the one hand, and man's passional urges on the other) may be discreetly but fruitfully channelled. It eschews the manifestly ungoverned, for the Elysée is no more a friend to rampant growth than Saint-Preux proves to be when he refers to the uncontrolled wilderness of the landscaped garden at Stowe.[9] But in the restrained diversity of its natural constituents it achieves a unity which first impressions belie. Like Addison's garden, it shows a preference for native over exotic plants; and in the temperate climate of Clarens it offers the very image of fertility, being irrigated by a stream that had previously supplied an unproductive fountain.

The Elysée appears to respect essential liberties, for even its aviary allows birds to come and go as they please. But as Wolmar's admonition to Saint-Preux makes clear, the natural freedom of the garden is subordinated to a supervening condition of constraint. Just as intimate access to Julie's person is now prohibited to all but her husband, so entry to her garden is restricted. Julie holds the key, and her domain is jealously guarded. Like her self, the

---

[8] p. 485. See also Jean Gillet, *Le Paradis perdu dans la littérature française de Voltaire à Chateaubriand* (Paris, 1975), p. 459.   [9] p. 484.

Elysée exists under a control which is no less real for being scarcely apparent; and if the garden is fertile, its fertility is exploited according to the will of its owner. The balance on which its flourishing depends is a precarious one, mirroring the uncertainty of the equilibrium attained by the novel's two main characters. However fruitful this paradise appears, its name of Elyseum reveals that it is subject to the ultimate power of death, like everything in nature.[10] When Julie makes it over to her children, she lets it be known, she will be yielding up the control under which alone it has subsisted until that time. On learning subsequently that Julie dies (ostensibly as a result of saving one of her children from drowning) to gain eternal happiness with Saint-Preux, we see how the garden at Clarens symbolizes both the repression of her 'natural' love and the inevitable dismantling of the moral edifice which she and Wolmar have so patiently constructed. Although Julie's Elysée spurned in its conception the safety and unadventurousness of her husband's beloved rules—for his own taste inclined towards the symmetry associated with the seventeenth-century French horticultural style—its controlled freedom proves to be as representative of the perils attending human love as does rampant nature.

While appearing to offer the haven from troubled passion of Horace's 'locus amoenus', then, the garden actually symbolizes the uncertainty of such a retreat. It suggests that even the most careful of efforts to divert the course of nature may be in vain, for trying to redirect passion as Julie and Wolmar rechannelled the stream flowing through the Elysée is to do no more than transpose it. Julie writes in her posthumous letter to Saint-Preux that 'le premier sentiment qui m'a fait vivre . . . s'est concentré dans mon cœur',[11] and that she cannot stem it, merely remove it to the other world. To struggle any longer against her passion is, quite literally, to go against nature. She prefers instead to immerse herself in the life-force: the act of plunging into the water after her child, although apparently fatal, is in truth the way to renewal. It is significant that Wolmar should be afraid of water,[12] for here as throughout literature it stands for the power of eros, which he above all others must fear.

In *La Fin des amours de Faublas* there is a return to the mood captured in Horace's ode. Here the vast English garden acts as a restorative for the man maddened by love. It is a 'séjour des regrets'

[10] p. 486.   [11] p. 741.   [12] p. 504.

peopled with cypress-trees and weeping willows, but its beauty and tranquillity, along with the purity of its air, promise to dispel violent passion and dispose the patient's soul towards tender melancholy. Its groves are less enticing to a lover than Horace's, or the 'bosquet fatal' of *La Nouvelle Héloïse*, but the most sombre among them plays an essential part in Faublas's restoration to sanity. The block of marble which Dr Willis has transported there to act as a *memento mori*, a constant reminder to Faublas of the Comtesse de Lignolle's death, gradually makes him reconciled to her fate; and when the thoughtful Willis has mattresses carried to the tranquil spot Faublas's treatment even comes to resemble a rest-cure.[13] The grove nevertheless remains a focus of the patient's melancholy thoughts, particularly when the doctor has the Marquise de B***'s tombstone erected there as well; and after his recovery Faublas still retires thither every evening to weep. Its association with the unhappy course of passion is indelible, but it becomes the 'pius lucus' that permits devotion more elevating than the love Faublas had earlier shown the mistresses of his pleasure.

The garden and its products also constitute a rich source of erotic symbolism in *Paul et Virginie*, which was published a year after the first volume of *Faublas*. Paul makes the garden for his beloved's delight, and as long as she remains free to stay on Mauritius its fertility continues, emblematic of the fruitful love that joins the children to each other. Once her mother's plan to send her to Europe has been conceived, however, and after Virginie herself has acceded to the proposal, nature seems to work against the horticultural efforts of the little community. Shortly before her departure from the island, a hurricane lays waste the little domain. Nothing in the garden survives; nature has cruelly demonstrated its changeability and man's impotence in the face of cosmic forces. 'Tout périt sur la terre,' Virginie sadly concludes; 'il n'y a que le ciel qui ne change point.'[14]

Nature's mutability is a leitmotiv of *Paul et Virginie*, and may readily be translated into the passional terms which the image of horticultural fertility and barrenness suggests. If Paul's garden represents the effort to control nature—as does Mme de la Tour's plan to send her daughter to Europe, pending marriage to Paul—its destruction by the hurricane shows the limitations which nature

---

[13] p. 1209.
[14] *Paul et Virginie*, ed. Pierre Trahard (Paris, 1958), p. 137.

imposes on such human effort. Virginie sends from Europe some seeds:[15] the symbolic violet, shrinking like her modest self, but exuding a delicious scent which causes it to be sought out, and the scabious, whose appearance of mourning has given it the folk-name of widow's flower—an apt reminder of Virginie's desolate state away from Paul. Paul plants them, taking care to follow Virginie's instructions to sow the violet in the 'repos de Virginie' at the foot of her coconut-tree (one of two planted at the children's birth, and which grow according to the measure of their own growth), and the scabious on the Rocher des Adieux, where they said their last goodbyes. But even in these symbolic positions most of the seeds fail to germinate. They may, the reader is told, have been spoiled during the crossing, or been unsuited to the climate of Mauritius.[16] However, the poetic reason is surely that Virginie's move from the Indian Ocean to the civilized West is tantamount to a loss of fruitfulness in her relationship with Paul. Having cut her ties with Mauritius, and therefore with him, she cannot restore fertility from afar, although Paul writes to her that he intends to unite Europe and Africa in sowing the seeds. With his letter he sends some coconuts from their two trees, 'parvenus à une maturité parfaite'.[17] But although, as this gift makes clear, he is now ready to be joined with his beloved, their union can never mature on earth.

Further erotic symbolism is provided by the apple-pips which Virginie sends along with seeds of other fruits in her letter from France, pointedly adding that the family will take more pleasure in them than in the bags of money which had been the original cause of their separation. Her Eden in Mauritius had previously contained no forbidden fruit.[18] But the apple-pips 'take' no more successfully than anything else she sends, and suggest that Saint-Pierre intends the consignment of seeds to signify that the innocence of childhood and of life in the natural paradise of Mauritius has ended. His contention in the fourteenth *Etude de la nature* that Europeans neither like nor understand the young is a further indication of why this childhood idyll had to be played out against a non-European background. (Other contemporary writers of pastoral seem to have disagreed, however: Florian's *Estelle*, for example, avowedly written

---

[15] p. 164.   [16] p. 165.   [17] Ibid.
[18] See Roseann Runte, '*La Chaumière indienne*: Counterpart and Complement to *Paul et Virginie*', MLR, 75 (1980), p. 777.

in celebration of the author's own homeland, describes the growth of the child-love between Estelle and Némorin and its eventual prospering in an embattled Languedoc, after the couple have passed through adolescence and retained their devotion in adult life.)

Saint-Pierre is the first French writer to attempt extended description of children in and for themselves, and to make nature the principal 'character' of narrative fiction.[19] In doing so he makes it clear that Paul and Virginie's love cannot be understood in isolation from the natural world in which they have been brought up. As they are thoroughly attuned to nature, so it appears to feel, rejoice, and suffer with them: it would have been unusual for a writer close to Rousseau, as Saint-Pierre was, not to have been influenced by this Rousseauist sense of oneness. But to live in such a world is necessarily to be impressed by man's inevitable subjection to nature's laws, such as the law of mutability and the necessary ending of all things in death. Saint-Pierre's treatment of this theme is muddled, but tells his reader something about the author's conception of love in relation to nature as a cosmic force. Bernardin rejects the optimistic portrayal of idyllic love in a pastoral world which was familiar from the contemporary work of Gessner (whom he, like Rousseau, is known to have admired),[20] Florian, and Marmontel, showing how an upbringing against such a background may lay man open to inconveniences as considerable as those faced by townspeople.

Nature, then, seems not reliably to promote the cause of love. It destroys Paul's garden and the 'repos' of Virginie, both literal and figurative. The fact of universal impermanence on earth, which is borne in upon Virginie following this event, is also impressed on Paul by their elderly neighbour in the face of the boy's desolation after her death. The Vieillard tells Paul that his beloved died before the corroding touch of nature and time could make itself felt on her—before the travail of motherhood could weaken her body as it did Mme de la Tour's, grown old prematurely as do all physical organisms in a tropical climate.[21] The 'civilized' West, besides returning Virginie to the island in a more delicate state than she left

---

[19] See D. G. Charlton, *New Images of the Natural in France* (Cambridge, 1984), p. 149; Paul van Tieghem, *Le Sentiment de la nature dans le préromantisme européen* (Paris, 1960), p. 95.

[20] See Daniel Mornet, *Le Sentiment de la nature en France de Jean-Jacques Rousseau à Bernardin de Saint-Pierre* (Paris, 1907), pp. 152, 154.

[21] p. 138.

it in, might have preyed on her once she had settled back on Mauritius, forcing the attentions of odious male settlers on her and destroying her virginal purity. (In roughly similar fashion, Diderot dramatically, if perhaps inaccurately, describes the importation of venereal disease into Tahiti by travellers from the West in his *Supplément au Voyage de Bougainville.*) Had she been able to retain her virtue, the difficulty of life in this natural paradise would have ensured that she, Paul, and their children remained poor and eternally hard-pressed.[22] As it is, we know that nature has already done her worst. Virginie has died in the shipwreck, although she need not have done had she sailed at a more clement time of year, or forgotten her 'pudeur' and stripped naked in front of the sailor who tried to save her from drowning.

This catastrophe draws attention to a major disadvantage of life in the unworldly setting of pastoral. As the old man reflects, a bucolic paradise provides too little with which to test man's fibre. 'Natural' events are rarely of alarming proportions, perhaps; but when they are, they threaten to destroy even those who are not materially struck by them. Thus 'les malheurs du premier âge préparent l'homme à entrer dans la vie, et Paul n'en avait jamais éprouvé.'[23] As a result, he dies.

In other words, Paul's experience need no more have been fatal than Virginie's at sea. In the former case it was so because a human nature had been insufficiently formed by life to withstand sudden shock; in the latter, because an individual had been removed from her 'natural' setting and imbued with foreign notions, unnatural in their very foreignness. Neither of these themes is very effectively worked out by Bernardin, and as a consequence it is hard to be certain whether he is really attacking the natural upbringing which Paul and Virginie have enjoyed or not. It is difficult, too, to see why Saint-Pierre, a writer who seems to have believed celibacy to be an unnatural and melancholy state (he remarks in the *Voyages de Codrus* that 'tous les célibataires sont tristes,'[24] and has the old man say in *Paul et Virginie* that having a life-companion is preferable to leading a solitary existence),[25] should apparently wish Virginie to be worshipped as a virgin-goddess. He clearly sanctions the old man's

---

[22] p. 216.   [23] p. 223.

[24] *Voyages de Codrus*, *Œuvres complètes* (Paris, 1820), xviii. 437.

[25] p. 166.

consoling words to Paul that his beloved has died 'pure et inaltérable comme une particule de lumière',[26] as though to leave life in the state of virginity was better than to die after a lifetime of natural fertility, like the mother-figures extolled in other eighteenth-century works of fiction. But what does emerge clearly from the end of *Paul et Virginie*, where Paul, Marguerite, the devoted slaves Domingue and Marie, and the dog Fidèle die in quick succession, followed a month later by Mme de la Tour, is his desire to show how the natural law of death holds all in its sway.

The operettas based on Bernardin's novel, by Favières and Kreutzer, Dubreuil and Lesueur, save Virginie from drowning.[27] In the pastoral romance *Daphnis and Chloë* Longus, like other writers of the Hellenistic period, describes the foundering of a ship at sea, but uses it to advance the happiness of his principal characters rather than frustrate it. When pirates abduct Daphnis, Chloë plays her pipe to summon back the cattle they have also taken on board ship: the excited trampling of the animals causes the ship to sink, and permits the light-footed Daphnis (who, unlike Virginie, takes off his clothes) to make his escape on the back of the oxen, who carry him ashore. In the seventeenth-century 'heroic' novels influenced by ancient Greek and Roman romances, the topos of shipwreck is an incidental occurrence which adds drama to the proceedings but does not lastingly affect the fortunes of the principal characters.[28] But for Bernardin it is of central importance, however ineptly conceived and executed. Virginie's death by drowning is meant to seem both tragic—in that it appears a cruel waste—and edifying.

The edification comes, if Virginie is to be believed, from the fact that she has been found faithful to the laws of nature, love, and virtue. She went to Europe at her mother's prompting and returned at the command of her aunt; she has therefore done her 'natural' duty towards her family. She renounced riches to remain true to Paul, and has thus stayed faithful to the laws of the heart. And she preferred death to the violation of her modesty, so preserving her virtue intact.[29] Perhaps to the modern reader these achievements are not very considerable; but to Saint-Pierre and his contemporaries they

---

[26] p. 221.
[27] See Jean Fabre, '*Paul et Virginie* pastorale', *Lumières et romantisme*, new ed. (Paris, 1980).
[28] See M. Magendie, *Le Roman français au XVIIe siècle* (Paris, 1932), *passim*.
[29] p. 221.

evidently were. Virginie's drowning, in particular, was rated one of the great death-scenes of eighteenth-century literature, to be ranked alongside Manon's in the American desert and Julie's at Clarens.[30] When later adaptations of Saint-Pierre's story avoided its tragic conclusion, it seems safe to say, they did so less for aesthetic reasons than for sentimental ones. For Bernardin, on the other hand, Virginie's death is the natural climax of the novel. It gives greater magnitude to the action than could have been procured by a conventional happy ending. It provides an element of drama which the idyll generally lacks, or which it possesses only intermittently,[31] and thus has artistic point. Above all, it emphasizes a notion on which the old man dwells at length in his conversation with Paul: that love necessarily perishes, for the natural law of death is a final one against which there can be no appeal. Or, if there can, it is only in the afterlife that love may continue changelessly, as Julie also realizes in *La Nouvelle Héloïse*. For Virginie this awareness came after the destruction of Paul's garden, the symbol of his love for her: 'il n'y a que le ciel qui ne change point.' The old man imagines Virginie telling her desolate lover:

Soutiens . . . l'épreuve qui t'est donnée, afin d'accroître le bonheur de ta Virginie par des amours qui n'auront plus de terme, par un hymen dont les flambeaux ne pourront plus s'éteindre. Là j'apaiserai tes regrets; là j'essuierai tes larmes. O mon ami! mon jeune époux! élève ton âme vers l'infini pour supporter des peines d'un moment.[32]

This, surely, was Saint-Pierre's reason for writing a Christian pastoral; and the artistically puzzling or clumsily handled features of his novel, such as the worshipping of Virginie's purity, may be an essential part of that aim. Virginie can be changeless *because* she is pure; she has done nothing that calls for alteration, and so she may have eternity.

Bernardin's desire to have nature speak symbolically in *Paul et Virginie*, and to do so in a way which suggests the close connection between the functioning of the natural world and that of its human inhabitants, is evident throughout the novel. Often, although not invariably, nature's workings evoke the development or suppression

---

[30] See John McManners, *Death and the Enlightenment* (Oxford, 1981).

[31] Apropos of Florian's *Estelle*, the Vicomte de Ségur allegedly remarked: 'Ces bergeries sont charmantes; mais elles le seraient bien davantage si de temps en temps on y rencontrait quelques loups.' See *Fables de Florian* (Paris, 1858), p. 3.

[32] p. 222.

of emotion in the characters. Thus Virginie's pubescence, whose symptoms drive her to seek refreshment in the pool where their mothers had bathed her and Paul in their infancy,[33] and make her burn with ardour as she thinks of him, occurs during a summer of unusual drought in Mauritius. When a hurricane ends the weeks of dryness, the pool is found to have been silted with mounds of sand.[34] But this obstacle, if such it be, does not signify the termination of Paul and Virginie's love. Although the source of fertility has been dried up, the coconut trees remain standing with their tops still intertwined.

Nature wears a dual aspect in *Paul et Virginie*, sometimes sympathetic and sometimes hostile towards man. Yet the evidence of natural destructiveness is not always straightforward. The storm that sinks the *Saint-Géran*, for instance, would not have been fatal had humans respected nature's rhythm: no native of Mauritius, we are told, would have set sail during the hurricane season. But the destruction of Paul's garden shows only man's impotence in the face of cosmic forces, which can scarcely be seen here to work in his favour. Indeed, nature appears on this occasion to mock the human effort to govern its variability by imposing the order of art. The seventh of the *Etudes de la nature* similarly highlights nature's tendency to punish man's self-assertiveness.

Si les orages détruisent quelquefois ses vergers et moissons, c'est qu'il les place souvent dans des lieux où la nature ne les a pas destinés à croître. Les orages ne ravagent guère que les cultures de l'homme. Ils ne font aucun tort aux forêts et aux prairies naturelles.[35]

If this is true, it must be concluded that nature is decidedly unsympathetic to man. Confounding the lessons professed elsewhere in the *Etudes* and in the *Harmonies de la nature*, Bernardin seems to be arguing that the old oneness between nature and man which held in past golden ages, and in the world of pastoral generally, no longer exists. The other harmonies which accompanied that union, and particularly the harmony of love, must likewise vanish, leaving a sense of anguished incompleteness behind. The unfulfilled yearning for union,[36] which Goethe's Werther epitomizes, is characteristic of literature that deals with the natural world and human passion.

---

[33] p. 134.   [34] p. 136.
[35] *Etudes de la nature*, *Œuvres complètes* (Paris, 1820), ii. 88.
[36] See Richmond Y. Hathorn, 'The Ritual Origin of Pastoral', *TAPA*, 92 (1961), p. 237.

It is undeniable that the literary depiction of fulfilled love in nature can appear anodyne. The pastoral framework in its traditionally one-sided presentation of nature as peaceful and tame contains a story essentially constructed for urban man. Nature is described in the bucolic poetry of Theocritus and Virgil as if it were a sentient being attuned and sympathetic to humans; and in a sense this is clearly a fanciful interpretation, not far from Ruskin's 'pathetic fallacy'. The humorous eroticism of Longus's narrative saves it from the flatness which can accompany calm portrayals of pastoral life. Such sensuality as informs *Daphnis and Chloë* was, however, later deemed reprehensible by theorists of pastoral: Florian, for example, claimed that the love of shepherds should be as pure as the crystal water of streams.[37] Yet it gives considerable artistic, if not moral, force to Longus's romance.

Many other novels of the eighteenth century link love with the natural world, whether the latter be the confined and cultivated space of the garden or the vast expanse of meadow, moor, or mountain. *La Nouvelle Héloïse* and the novel it influenced, *Die Leiden des jungen Werther*, give perhaps the most celebrated novelistic expression to the eighteenth century's fascination with the union of nature and man's passions. In both of them it has the dual aspect of blessing and curse: nature uplifts and enlightens, but also casts down and threatens. In both Rousseau and Goethe it seems to feel with the suffering victims of love. When Saint-Preux is exiled by Julie, he makes for the rocky Alpine crags whose ungratefulness matches the barrenness of his emotional state; when Werther experiences the torment of unrequited love, he identifies his inward desolation with the gloom of Ossian's landscape. If nature brings maternal comfort and warmth, it may also have the stepmotherly aspect to which Kant refers, or exhibit the miserliness which Schopenhauer finds in it. Because it both provides and, through the trick of time, removes sustenance, it can mirror the plenitude, mutability, and poverty of the lover's existence. The most poignant image of its infertility is presented in the closing pages of *Manon Lescaut*, where the dry desert in which Manon perishes symbolizes the end of the richness associated with love. For des Grieux and Manon, the erstwhile profligates, the source of bounty has dried up. If the tamed nature

---

[37] See Winfried Engler, 'Jean-Pierre Claris de Florian, *Estelle*, roman pastoral?', *ZfSL*, 78 (1968), p. 315.

within the Louisiana settlement illustrates the responsible emotional maturity gained by the couple and the new orderliness of their habits, the aridness of the desert signifies the unwelcoming hostility that surrounds them even in the New World. Nature's sterility, marked by the extinguishing of Manon's life, is subsequently echoed in the emptiness of des Grieux's existence, bereft as he is of the woman who gave it meaning.

Prévost's work reminds us of a further sense in which nature and eighteenth-century literature are intimately linked. Rousseau's name is most often invoked in this connection, and the meaning of 'nature' is evinced in its opposition to 'culture' or 'civilization'. As Diderot noted in his *Essai sur Sénèque*, however, neither Rousseau nor his age invented the contrast between the primitive and the artificial on which the *Discours sur les sciences et les arts* and the *Discours sur l'origine de l'inégalité entre les hommes* waxed eloquent. It is found in classical antiquity, both in the form of pastoral literature and in other speculations on the mythic age of gold which was popularly believed to have preceded the artificial present. The essence of the opposition is present in the contrasting *téchnē* and *phýsis* of the Greeks. In this sense *téchnē*, art, is what is practised in towns, or places where men love 'artfully' rather than according to the laws of nature (*phýsis*). *Emile et Sophie* shows how the *téchnē* encouraged by urban living can be inimical to the flourishing of responsible love, particularly that between husband and wife. Emile's mentor, preparing his pupil in country fastness for eventual marriage, feared the dangers of *urbanitas* for reasons whose validity Emile and Sophie will later acknowledge, like des Grieux and Manon before them. The temptations of the city lead to a denaturing of the personality, which in its most acute form culminates in moral bankruptcy. The simplicity of a rural existence, on the other hand, means that there is little encouragement for man to appear other than he truly is, with the consequence that his moral character has ample opportunity to develop. Genuineness of emotion, on such an interpretation, is a 'natural' product of a person's mode of living.

The logic behind such reasoning as this will not commend itself to everyone, but it satisfied the many eighteenth-century writers who incorporated it in some form into their imaginative works. There were celebrated dissenters, however. Marivaux, an agnostic rather than an out-and-out enemy of what later came to be called the school of Rousseau, at least considered the possibility in his novels and plays that comparative civilization may be no less favourable to the growth of stable emotional relationships than rustic simplicity. But the works in which he explores this idea are either too stylized in form to carry much weight—as witness the part-pastoral comedy *La Double Inconstance*—or too equivocal in their treatment of the question to offer much comfort to supporters of either persuasion. While conveying none of the pessimism about the harmful effects of urban life that marks Restif's *Le Paysan perverti* and *La Paysanne pervertie*, Marivaux's *Le Paysan parvenu* fails to demonstrate that love can acquire much depth in a world of sociability. Rather, Jacob's removal from the comparative isolation and simplicity of the country to the French capital serves initially only to increase the number, but not necessarily the quality, of the sexual opportunities available to him. We never discover whether or not more has meant worse, as Jacob tells us little about his erotic experiences before coming to Paris. His emotions, however, are not evidently perverted by his removal there: his opportunism, which is always tempered by a degree of authentic feeling, does not extend to conscious exploitation of the women with whom he becomes involved; and if he displays no superhuman virtue in his scaling of the social ladder, he is at no time tempted into vice. He knows that his air of rustic ingenuousness will stand him in good stead in his relations with society ladies, but the amused irony with which he lets this become apparent prevents Marivaux's reader from judging him adversely. In the case of Marianne it is no clearer that the city poses a threat to the individual who seeks a lasting emotional relationship. Certainly it encourages in Marianne something of the superficiality to which the Silvia of *La Double Inconstance* is also prey in the court environment. In particular, town life develops feminine vanity, which is constantly fed by luxuries of the kind Climal offers to the coquettish Marianne. Yet the girl's moral character does not appear to have been fundamentally weakened by such attentions; it is Valville, not Marianne herself, who causes their relationship to be broken off. No more than in *La Double Inconstance*, where Silvia remains true to her 'jeune officier' until she realizes that he and the Prince are one and the same, do the temptations of refined life appear to have a destructive effect on sentiment.

But other 'worldly' novels of the eighteenth century—*Manon Lescaut*, *Les Egarements du cœur et de l'esprit*, Restif's *Paysan* and *Paysanne pervertis*, *Les Liaisons dangereuses*—suggest the prevalence of such corruption. It is true that some make no explicit comparison with a different order of things in the countryside such as might indicate that the answer to urban degeneracy is to be found in the rustic world. In *Les Egarements* we learn that Hortense has no liking for the latter, and that her mother has therefore been obliged to return with her daughter to the capital. If Crébillon believed purity and rusticity to go together, this might imply that Hortense is not to be the saviour of Meilcour's virtue to whom Crébillon alludes in his preface. Equally, it might mean that he thought the town to be as capable as the country of promoting man's moral growth. A third possibility is that he considered city life to be a more interesting background for his novel of sentimental education; a fourth, that he intends no special significance to be attached to Hortense's stated preference. As the book is unfinished, we have no way of knowing. What can be said with confidence is that, on the evidence of *Les Egarements*, the metropolis encourages modes of living, at least in its wealthy and leisured inhabitants, which make stable love-affairs unlikely to prosper there.

In *Le Paysan perverti* and *La Paysanne pervertie* Restif describes the city's effect on erstwhile innocents in terms which point to his affinity with Rousseau. Just as *Emile et Sophie* unambiguously declares that the emotions are corrupted by urban life, so Restif leaves no doubt that the cause of his peasants' perversion is their exposure to the town's wicked ways. It is significant that Edmond's dream anticipating Ursule's ravishment in the city should take the pastoral form it does. He sees his sister tending her sheep in the country, and a wolf carrying off the prettiest ewe; Ursule tries to save the sheep, but is carried off herself in the process. The wolf then turns into a man, and Ursule caresses him. Edmond warns her against her action, she fails to hear the warning, and the man devours her on becoming a wolf again.[38]

Although Crébillon offers in *Les Egarements du cœur et de l'esprit* no direct hint as to whether a rural environment would be more likely to foster stable relationships than an urban one, it is clear that many of the circumstances which militate against settled love in the city are absent from country living. The rustic world knows no such regular gatherings of the *beau monde* as occur in the capital, and consequently, we may guess, puts people under no such obligation as is felt in Paris to cultivate the insincerity in emotional relationships

---

[38] *La Paysanne pervertie*, pp. 48–9.

which so shocks Meilcour—the prescribed need to abandon a loved person immediately he or she has been sexually enjoyed, for example, purely because etiquette deems lasting love-affairs to be bad form. Away from the city, as Laclos's Valmont reveals, a different rhythm is observed.

For some the countryside offered, not a place in which to pursue love, but a refuge from it. In the eighteenth-century *libertin* novel, as well as in works like *Paul et Virginie* which stood as foils to it, the city often represents the dangers of passion and the country their antidote. *Le Paysan perverti* and *La Paysanne pervertie* set against the metropolis that corrupts Edmond and Ursule the honest rustic milieu from which they come. It is to this safe retreat that the regenerate Ursule is welcomed back by her parents and their brood, in a scene worthy of Greuze at his most sentimental.[39] If Paris has led her and her brother astray, the village with its genuine sense of community will bring them back to moral health, and inspire them anew with the decent values which urban debauchery has corroded.

For Faublas too, removal from Paris is effectively removal from the source of love-madness, and his rebirth occurs in the bosom of nature. Where the temptation to indulge in frenetic sociability no longer exists, the sufferer can re-establish contact with the life force of regular love. Safe in the calm of his English garden, Faublas perceives the dangers he has escaped with the clarity that had attended his initial perception of Paris, as a sixteen-year-old boy:

Je cherchai cette ville superbe dont j'avais lu de si brillantes descriptions. Je voyais de laides chaumières très hautes, de longues rues très étroites, des malheureux couverts de haillons, une foule d'enfants presque nus; je voyais la population nombreuse et l'horrible misère. Je demandai à mon père si c'était là Paris.[40]

His tones echo those of the young Rousseau, disabused by his first encounter with Paris:

Je m'étais figuré une ville aussi belle que grande, de l'aspect le plus imposant, où l'on ne voyait que de superbes rues, des palais de marbre et d'or. En entrant par le faubourg Saint-Marceau, je ne vis que de petites rues sales et puantes, de vilaines maisons noires, l'air de la malpropreté, de la pauvreté, des mendiants, des charretiers, des ravaudeuses, des crieuses de tisanes et de vieux chapeaux.[41]

---

[39] Ibid. 373–4.   [40] p. 419.   [41] *Les Confessions*, p. 159.

All this made such an adverse impression on him, Rousseau writes, that the real magnificence he subsequently saw there was powerless to destroy its effect. Consequently he has always had a lingering dislike for living in the capital.

In Louvet's novel the Baron de Faublas gives a chilly response to his son's dismay, and remarks that he will soon be seeing more elegance than he initially meets with. The consequences of Faublas's discovering this refined milieu and the *beau monde* that inhabits it are related over the ensuing 800 pages of the novel. It is Faublas's weakness for women (as well as for dressing like them), which announces itself soon after his arrival, that leads to his eventual confinement in the country; and Dr Willis tells the Baron that his son must never set foot in the city again, lest he should encounter unsettling reminders of his unfortunate experiences there.

The 'civilized' values associated with urban living also pose a threat to young lovers in *Paul et Virginie*. It is less the antithesis of *Une année de la vie du chevalier de Faublas* than its complement, as *Les Liaisons dangereuses* is the complement to *La Nouvelle Héloïse* (whose epigraph, 'J'ai vu les mœurs de mon temps, et j'ai publié ces lettres', Laclos borrowed for his own work). In *Paul et Virginie* Saint-Pierre uses the familiar theme of escape from civilization to an uncultured retreat as the basis for a book which he described as 'une espèce de pastorale', one that would rival the literary bucolics of antiquity.[42] Virginie's unhappy experiences in the French capital, her refusal to marry a wealthy elderly Frenchman for the sake of keeping faith with Paul, her rejection by the aunt, her enforced voyage back to Mauritius during the hurricane season, all demonstrate the perils to which those who cease living according to nature's rhythm are exposed. It seems clear that Bernardin meant to argue the purity of life lived away from the 'civilized' world, where marriages are concluded as a matter of financial convenience between parties of disparate age,[43] and where love-matches such as that between Mme de la Tour and her husband, or between the illegitimate and penniless Paul and the better-born but equally penniless Virginie, are deemed unsuitable.

[42] ed. cit., p. cxlvi.
[43] Bernardin himself, however, was successively married to two child-wives. See Maurice Souriau, *Bernardin de Saint-Pierre d'après ses manuscrits* (Paris, 1905), p. 318 ff., and Fernand Maury, *Etude sur la vie et les oeuvres de Bernardin de Saint-Pierre* (Paris, 1892), pp. 187 ff.

Not even Mauritius, however, is an unsullied paradise. Although nature extends its benevolent protection to the children when they lose their way in the wild far from home, and although the island Saint-Pierre describes in his novel is a greatly idealized version of the Mauritius on which he had himself spent two dissatisfied years,[44] he does not portray this refuge from civilization in blandly eulogistic terms. Quite apart from the social inequalities Mauritian society preserves (Paul and Virginie get lost in the jungle after they have been to plead with the cruel owner of a runaway slave to show clemency towards the fugitive), nature itself proves a threat even to those who adapt themselves to its seasonal changes. The destruction of the garden which Paul created for his beloved is not the work of a jealous fellow-human, as a similar event is shown to be in *Daphnis and Chloë* (where a rival for Chloë's affections lays waste the flower-beds and plants which Daphnis has tended for his love); it is the unavoidable, if not wholly unpredictable, consequence of a meteorological phenomenon.

As the story of the slave suggests, a given environment may present to one person an aspect quite different from that which it presents to another. And although the little community formed by Mme de la Tour, Marguerite, and their respective children is a comparatively privileged one (for the families possess the two faithful slaves, Domingue and Marie), the life they lead is still far from easy. Mme de la Tour, as we saw, has been prematurely aged by her unrelenting labours and the climate of the tropics. The living to be made from nature is sufficient, but never easy. Only townspeople, it seems, view the natural world as straightforwardly fruitful, an Eden of pastoral simplicity quite without privation.

If the countryside was sometimes host to the amorous encounters of leisured people in eighteenth-century France, it was rarely a countryside which its permanent inhabitants would have recognized. The workaday world of ordinary country folk had little in common with the escapist paradise of the 'fête galante', where men and women had nothing to do but indulge in sentimental amusement.[45] The joys of the fête, a refined blend of music-making, dancing, promenading, and love-dalliance, could be indulged in only by those

[44] See *Voyage à l'Ile de France*, ed. Yves Benot (Paris, 1983), p. 114.
[45] See Robert Tomlinson, *La Fête galante: Watteau et Marivaux* (Geneva, 1981), p. 157.

with time on their hands, just as its setting (typically, parkland or extensive wooded garden) was available only to the well-off.

Although the fairy-tale allegory of Edmond's dream in *La Paysanne pervertie* is obviously intended as a caution against the dangerous pleasures of the town, the terms in which it is cast give a salutary indication of the non-ideal aspects of country life (which in Restif's two novels appear principally in its prevailing poverty, the original reason for Edmond and Ursule's departure for the city). This suggestion that rustic existences have their perilous or unromantic side finds confirmation in other novels published in the last two decades of the eighteenth century. Laclos and Louvet, as well as Saint-Pierre and Restif, show the lives of ordinary country-dwellers as far harsher than the idealizing pastoral makes them appear. For civilized eighteenth-century man, rural domesticity offered an image of tranquil settledness which he aspired to emulate in his emotional life. This is how the lyrical Léonard portrays the comfort of family community in the bosom of nature:

> Que n'ai-je le destin du laboureur tranquille!
> Dans sa cabane étroite, au déclin de ses ans,
> Il repose entouré de ses nombreux enfants;
> L'un garde les troupeaux, l'autre porte à la ville
> Le lait de son étable, ou les fruits de ses champs,
>   Et de son épouse qui file
>   Il entend les folâtres chants.[46]

These lines recall one of Greuze's idealizing family interiors, like the rustic scene of contentment depicted in *L'Accordée de village*. But other pictures by Greuze, like *Le Mauvas Fils puni*, tell a different story.

Cultivating a rural love away from Paris, Valmont desires to impress the 'sensible' Mme de Tourvel with a show of the philanthropy he knows she admires; and in so doing, he demonstrates that country idylls are rarely available to real country folk. Letter XXI, to the Marquise de Merteuil, describes how he saves a village family from having its effects confiscated for non-payment of taxes.[47] The scene, whose outcome Valmont engineers by settling the family's debt, reveals nothing so clearly as the true harshness of the existences so frequently romanticized by 'civilized' men. They alone, it seems, can

---

[46] *Les Regrets*, ll. 45–51.   [47] pp. 45–7.

afford to be the 'beaux bergers' of pastoral. Danceny, to whom the contemptuous Marquise gives this name as well as that of 'Céladon', is privileged enough to have no need of earning his living, and can therefore cultivate his love for Cécile in a world of music-filled leisure (he sings with her in her mother's house) worthy of *L'Astrée*'s amorous shepherds.[48] Similarly, *Faublas* offsets the idyllic imaginings of the romantic aristocrat, the Comtesse de Lignolle, against the actual sufferings that attend peasant life. If the Comtesse can lose herself in an invented pastoral world in which she and Faublas enjoy a paradisiacal rustic existence, as châtelaine of a country estate she is forced to acknowledge that the simple people she professes to envy undergo privations of which the urban rich often know nothing. To her credit, she shows herself at the age of fifteen to be a humane and conscientious mistress, freeing families from the threat of crippling rent increases and exercising a benevolent paternalism which recalls that of the Wolmars at Clarens. But the limitation in her understanding of the peasants' existence is sharply brought out by the fanciful scenario of pastoral bliss she constructs for her lover and herself.

In *Faublas* it is both the enforced conviviality of life in the capital and the sophistication (in the old sense of 'adulteration' or 'falsi-fication' as well as the modern one of 'refined smartness') of its ways that makes the Comtesse de Lignolle desire a country retreat for herself and her lover, where they can live untroubled by the claims of society (including Faublas's wife Sophie). The Comtesse imagines a state of bliss similar to that detailed by Rousseau in the fifth *Rêverie du promeneur solitaire* and the third letter to Malesherbes, where the complete happiness procured by surroundings of natural beauty combined with the absence of importunate company and a simple mode of living is described. The Comtesse's plan is that she and her lover shall escape the city with a sum of money realized by the sale of her jewels[49] and settle in a 'cabane petite et tranquille' in the country, living without maid or manservant and subsisting on the produce of their garden (which an obliging peasant couple will cultivate for them). Like Rousseau, they will live on eggs, dairy produce, and fruit; but if, like Rousseau, they have a desire for sweet things, the Comtesse will turn pastrycook: 'j'ai appris la pâtisserie, je te ferai des brioches, des galettes, et de temps en temps de bonnes petites crèmes.'

<hr />

[48] pp. 105–6.　　　[49] p. 1110.

Not all ties with society and its financial transactions will be severed, for the money remaining after the purchase of their cabin will be invested to yield a small but adequate income. Loving their fellows provided that they respect the lovers' privacy, they will devote half of their disposable income to succouring the poor and needy. And as a confirmed Rousseauist, the Comtesse will breast-feed her children[50] and teach them (irrespective of sex) to embroider and play the piano, as well as to make milk puddings. Unlike Paul and Virginie in Mauritius, the infants will be taught to read and write; they will not, however, learn the handling of arms, for '[dans] cette campagne, où nous ne serons environnés que de bonnes gens qui nous voudront du bien, qu'a-t-il besoin de tuer quelqu'un?' Mindful of what they owe to those who had charge of their upbringing, they will construct a second cabin adjacent to their own and install in it Faublas's widower father and the Comtesse's aunt Mme d'Armincour. Even the presence of Faublas's sister will be tolerated, for 'nous la marierons à quelque bon laboureur, à quelque honnête homme qui n'épousera pas son bien, mais sa personne, et qui la rendra plus heureuse.'[51] Gently, Faublas advises his mistress that her scheme is unworkable. Not only is he himself married and, unlike the Comtesse, in love with his spouse, but the Comtesse's money is necessary to support the legions of needy peasants who inhabit her estates.[52] Moreover, as married individuals who are both minors, they are under the triple authority of family, sovereign, and state.[53]

Other novels of the late eighteenth century refuse the simple identification of country living with the enjoyment of unsullied love-relationships, or with freedom from the toils of love set by city-dwellers. In the cautionary tale of Justine, Sade shows how innocence may be beset in rural isolation as well as amidst the artifices of the town. Indeed, the very solitude of the country may facilitate such assaults on virtue as Justine suffers. The fact that the monastery of Sainte-Marie-des-Bois, to which she is directed after escaping from the household of the evil Dr Rodin, is far from all human habitation means that the vile practices of its four monks pass unnoticed by the rest of the world.[54] The sexual horrors perpetrated against the

---

[50] Rousseau strongly disapproved of mothers refusing to breast-feed their own babies. On the 18th-century use of wet-nurses, see Charlton, op. cit. 146–7.

[51] p. 1112.   [52] pp. 904 ff.   [53] p. 1113.

[54] Sade, *Les Infortunes de la vertu* (Paris, 1969), p. 109.

girl by Raphaël and the others end only when a troupe of saintlier men take over the 'cloaque d'impureté et de souillure' which she has unwittingly entered.[55] Before that time she is subjected, along with the other females imprisoned there, to rape, sodomy, and all manner of indecent practices by the monks. Later, on the road to Vienne, Justine saves a traveller whom she sees being set upon by two men on horseback, and accompanies him back to his château on the frontier of Dauphiné.[56] Her reward is to be kept prisoner by her beneficiary, chained for twelve hours a day to a water-wheel in order to assist his counterfeiting operations. In this prison too she becomes the victim of male lust, raped by the very man to whom she had earlier performed an act of mercy.[57]

A more subtle, but scarcely less chilling, illustration of the same idea, that the countryside offers scant protection against sexual persecution, is provided by *Les Liaisons dangereuses*. So unsafe is Mme de Rosemonde's château that the Présidente finds flight to Paris her only resource in the face of Valmont's pressing attentions.[58] According to Mme de Merteuil, the Vicomte's own absence from the capital is regarded with contempt by the Parisian *beau monde*, who assume him to be engaged in an 'amour champêtre' with some simple country wench.[59] But Valmont has earlier informed her of the intensity which absence from the city and its multiple distractions confers on amorous affairs: 'Je n'ai plus qu'une idée; j'y pense le jour, et j'y rêve la nuit.'[60] The Présidente, meanwhile, plainly shares the widespread belief that country air and removal from town vices have a salutary effect even on hardened libertines. When Mme de Volanges warns her that Valmont is a 'liaison dangereuse' for one so pure as the Présidente, the younger woman replies that the erstwhile rake evidently laid down his deadly weapons before entering his aunt's house, and now seems merely a 'bon enfant'. 'C'est apparemment l'air de la campagne qui a produit ce miracle',[61] for Valmont has uttered not a word 'qui ressemble à l'amour' since arriving there. The most she is prepared to suspect is that he is sentimentally involved with a local girl, since '[il] y a bien quelques femmes aimables à la ronde.' But this is perhaps disproved by the fact that

---

[55] p. 115.   [56] p. 150.   [57] p. 154.
[58] A country château is also the background to scenes of sexual libertinage in Crébillon's *conte dialogué*, *La Nuit et le moment*.
[59] p. 259.   [60] p. 18.   [61] p. 32.

he goes out little save in the morning, allegedly to hunt (although 'il rapporte rarement du gibier'). Once Valmont has departed for Paris, she believes her persecution to be over; but he continues to woo her by letter.

The next change of location involves Mme de Volanges and Cécile, for the former has decided that only removal to Mme de Rosemonde's château will ensure her daughter's safety from the inconvenient attentions of Danceny.[62] Since Valmont has by this time discovered that Mme de Volanges was responsible for the blackening of his name with the Présidente, and is therefore willing at last to undertake the seduction of Cécile, as Mme de Merteuil has long urged him to do,[63] he too makes once more for his aunt's country seat.[64] So completely has this rural refuge replaced Paris in his eyes as the hub of erotic activity that he can write to his confidante 'que dans le triste château de ma vieille tante je n'ai pas éprouvé un moment d'ennui. Au fait, n'y ai-je pas jouissances, privations, espoir, incertitude?'[65] Later on, however, he is to turn the tables on the very Mme de Merteuil who earlier chided him for his obstinate confinement to the country. Now it is she whom a lover has transported away from Paris (but a lover of whom the Marquise is tired); hence Valmont's triumphant cry in letter CXV, 'Vous voilà à la campagne, ennuyeuse comme le sentiment, et triste comme la fidélité!'[66] And Mme de Merteuil intends to use for the purposes of breaking with Belleroche the very property of country life which Valmont found a spur to his pursuit of the Présidente—its utter monotony. Her party, she tells the Vicomte, will consist of Belleroche, herself, and

quelques personnes désintéressées et peu clairvoyantes, et nous y aurons presque autant de liberté que si nous étions seuls. Là, je le surchargerai à tel point d'amour et de caresses, nous y vivrons si bien l'un pour l'autre uniquement, que je parie bien qu'il désirera plus que moi la fin de ce voyage dont il se fait un si grand bonheur; et s'il n'en revient pas plus ennuyé de moi que je ne le suis de lui, dites, j'y consens, que je n'en sais pas plus que vous.[67]

In the meantime, Valmont professes disbelief in her intention to switch affections from Belleroche to Danceny, a child in comparison with her other lovers,[68] and attributes her sudden whim to the fact

---

[62] pp. 126, 217. [63] pp. 133, 227. [64] p. 152.
[65] p. 220. [66] p. 268. [67] pp. 263–4. [68] p. 268.

that Paris is languishing in the 'dead' season.[69] He is later to use the
same circumstance as part-explanation of his own involvement with
Mme de Tourvel:

Peut-être aussi la saison morte [the summer months] dans laquelle est venue
cette aventure m'a fait m'y livrer davantage; et encore à présent [November],
qu'à peine le grand courant commence à reprendre, il n'est pas étonnant
qu'elle m'occupe presque en entier.[70]

The *beau monde* generally reserved its love-campaigns for the winter,
when Paris society picked up after the recess. It is a pleasant irony
that the love-lives of creatures so out of touch with the world of
nature should nevertheless be governed by seasonal variations.
It scarcely needs saying, however, that Valmont's excuse to the
contemptuous Marquise is the merest pretext.

What country living offers Valmont is time—time for a relationship
to develop away from the rush of the city, where liaisons are quickly
formed and as quickly broken off. The countryside follows a slower
rhythm. It resists the attempt to force growth for the sake of swift
enjoyment, regarding such a procedure as unnatural. Nature has no
need of the artifice employed by those who chafe at its leisurely pace.
It allows things to ripen of their own accord rather than speeding
their growth by extraneous means. Like the gardener Addison, the
true nature-lover of the eighteenth century spurns the glasshouse.
For the *libertin*, on the other hand, speed is of the essence. He prefers
the fruits of cultivation to those which ripen slowly in the wild; and
the likelihood that forced growth will wither more quickly than
unforced worries him little, for his erotic life is devoted to the
ephemeral.

As we have seen, this creed is a central one in Laclos's novel.
Valmont senses that he will be ill-served by attempting to use society's
methods of seduction on the Présidente, that he can never win her
love except by allowing time to run its course. And love is what he
desires from Mme de Tourvel. He knows that this emotion is rooted
in time, unlike the sensations that alone preoccupy the *libertin*, and
which have no permanence. Love requires patient husbandry. Hence
his distaste—here Valmont borrows a metaphor from another
countryside activity—for the huntsman's quick kill. The fact, too,
that he loves the Présidente as well as wanting her love for himself

[69] p. 269.   [70] p. 309.

means that he luxuriates in time when he is with her. Like Danceny, he feels no impatient need for advance, for in idylls time stands still. However, such immobility may be artistically regrettable, for it precludes dramatic development; hence the kind of counter-idyllic peripateia met with in *Les Liaisons dangereuses*, *Paul et Virginie*, and *Faublas*.

'Natural' impulses, according to the authors of these three novels, may be freely indulged neither by the inhabitants of cities nor by country-dwellers. There is a time and a place for sexual activity. Limits may be set by age or marital status, or by a consciousness of what is owed to or expected by the society in which the individual lives. Parents have a certain influence over the emotional lives of their offspring, as both *Paul et Virginie* and *Faublas* demonstrate, even if, on the evidence of the latter work and *Les Liaisons dangereuses*, their begetters seem strangely neglectful of the young's education in crucial areas of erotic experience.

Many of the idylls described in the eighteenth-century French novel run foul of the constraints imposed by parental desire. Des Grieux calls his father a 'père barbare et dénaturé' for refusing to sanction his relationship with Manon.[71] When the couple finally attempt to regularize their union, it is in the still primitive culture of the New World where such prejudices as those of des Grieux's father should hold no sway. (The discovery by the Governor of the colony that the pair are not married still has adverse consequences, but less because their relationship is felt to be improper than because the Governor's nephew Synnelet wants Manon for his own wife.)

Faublas's case is somewhat different. When he first encounters Sophie in his sister's convent, he falls as spontaneously in love with her as des Grieux had done with Manon. Sophie is the same age as the latter, and Faublas a little younger than des Grieux. In the eyes of the Baron de Faublas this makes his son unripe for matrimony; but he is merely requested to postpone the fulfilment of his desires, not suppress the thought of wedding his beloved altogether. Like des Grieux, Faublas eventually unites himself with his mistress informally, taking advantage of her unconsciousness as she swoons on hearing that he is about to fight a duel.[72] But if Faublas is right in assuming that his sexual possession of Sophie makes it impossible for his or her father to prohibit their marriage,[73] the remaining two

[71] p. 172.   [72] p. 706.   [73] p. 707.

books of Louvet's novel reveal that the social obstacles to their full union are, if not insuperable, still considerable. One of them disappears when Sophie is discovered to be the long-lost daughter of the Baron's friend M. de Portail, to whom the Baron had earlier promised to marry his son. Another, Faublas's 'natural' promiscuity, is the cause of M. de Portail's abducting Sophie after the couple's wedding, and also of Faublas's unwillingness to ensure recovery of his wife as expeditiously as possible; for he has meanwhile fallen deeply in love with two other women.

The loudest complaint against parental contempt for the natural impulses of their offspring, however, come from a different quarter. At the origin of high society's pervasive 'galanterie', according to Louvet's Marquise de B***, is the 'civilized' habit of marrying off children for reasons other than those of natural attraction. Her moving speech to Faublas in the second book of the novel bitterly criticizes the institution of the 'mariage de convenance' which, for the sake of ensuring social prestige and financial stability, causes a girl of fifteen to wed a man of fifty. (In the Comtesse de Lignolle's case, such ambitions have resulted in the concluding of a marriage that can have no issue, since the Comte is impotent as well as elderly.) The Marquise bitterly indicts a system which prevents love-matches, and which instead pairs individuals for reasons that have nothing to do with affection:

Une fille qui s'ignore elle-même tombe à quinze ans dans les bras d'un homme qu'elle ne connaît pas. Ses parents lui ont dit: 'La naissance, le rang et l'or constituent le bonheur; tu ne peux manquer d'être heureuse, puisque sans cesser d'être noble tu deviens plus riche, ton mari ne peut être qu'un homme de mérite, puisqu'il est homme de qualité.'[74]

Mme de Lignolle similarly upbraids those who, as adults, should have used their experience of the world to warn the innocents about to be sacrificed of what awaited them, and where possible avert their fates, rather than encourage unions that are deemed suitable for purely materialistic reasons. The voice of maturity, she tells her aunt, ought to be the counsellor of innocence; but Mme d'Armincour resisted the desires of her niece's parents too weakly to prevail over their avarice. She failed to warn the Comtesse that she was about to forfeit the inalienable freedom of natural creatures to dispose of

---

[74] p. 747.

their affections as they please, and allowed her to be deceived by the empty promise of social advantage and wealth.[75]

Before her 'education' by Faublas, the unhappy Comtesse remained in childish ignorance of what her natural urges meant, despite her married state. Knowing nothing of the art of love-making, like the children of nature in Longus's *Daphnis and Chloë*, she understood neither what sexual desire was nor how it might be assuaged. What Mme d'Armincour in her criminal negligence failed to alert her niece to was the significance of the natural process of change within her, and to which a well-assorted marriage would have been a proper response:

quand une fille naturellement vive se montre au printemps émue du spectacle de la nature, est surprise dans de fréquentes rêveries, avoue des inquiétudes secrètes, se plaint d'un mal qu'elle ignore, on dit communément qu'il lui faut un mari . . . [mais] je crois qu'il lui faut un amant.

The nature-metaphor continues as the imagined Mme d'Armincour considers the valetudinarian husband which it is proposed to give the girl: 'ta jeunesse à peine commencera, que son automne sera fini.'[76]

On reading the veiled description of the Comtesse's sexual awakening, we may be reminded of the similar passage in *Paul et Virginie*, published two years before *La Fin des amours de Faublas*. The 'mal inconnu' which strikes Virginie, the shadowing of her blue eyes, the yellowing of her flesh-tone, and the general languor of her body signal the onset of sexual maturity.[77] Like the Comtesse she is overcome by melancholy, and develops a taste for solitude. The natural response to this change, but one which is not forthcoming from her mother, would have been to find her a husband, or at least to explain to the innocent girl the meaning of the changes affecting her organism. Mme de la Tour does neither of these things, despite the fact that the family's life away from the etiquette-bound society of eighteenth-century France has allowed a much closer bond to develop between mother and daughter than would have been usual in the 'civilized' world of the aristocracy. One reason for her failure to arrange the marriage that would 'cure' Virginie is the fact that Paul, for whom she seems destined, lags behind her in sexual development, being still at the pre-pubertal stage. But Virginie's mother is also unwilling to permit her daughter to marry young.

[75] p. 1007.   [76] p. 1097.   [77] p. 132.

Neither Mme de la Tour nor Marguerite questions the natural rightness of the children's relationship, or doubts that it should end in marriage. Nor is Virginie's mother so marked by her own upbringing as to believe that union with some well-born Frenchman should be preferred to becoming the wife of a peasant-woman's son, although she impresses on Paul the need to earn financial independence before marrying Virginie. Mme de la Tour's view remains the one that Mme de Volanges eventually adopts in Cécile's case, namely that her daughter should be allowed to have the husband of her choice. But the obstacle of Paul and Virginie's youth remains a considerable one. This is perhaps the more surprising in that, at least on the evidence of other late-eighteenth-century novels, the *beau monde* does not itself see youth as a barrier to marriage. The Comtesse de Lignolle is a child-wife, as Cécile is destined to be (both of men considerably older than themselves). In *La Fin des amours de Faublas* the Comtesse's cousin, Mlle de Mésanges, weds Faublas's friend Rosambert at the age of fifteen, but not before she has learnt to 'causer ensemble' with Faublas (disguised as a woman) in bed. When she appears surprised that Faublas possesses a male organ, she is reassured by him that this endowment comes when one is 'bonne à marier'—in Faublas's own case, he tells her, at the age of sixteen.[78] She has therefore lost her virginity by the time of her wedding; and Rosambert, whose ambition (like Gercourt's in *Les Liaisons dangereuses*) was to marry a virgin, is chagrined to find himself the possessor of second-hand goods.[79] Unlike Gercourt, however, he seems to have cherished this ambition solely for the sake of the pleasure Valmont also enjoys with Cécile, that of observing the 'airs de lendemain' after a girl's first night of love.

The question whether it is natural to prolong innocence, or conversely whether the sexual induction of youth is 'in nature', is a recurrent theme of the eighteenth-century novel. Prévost leaves his reader uncertain what, precisely, were the 'penchants au plaisir'[80] which decide Manon's parents to commit her to a convent, but des Grieux's recognition at the start of their affair that 'Manon était beaucoup plus expérimentée que moi'[81] suggests that even at the age of fifteen she has acquired some sort of sexual knowledge. Throughout the story des Grieux emphasizes the youth of the lovers, but less in order to suggest the imprudence of forming a sexual

---

[78] p. 1080.   [79] p. 1143.   [80] p. 20.   [81] Ibid.

relationship at their age than as an extenuating circumstance for the follies that love leads them to commit. Struck by the lightning-bolt of passion, he cannot but see its swift consummation as natural. Indeed, the description he gives of his life before meeting Manon makes its single-minded devotion to scholarship, which entailed complete ignorance of the opposite sex, sound contrary to the natural order of things. His father recognizes that des Grieux is temperamentally unfitted for celibacy, and seems to respond favourably to his son's argument for taking a mistress; and implicit in the debate between des Grieux and Tiberge at Saint-Lazare about the good life is the assumption (a common one in Enlightenment polemic) that celibacy is not just difficult to practise, but may actually be pointless. In retrospect, des Grieux cannot see the ignorance of his ways before the encounter with Manon as blissful. Rather, it seems a state of incomprehension whose termination he welcomed (at least until the time of Manon's first betrayal) because it gave him access to the pleasure that most conduces to human happiness.

To be young in the state of nature, however, may be to lack the social encouragements to sexual experience which the city environment provides, and which appear normal in civilized milieux. When Rousseau claimed that children were corrupted by urban life, and had Emile's emotional education take place in the country, it was sexual as well as material corruption he was thinking of. For Rousseau, the integrity of the child was inevitably compromised by his entry into the adult world; and to judge by what he tells us in the *Confessions*, Rousseau's own sexuality retained an element of the childlike throughout his life. (In abandoning his own offspring on the steps of the foundling's home he was in a sense denying his responsibility for their existence. Being but a child, he could not have begotten children.) Julie and Saint-Preux are driven from the paradise-garden of Eden as a result of their sexual fall, and even the Elysée Julie constructs is incapable of restoring her and her lover to their old purity. Although nature itself is the image and archetype of fruitfulness, it punishes those who, in Rousseau's world, imitate its progenitive drives. Julie thus loses the child she has conceived, for only marriage legitimizes the production of human lives (which may be another reason for Rousseau's abandoning the children of his common-law wife). And for those who remain children at heart, generation is best avoided. Rousseau never sought possession of the women he truly loved, and felt overt sexual activity to be degrading.

The empirical reality of natural sexual urges, and still more what he regarded as unnatural (such as the homosexual advances to him which he details in the *Confessions*), is repellent or ludicrous in comparison with the creation of life which took place in his literary imagination. Can men wonder that women discourage their brute attentions, he asks apropos of the experiences in the Turin catachumen's seminary described in the *Confessions*, when male physical arousal wears such a gross aspect?[82]

The innocence of children in the state of nature is a recurrent theme in literature. Schiller writes in *Über naive und sentimentalische Dichtung* that the child embodies an ideal which adults have lost, and consequently possesses pure and unlimited strength.[83] The child's integrity is like the uncorruptedness and fulfilment of nature, as opposed to the compromised state in which the 'sentimental' adult subsists. The man of morals and feeling therefore regards children as holy objects, not as representatives of limitation and need. Pastoral literature, which in the eighteenth century often concerned itself with purer loves than seemed possible in an urban environment, is naturally inclined to choose children as its subjects. Experience, which forms the personality, has yet to mark them. They thus body forth an unsulliedness to which the adult aspires, but which he has inevitably left behind him.

But the child is father to the man. Without necessarily holding such beliefs about child-development as modern theorists have espoused, eighteenth-century writers often implied that his sexual character was inherent as well as externally governable, and as such natural. In that respect efforts like those made by the Baron de Faublas and Mme de la Tour to check spontaneous impulses run counter to the preordained order of things. The attempt to guide such impulses may be regarded as laudable—particularly if man's superiority to beasts is seen as residing in his ability to constrain brute impulses—but the effort to quash them altogether perhaps as less so. The responsible course may rather be to teach nature how best to realize itself. In *Une année de la vie du chevalier de Faublas* Louvet describes the young Faublas's ignorance of the meaning of his natural impulses in Sophie's company: 'Aveuglément livré aux premières

---

[82] p. 67.
[83] *Über naive und sentimentalische Dichtung*, ed. W. F. Maitland (Oxford, 1951), p. 4.

impulsions de la nature, j'étais loin de soupçonner son but secret.'[84] When he kisses his beloved's hand, he occasionally wants more, but would have been at a loss to say what that 'more' was. Later on, installed in the Marquise de B***'s bed, he still fails to suspect the truth. Mlle de Mésanges shares that ignorance until she has her bed-time 'conversation' with the *soi-disant* Mlle de Brumont.

All this may recall the situation described in *Daphnis and Chloë*, which presents a teasing succession of attempts on the children's part to discover what the ultimate purpose of lying together might be[85] (and incidentally reveals that the model of the natural world is of limited usefulness to human beings in search of sexual fulfilment). The old man Philetas informs Daphnis and Chloë in book two that for the torment of love—which has induced in Chloë the state of restless-ness and changeability shared by many of literature's pubescent females—there is no remedy but the act of kissing, embracing, and lying together naked. But the natural conclusion to the last-named cannot come about until, in book three, Daphnis has been taken under the wing of the sophisticated older woman Lykainion.[86] Although the coming of spring merely increases Daphnis and Chloë's ardour, and although the still virgin Daphnis suggests to his beloved that they should try doing what animals do with one another, the observation and imitation of beasts copulating is inevitably frustrating. The lovers learn to sing, dance, and gather flowers by copying the creatures of the natural world—birds, lambs, bees, and so on—but this education is inadequate where sexual love is concerned.[87] Even when Lykainion has taught Daphnis the technique of love-making, he is loath to follow her example with Chloë for fear of hurting her. Although there is some crudeness in Longus's account,[88] this is a delicate touch; and the fact that Lykainion lacks the impure motives of some eighteenth-century teachers of sex gives the third-century work an innocence which novels like *Les Liaisons* lack. If Daphnis and Chloë seem young to

---

[84] p. 427.

[85] See S. L. Wolff, *The Greek Romances in Elizabethan Prose Fiction* (New York, 1912), pp. 130–1. *Daphnis and Chloë* was much re-edited in the 18th century, and two new French translations were published shortly before Bernardin's novel.

[86] See G. Rohde, 'Longus und die Bukolik', *RhM*, 86 (1937), p. 39.

[87] See R. L. Hunter, *A Study of 'Daphnis and Chloë'* (Cambridge, 1983), p. 20.

[88] Erwin Rohde, *Der griechische Roman und seine Vorläufer*, 3rd ed. (Leipzig, 1914), calls Longus's treatment of love an 'abscheuliches muckerhaftes Raffinement' (p. 549).

be schooled in nature's ways, they are in no sense untypical of heroes and heroines of Greek romance, whose acquaintance with sexuality, practical as well as theoretical, usually begins between the ages of fourteen and sixteen.[89]

Before his description of Virginie's sexual awakening, Saint-Pierre alludes to both her and Paul's swift physical growth, raised as they have been in an environment where bodies are purely and healthily nourished. He remarks, in *Voyage à l'Ile de France*, that physical maturity comes early in the Tropics,[90] and in the fourteenth of his *Etudes de la nature* he dilates on the desirability of loving relationships starting young. In *Paul et Virginie* Marguerite observes that, given the children's evident passion for one another (and despite Paul's ignorance of the fact that he loves Virginie in a more than brotherly way), their mothers would do well to marry them forthwith; for '[lorsque] la nature lui [i.e. Paul] aura parlé, en vain nous veillerons sur eux, tout est à craindre.'[91] Mme de la Tour's view that their marriage should be postponed is none the less allowed to prevail.

Saint-Pierre's desire to Christianize a literary type which, from Longus onwards, had been marked by a very pagan sensuousness helps explain why this had to be so. The heroine's name has an obvious significance, and it is plain at the end of the novel that she is worshipped for a sexual purity that marriage would have compromised, as well as for her determined faithfulness to Paul. When at an earlier age in the novel Saint-Pierre likens the children to Adam and Eve, conversing in their Edenic setting like brother and sister,[92] he underlines the parallel between their story and that of pre-lapsarian man. But his failure to develop this theme coherently must be counted a major weakness of the novel. If he intends Virginie's death to represent an extension of the state described in Genesis, where man and woman, having eaten the fruit of the tree of knowledge, saw that they were naked and formed an idea of evil, it is unfortunate that he should strike a note of such bathos in the shipwreck scene, where Virginie refuses to strip naked and so goes to the bottom of the ocean with the *Saint-Géran*. In their pre-pubescent days, and before Virginie absorbs different ideas about

---

[89] See introduction to *Daphnis and Chloë*, ed. Otto Schönberger (Berlin, 1960), p. 113.

[90] p. 114.   [91] p. 138.   [92] p. 130.

modesty in France, Saint-Pierre shows the children splashing naked in the pool of their garden as happily as had Longus's Daphnis and Chloë. Indeed, he turns the traditional erotic motif of water[93]—the fountain, stream, or pool—as effectively to account as his Greek predecessor, a fact which adds no more philosophical coherence to the book than does Rousseau's similar procedure in *La Nouvelle Héloïse*.

The physical change which is a part of childhood and adolescence is memorably described in *Daphnis and Chloë*, Florian's *Estelle*, and *Paul et Virginie*. But these same works depict moral and emotional constancy in their pairs of young lovers: the growth towards adulthood is accompanied only by a maturing of the children's love, not by a change in the object of their affections. This kind of steadiness is a common feature of pastoral literature, and it is necessarily bound up with an aspiration to move beyond the merely temporal. The effort to capture the moment and make it eternal, which confers their elegiacally moving quality on Watteau's 'fêtes galantes', is at the very heart of the pastoral ideal. The Faustian invocation, 'Verweile doch, du bist so schön', represents such a desire to achieve timelessness. Its association with the sentiment of love arises from the human desire to foster and preserve what is best, but which has no material existence. Hence the pursuit of permanence as it can be procured: the written record which des Grieux's story of his love for Manon eventually becomes (but unbeknown to him), or the heavenly eternity to which Julie and Virginie aspire, and which we are to assume they gain.

Yet this human desire, against which Pascal cautioned, to find an 'assiette ferme' and use it as a tower reaching up to eternity may lead only to the unsettledness of 'inquiétude'.[94] Léonard's *Les Regrets* suggests that mortal happiness is always subject to the threatening touch of time:

> Mais le temps même à qui tout cède
> Dans les plus doux abris n'a pu fixer mes pas!
> Aussi léger que lui, l'homme est toujours, hélas!
> Mécontent de ce qu'il possède
> Et jaloux de ce qu'il n'a pas.
> Dans cette triste inquiétude

---

[93] See Philip Stewart, 'Décence et dessein', in Viallaneix and Ehrard (eds), *Aimer en France*, i. 38.

[94] See also Tilo Schabert, *Natur und Revolution* (Munich, 1969), p. 9.

On passe ainsi la vie à chercher le bonheur.
A quoi sert de changer de lieux et d'habitude
Quand on ne peut changer son cœur?[95]

Rousseau's more limited goal was 'une assiette assez solide' where the soul might rest and enjoy its own existence, unaffected by the passage of time.[96] It is this fixity which many lovers seek, and it involves mastering the fates that threaten permanence. Longus's pastoral romance differs from other ancient romances in replacing chance with the settled rhythm of nature,[97] and in that respect looks forward to *Paul et Virginie*.

Natural man, in Rousseau's conception of him, cannot be concerned by the passage of time: he lives only in and for the instant, and consequently must lack memory and anticipation. Not knowing how to draw comparisons, he must be without aspiration, as he is necessarily without nostalgia—both sentiments which lie behind the idea of pastoral life conceived by the sophisticated. According to Rousseau, happiness—the supreme form of which may be found in love—is not the pleasure of a moment: 'il ne consiste pas dans une modification passagère de l'ame, mais dans un sentiment permanent et tout intérieur dont nul ne peut juger que celui qui l'éprouve.'[98] But then, and crucially, the savage is incapable of feeling happiness, for his life is a succession of non-identical 'presents' which cannot be joined together to give duration. The desire for what is unchanging is necessarily a conscious, intellectual desire that is not available to the purely sentient being. It may be a desire born of revulsion from things intellectual, but its origin is itself in the intellect.

As Virginie reflects after the storm has laid waste Paul's garden, the only truly changeless element is the divine. Nature's seasons embody mutability, and may even reflect the changeableness of man's physical being. Yet seasonal change is subject to natural law, and exhibits a dependable regularity. Paul and Virginie have no clocks, almanacs, or books of chronology, because the periods of their lives followed those of nature. They know the hours of the day by the shadows of trees, the seasons of the year by the time at which the

---

[95] ll. 52–60.

[96] *Les Rêveries du promeneur solitaire. Œuvres complètes*, i. 1046.

[97] See B. P. Reardon, *Courants littéraires grecs des IIe et IIIe siècles après Jésus-Christ* (Paris, 1971), p. 377.

[98] *Du bonheur public, Œuvres complètes*, iii. 510; also *Rêveries*, pp. 1085, 1099.

trees blossom or bear fruit, and the year by the number of harvests. 'Quand viendrez-vous nous visiter?' some neighbours ask, and Virginie replies: 'Aux cannes de sucre.' She measures her age, like Paul's, by that of the coconut-trees beside the fountain, and the number of times the mango-trees has borne fruit or the orange-tree blossomed.[99]

To link seasonal changes with changes in man's emotional life is not to posit the same regularity in the latter's rhythm as in nature's. *Daphnis and Chloë* is a special case, for the love between the two protagonists is directly and emblematically bound up with a yearly celebration linking love with the fixed character of the seasons. This gives the story a poetic quality; the development of the children's love is so stylized as to make realist interpretation inappropriate. We are not invited to speculate as to whether the couple will annually see their love decline like winter, burgeon like spring, and flourish like summer. Only to the inconstant lovers of eighteenth-century fiction might such a pattern be applicable (and then only partially, for the regular pattern of seasonal changes would doubtless seem excessively ordered to them): Valmont's love for the Présidente, conceived in the dead heat of summer, withers and dies as autumn turns into winter. For those who, like Saint-Pierre, want to show the virtue of fidelity, nature offers an imperfect analogue.

Although both Longus and Bernardin perceive a divine essence in nature, the conclusions they draw from this perception are sharply distinct. The former tells his reader that those who live in harmony with the natural world live also in harmony with the gods. *Daphnis and Chloë* is an oblation to Pan, the god of shepherds, the countryside, and song; and the love of Daphnis and Chloë, which sprang up in the rural paradise of Lesbos, is finally expressed in their serving as their deities the nymphs, Cupid, and Pan. Daphnis observes that his foster-mother, like Jove's, was a goat; he himself plays the syrinx like a god; and Longus compares him with Apollo. The beauty of both children raises them above ordinary mortals, but in the course of the story they themselves are led more and more to recognition of the gods. The joyful paganism of the romance, however, excludes all possibility of the tragic ending which Saint-Pierre's Christian conception of pastoral suggested to him. The sensuous richness of Longus's natural world does not stand in contrast to the love that

[99] p. 129.

develops in its midst, but is a part of that love. The ascetic Christian tradition, on the other hand, means that such sensuousness is ultimately subordinated in *Paul et Virginie* to the purity of Virginie's sublimated passion: in determining to remain pure and changeless as a ray of light, she has forever cut her ties with the living but flawed world of Mauritius. To Bernardin, joy in nature such as Longus portrayed comes to appear excessively secular.

As the *Rêveries du promeneur solitaire* and *Lettres à Malesherbes* reveal, Rousseau was able to find God in nature, and hence to find permanence in the transitory. His intoxicating communion with the natural world on the Ile de Saint-Pierre led the communicant to feel a closeness to the divinity which town life could not procure. This transported state robbed him of the power to think or speak (save for uttering the ecstatic cry 'O grand Etre! ô grand Etre'), so that all he could do was feel.[100] In *Paul et Virginie* such devotion is clearly felt to be insufficient. Although the children's religion initially appears to be the informal one that existence in nature and the enjoyment of a happy family life might foster, it finally acquires an other-worldly aspect.

With the loss of its linchpin, Virginie, the small community turns its attention from this world to the afterlife, and so falls into what Rousseau's *Du contrat social* called the trap of Christianity, namely that it deflects the believer's attention from the here and now and focuses it, to the detriment of the social fabric, on the hereafter.[101] Once the two families have sustained their terrible loss at the hands of nature, they must fix their minds and hearts on the intangible world to come. The fact that the sufferings of Paul and Virginie, and particularly the frustration of their love, cannot be straightforwardly attributed to nature perhaps makes their successive deaths ironic; but Bernardin himself, if he perceived the irony, does not make them privy to it.

This might give further substance to the notion which the old man voices in the novel that the country does not provide humans with all they need for a balanced life. On this interpretation, innocence is tantamount to ignorance of certain truths that matter, and the habit of compromise that ordinary social living engenders may be of life-preserving significance when set against such intemperate extremism

---

[100] *Lettres à M. de Malesherbes*, *Œuvres complètes*, i. 1141.
[101] *Du contrat social*, *Œuvres complètes*, iii. 465.

as Virginie exhibits in her desire for death. Perhaps the favour which Saint-Pierre occasionally showed towards civilization in his writings indicates that he realized this. If so, it would bring him closer to Longus than a preliminary comparison of their respective pastorals indicates. In his projected Utopian novel *L'Amazone* Saint-Pierre imagines a post-Revolutionary idyll where men can recover from the shocks of political upheaval and find tranquillity. But they do not live in an isolated state of nature: rather, their new world is revealed to be one of perfect civilization. The valley of the Amazon shelters a Republic of Friends devoted to the principles of equality and humanitarianism, and guided by virtue.[102]

The fixed happiness which Rousseau describes in the fifth *Rêverie* can be gained only in the world of timelessness. Nor can the joys of love, whose fleeting quality is captured in the lightness of Watteau's brushstrokes, match nature's eternal return. What gives permanence to the moment may be what kills it: as Versac tries to impress on Meilcour, only a mistress who can appear ever different to her lover (as Mme de Merteuil prides herself on doing) may hope to fix him. The seal of death preserves Virginie's love for Paul and makes it unalterable, just as Julie's *Liebestod* gains for her the eternity in which alone she can love Saint-Preux. Only in the pastoral world can the assurance of changeless love on earth be given, as Rousseau's Daphnis gives it to Chloé:

—Dans un nouveau parentage,
Te souviendras-tu de moi?
—Ah! je te laisse pour gage
Mon serment, mon cœur, ma foi.

—Me reviendras-tu fidèle?
Seras-tu toujours mon berger?
—Quelque destin qui m'appelle,
Mon cœur ne saurait changer.

—Ah! sois-moi toujours fidèle!
—Je serais toujours ton berger.[103]

In the real world man can only assert his moral, not his physical, power over the fate that tries to refuse him permanence. Virginie,

---

[102] *Fragments de l'Amazone, Œuvres posthumes*, ed. L. Aimé-Martin (Paris, 1840), pp. 512 ff.
[103] *Daphnis et Chloé, Œuvres complètes*, ii. 1165.

the redeemer through love, wins out against the material death which appears to end her ties with the human, and particularly her union with Paul. She rises too above the power that ordains fluctuation, epitomized by the changing seasons which signify both death and renewal. Although 'tout périt sur la terre,' the non-material cannot be subject to mutability. But the price which must be paid for it is infertility: a nature without substance cannot bring forth substantial life formed in its own image. It is a price which Daphnis and Chloë do not have to pay. The ascetic strain in Saint-Pierre's 'espèce de pastorale' means that the affinity of its characters with nature is limited as that of Longus's is not; the forfeiture of sensuousness entailed by Bernardin's Christian conception of pastoral finally amounts to an exclusion of the natural world.

5

Beauty and its Trappings

LA *Nouvelle Héloïse* contains an episode describing Saint-Preux's feelings about a portrait of his beloved Julie, feelings which seem to qualify the familiar notion of beauty's residence in the eye of the beholder. Julie has sent her lover, now living in Paris, a present to which an advance letter alludes in secretive terms. According to this arch missive, Saint-Preux must postpone opening the parcel until he is alone in his bedroom, apparently lest its effect on him should prove an embarrassment in a public place. Pending that time he has Julie's mysterious description of its contents to whet his appetite:

C'est une espèce d'amulette que les amants portent volontiers. La manière de s'en servir est bizarre. Il faut la contempler tous les matins un quart d'heure jusqu'à ce qu'on se sente pénétré d'un certain attendrissement. Alors on l'applique sur ses yeux, sur sa bouche, et sur son cœur; cela sert, dit-on, de préservatif durant la journée contre le mauvais air du pays galant. On attribue encore à ces sortes de talismans une vertu électrique très singulière, mais qui n'agit qu'entre les amants fidèles. C'est de communiquer à l'un l'impression des baisers de l'autre à plus de deux cents lieues de là. Je ne garantis pas le succès de l'expérience; je sais seulement qu'il ne tient qu'à toi de la faire.[1]

Days pass as the excited lover waits for his gift to arrive at a Paris depot. Eventually it does, and he rushes deliriously through the city streets with it to regain his lodgings. The power of his *vade mecum* (presumably inoperative until he has unwrapped it) does not, however, prevent him from becoming hopelessly lost, so that he ends up at the opposite end of town from the one in which he lives. Bad-temperedly, he climbs into a cab; the packet burns in his hands as he waits impatiently for the journey to end. Once in the safety of his room he breaks the seal of the parcel with trembling hands and removes the layers of wrapping in which Julie has prudently swathed the amulet. What he uncovers is a miniature portrait of his beloved,

---

[1] p. 264.

whose effect on his senses is as immediate as (according to the philosophical orthodoxy of Rousseau's age) visual images are wont to be:

Julie! . . . O ma Julie! . . . le voile est déchiré . . . je te vois . . . je vois tes divins attraits! Ma bouche et mon cœur leur rendent le premier hommage, mes genoux fléchissent . . . charmes adorés, encore une fois vous aurez enchanté mes yeux.[2]

Soon, however, peevishness tempers the initial ecstasy. Why, he asks, should he be banished from his beloved's presence at all, and so be forced to find a substitute pleasure?—not the solitary pleasure of which Julie signals her extreme disapproval elsewhere in the novel,[3] but the less immediate one afforded by a painted simulacrum.[4] For a while the portrait's powerful charm does its work, and Saint-Preux feels restored by the acts of pious worship his goddess-mistress has enjoined on him. But mere obeisance palls with time. He begins to cavil, not at the monstrous egoism of his beloved, but simply at the fact that the miniature is not Julie in her more ordinary imperfections: the portrait is unsatisfactory because it presents an idealized version of its subject, although the phantom it conjures is too actual to permit the corrective functioning of his imagination. Why has the artist omitted the faults of Julie's physiognomy, the blemishes in her complexion, the irregular shape of her face, and the other features which constitute her individuality?[5] Inadequate as the image is, it none the less confers on her a perfection she does not possess, and in so doing it inhibits the communion with his loved one for which Saint-Preux yearns. Julie was mistaken, in other words, in sending him a visual image of herself, for visual likenesses are sufficiently close to reality to impede the work of fancy: since he feels, in virtue of the portrait, partially in Julie's presence, he is unable to imagine her.

Many eighteenth-century theorists would have agreed with Saint-Preux's observations on the inherent tendency of visual art to make concrete what the imagination might otherwise have formed according to its own whim, although they would not necessarily have condemned pictorial representations on those grounds. Images were generally thought to be more vivid than words, but words better able to stimulate the imagination. In his *Lettre sur les sourds et muets*

[2] p. 279.   [3] p. 237.   [4] p. 280.   [5] pp. 291–2.

Diderot reasons that imagination is less scrupulous than the sense of sight: literature allows us to enjoy images that would be unendurable in painting, because of the more directly mimetic quality of the latter.

En effet, qui pourrait supporter sur la toile la vue de Polyphème faisant craquer sous ses dents les os d'un des compagnes d'Ulysse? Qui verrait sans horreur un géant tenant un homme en travers dans sa bouche énorme, et le sang ruisselant sur sa barbe et sur sa poitrine? Ce tableau ne récréera que des cannibales. Cette nature sera admirable pour des anthropophages, mais détestable pour nous.[6]

Rousseau emphasizes the comparative indirectness both of verbal description and of visual depiction. In his notes on the engravings that accompanied the Duchesne edition of *La Nouvelle Héloïse* in 1764 he exhorted his reader to look beyond the necessarily inadequate images of Julie, Saint-Preux, and the rest, on the grounds that '[la] plupart de ces sujets sont détaillés pour les faire entendre, beaucoup plus qu'ils ne peuvent l'être dans l'exécution.'[7] To realize his subject successfully, Rousseau writes, the artist (and consequently the reader) must picture it not as it will appear in the engraving, but as it is in nature. Imagination will enable the observer to see hair as blonde or brown, although no such distinction of colour appears in the black-and-white plate; and in general the artist must use his skill to make the public visualize all manner of things which cannot appear on paper. Even where physical depiction is clear—a matter of line rather than colour—the onlooker is required to read into it qualities of the heart that cannot easily be suggested by the engraver's skill. Milord Edouard, for instance, is to exhibit '[un] air de grandeur qui vient de l'âme plus que du rang; l'empreinte du courage et de la vertu'.[8] In other words, nobility of character is not to be signalled by casual appurtenances: Edouard is to be well-dressed, but his true quality is to emerge from less tangible signs. However, Rousseau's remark that the Englishman's appearance is to be marked by '[un] maintien grave et stoïque, sous lequel il cache avec peine une extrême sensibilité',[9] reminds the reader just how difficult the non-verbal nature of his medium makes the artist's task.

---

[6] p. 97.
[8] p. 762.
[7] *La Nouvelle Héloïse*, p. 761.
[9] Ibid.

In *Manon Lescaut*, where the heroine's character is presented in the most shadowy terms, the writer perhaps makes a virtue of necessity. The eighteenth-century taste for the 'esquisse' or *non finito* was well served by artistic forms which gave scope for the play of the imagination through suggesting things rather than stating them outright. Diderot approved this quality even in visual art: 'Je vois dans le tableau une chose prononcée; combien dans l'esquisse y supposé-je des choses qui sont à peine annoncées!'[10] The interpretative share of the beholder or reader is enlarged in proportion as his imagination is activated.

But if human beauty can be, and in the eighteenth century certainly was, successfully captured in paint, why should writers in the age of Diderot so regularly have abandoned the attempt to convey it through language? Looks were rarely described in detail in the contemporary novel. On some occasions this perfunctoriness may simply reflect a lack of interest on the novelist's part; if so, it contrasts with the evident importance attached to beauty by the majority of male lovers who feature in the fiction of the age. (Women novelists, to judge by works like *Lettres d'une Péruvienne* and *L'Abailard supposé*, are no more concerned with this kind of description than men.) But sometimes the haziness of reference seems intentional— the product, perhaps, of a desire to add reverence to the view by implying that a certain person's beauty is beyond verbal evocation. Other types of visual appeal are more readily translated into words by writers of the time: the beauties of nature, for example, are exhaustively treated by Rousseau in *La Nouvelle Héloïse* and Saint-Pierre in *Paul et Virginie*, while the personal beauty of their respective heroines is described in the most general terms. For whatever reason, novelists as diverse as Prévost, Diderot, and Laclos were as a rule uninformative about a woman's looks, and this despite the interest of the eighteenth century both in the science (or pseudo-science) of physiognomy[11] and in the bodily nature of man, his purely physical as opposed to spiritual being. 'Literary' art criticism as pioneered by Diderot, however, showed an attention to the detail of human looks when these had been given plastic form which is nowhere matched by imaginative literature of the age, but which suggests that

---

[10] *Salons*, ed. Jean Seznec and Jean Adhémar, 4 vols (Oxford, 1957–67), ii. 154.
[11] See Graeme Tytler, *Physiognomy in the European Novel: Faces and Fortunes* (Princeton, 1982).

verbal resources were not lacking for conveying visual images. In the seventeenth-century *Caractères des passions* Cureau remarked on the French language's richness in words for describing the 'air' of a person, her or his grace, mien, bearing, attitude, and gestures.[12] On the evidence of Diderot's much more specific descriptions of human looks in the *Salons*, the poverty of reference in fiction may have been partly due to a failure of visual imagination on the writer's part. On the other hand, the contemporary novel paid considerable attention to apparel and the other trappings of beauty, an interest reflected in the development of fashion journalism during the period.[13]

Visual art seems to have given Diderot the moralist a freedom he welcomed to write explicitly about certain types of human beauty. Whereas in his fiction he does not, for whatever reason, write lengthily and analytically about aspects of the body usually associated with the erotic, when he was commenting on painted or sculpted nudes he had perfect licence to do so. The pornographic *Bijoux indiscrets*, for all its risqué preoccupations, gives few details about the female anatomy, and those supplied are often veiled and indirect (as in the discussion of the coyly named 'bijoux' themselves).

Nudity is not itself immodest in visual art, which has its own discretion.[14] That of literature is different. In *Manon Lescaut* there is complete reticence about the lovers' physical relationship, but also silence about physical details which the narrator could decently have supplied. Where Prévost's illustrators chose, like Desenne in 1818, to picture Manon as a Grecian-nosed and bejowled supplicant, or like Rossi in 1892 as a *fin-de-siècle* vamp of improbably slim proportions,[15] they were providing an interpretation where Prévost furnished none. Although often explicit about physical attitude and gesture (as in the scene at Saint-Sulpice which his illustrators have recurrently chosen to depict),[16] about the fine details of individual appearance he remains silent. Remembering that des Grieux's narrative is ostensibly to the Renoncour who has already beheld Manon, we may reflect that a detailed description would have been superfluous for the listener; but when we consider how flimsy is the

---

[12] Cureau de la Chambre, *Les Caractères des passions* (Paris, 1640), p. 16.

[13] See Annemarie Kleinert, *Die frühen Modejournale in Frankreich* (Berlin, 1980).

[14] On the distinction in English between the (shamefully) naked and the (decently) nude, see Kenneth Clark, *The Nude* (London, 1956), p. 1.

[15] See the plates in the edition cited.   [16] pp. 44–5.

pretext of an audience for des Grieux's story-telling, and that his words are really directed at the reader of Prévost's novel, that reasoning seems barely sufficient. Is it rather, then, that the author understood the imaginative appeal of the understated, and preferred to let each reader picture Manon as he would? There is much to recommend this view. Another possibility is that the narrator, if not the author, meant to suggest that Manon's looks belonged to the realm of the other-worldly, and that des Grieux's imprecision is the product of a reverence like that felt for divine beings (although one is bound to say that Prévost followed the same procedure when describing women in his other novels). At all events, the hagiographical tones in which des Grieux relates the last moments of his beloved mistress makes this a less fanciful interpretation than the revelation of Manon's all-too-human frailties in the earlier part of the narrative might lead the reader to suppose.

Whatever the case, such reticence contrasts strongly with other contemporary forms of exhibitionism—the tear-jerking appeals to sensibility characteristic of much eighteenth-century literature (including Prévost's novelistic *œuvre*), or the declamatory style of much painting in the period, with its theatrical emphases and highly coloured, flamboyant aspect. Renoncour himself matches des Grieux in understatement. Of his first sight of Manon the former simply tells the reader that her air and her face were wholly at odds with her situation, as she waited with the rest of the convoy of prostitutes at Pacy. Moving back in time, the narrative then takes us to the moment of des Grieux's initial encounter with Manon in Amiens, when the process he undergoes is allusively described in terms of bewitchment, the state of being under a charm; such detail as des Grieux furnishes about her looks, indeed, amounts to the statement that she appeared 'charmante'.[17] Nor does the narrative subsequently provide much more enlightenment. When Manon comes to visit des Grieux at Saint-Sulpice, he informs the reader that she appears enchanting, and that her charms exceeded 'tout ce qu'on peut décrire'. Her air is 'si fin, si doux, si engageant, l'air de l'Amour même'.[18] With this, des Grieux's attempt at explicitness peters out.

Although other novels of the period may not share the mystical perception of female beauty to which des Grieux's narrative seems to bear witness, most show a comparable imprecision about their

---

[17] p. 19.   [18] p. 44.

heroines' looks. Eighteenth-century literature gives us little indication of what contemporaries regarded as canonically beautiful in a woman's appearance. Despite the details about Julie contained in Saint-Preux's letter on the amulet, Rousseau is still as vague about her looks as he is about Saint-Preux's. We know that what her lover reveres in her is not beauty of any exceptional kind, but rather a collection of virtues to which her facial aspect is an inadequate guide. Precisely how her qualities of soul infuse her physiognomy is left tantalizingly imprecise, however. There is a comparable imprecision in Julie's later mentioning of her lover's own looks: she merely says that

d'autres jeunes gens m'ont paru plus beaux et mieux faits que vous, aucun ne m'a donné la moindre émotion, et mon cœur fut à vous dès la première vue. Je crus voir sur votre visage les traits de l'âme qu'il fallait à la mienne.[19]

What Saint-Preux and Julie appear to be saying about each other's appearance is that, although they recognize qualities of the soul in their beloved's visage, it is impossible to be specific about their physical nature; they are to be felt rather than analysed. As Julie says, 'j'aimai dans vous moins ce que j'y [in Saint-Preux's face] voyais que ce que je croyais sentir en moi-même.'[20]

Saint-Preux's rehearsing of Julie's facial defects in Letter XXV of Part Two would have been unthinkable in an age which set less store than the eighteenth century by the concept of individuality. The anti-rationalist bent of the period was against the belief, particularly favoured under the neo-classicism of the previous century, that the conditions governing beauty could be formalized and given institutional sanction. The prevalence of seventeenth-century treatises devoted to different arts was partly a product of the assumption that rules existed according to which their various perfections might be perceived and classified. The counter-systematic temper of the eighteenth century, by contrast, was manifested in the many works—philosophical, aesthetic, and literary—which approvingly discussed the phenomenon of irregularity, especially when it was perceived as a product of nature as opposed to art.

It would be wrong, none the less, to associate the praise of irregularity too closely with the eighteenth century. Bouhours's

[19] p. 340.   [20] Ibid.

Entretiens d'Ariste was eloquently proclaiming the pleasures of the elusive, indefinable, and non-rational in 1671; and as early as 1640 Cureau de la Chambre was writing in *Les Caractères des passions* of the piquancy which the 'je ne sais quoi' lent to facial beauty. Equally, however, eighteenth-century thinkers whose works may in other respects argue a devotion to the non-systematic associated themselves with the idea that beauty and order go together. In the *Eléments de physiologie* Diderot notes with apparent approval the possibility that fixed canons of beauty may make people less subject to the illusions of love than does appreciation of the irregular.[21] Even the sentimental Bernardin de Saint-Pierre, whose freedom from the encumbrances of rationalism is so plainly evident in the more preposterous hypotheses of the *Harmonies de la nature*, still attempts in the earlier *Etudes* to elaborate a reasoned theory of human beauty. Although the tenth *Etude* describes it as the fruit of virtue, it also states that beauty may be found in the fixed geometry of the countenance, the triangle of the nose according agreeably with the heart shape of the mouth and the circular orbits of the eyes.[22]

Despite such evidence, the age of sensibility was more inclined to heed the judgement of feeling than that of the rational faculty. Dubos's *Réflexions critiques sur la poésie et sur la peinture* (1719) argued that aesthetic judgement is the product of sensation, which all men possess, and that all are therefore qualified to pass judgement on the beauty or otherwise of things that arouse emotion. Nevertheless, it was clear to some theorists that analysing beauty is much less easy than perceiving it. Diderot remarked on this problem in his *Encyclopédie* article 'Beau':

Tout le monde raisonne du *beau*: on l'admire dans les ouvrages de la nature; on l'exige dans les productions des arts; on accorde ou l'on refuse cette qualité à tout moment . . . Comment se fait-il que presque tous les hommes soient d'accord qu'il y a un *beau*; qu'il y en ait tant entre eux qui le sentent vivement où il est, et que si peu sachent ce que c'est?

In his article Diderot does, all the same, try to establish objective rules governing the existence and therefore the perception of beauty, finding that it exists in the 'rapports' (relationships) which hold between the parts of a whole.

---

[21] *Eléments de physiologie*, ed. Jean Meyer (Paris, 1964), p. 277.
[22] *Etudes de la nature*, *Œuvres complètes* (Paris, 1820), iii. 83.

This theoretical notion of beauty finds no reflection in the imaginative writing of Diderot or others of his century, however, perhaps because it seems to exclude individuality. Their general unwillingness or inability to describe the essential beauty of a countenance led authors to explore other areas in which the humanly beautiful might be discerned. Sometimes this simply meant switching their attention from the face to other parts of the body. In erotico-pornographic works like *La Philosophie dans le boudoir* and *Le Rideau levé*, predictably enough, much attention is paid to the sexual organs, whose appearance and size are described in loving detail. In more modest works the limbs and their extremities often receive a degree of consideration which surprises the reader for whom such parts of the human frame, especially hands and feet, are less than arousing.

The erotic hand makes an appearance in *La Vie de Marianne*, where the heroine reflects on the excitement its uncovered state can occasion compared with the sight of a pretty face:

Ce n'est point une nudité qu'un visage, quelque aimable qu'il soit; nos yeux ne l'entendent pas ainsi: mais une belle main commence à en devenir une; et pour fixer de certaines gens, il est bien aussi sûr de les tenter que de leur plaire. Le goût de ces gens-là, comme vous le voyez, n'est pas le plus honnête; c'est pourtant, en général, le goût le mieux servi de la part des femmes, celui à qui leur coquetterie fait le plus d'avances.[23]

Marianne earlier remarked on the hand-kissing attentions of M. de Climal, precisely a man of the less refined tastes to which she here refers. Climal's tender kiss seems odd to her even amidst the transport of delight occasioned by his promise to buy her a dress; but at this stage she is apparently unable to see what it might signify beyond Climal's possession of a kind heart. For Saint-Preux, later in the century, a woman's uncovering of her hand and arm has lost none of its erotic power:

Ne vis-je pas quand tu te dégantais pour la collation l'effet que ce bras découvert produisit sur les spectateurs? Ne vis-je pas le jeune étranger qui releva ton gant vouloir baiser la main charmante qui le recevait?[24]

But despite Julie's liberality with her charms in this respect, he tells her, she never makes him jealous of other male witnesses to the

[23] p. 63.   [24] p. 107.

act. He knows her heart, which is incapable of loving more than one person.

Nothing better illustrates the shifting perception of beauty over the ages than the reaction of most modern readers to this rhapsodic account, or to Marianne's description of the episode involving her damaged foot. It is difficult now to share the excitement felt by Valville as the foot is uncovered, or grasp the point of the 'pud^ur' which makes Marianne blush at the need to display the injured part.[25] (Immediately after, it is true, she begins to take a coquettish delight in the situation.) The extreme delicacy of the proceedings, which necessitates Valville's and the surgeon's withdrawal while a maid removes Marianne's shoe and stocking, is matched only by the superfine sensibility revealed in her recounting of it, a prime example of the gossamer touch of Marivaudage. Valville's solicitude, manifested in his anxious imitation of the surgeon's close examination, gratifies Marianne, who is yet chary of revealing her gratification: 'il n'aurait pas été modeste de paraître soupçonner l'attrait qui l'attirait, et d'ailleurs j'aurais tout gâté si je lui avais laissé apercevoir que je comprenais ses petites façons.'[26] Something similar is described by Cécile Volanges in the letter to Sophie Carnay that opens *Les Liaisons dangereuses*; but in this case the imagined suitor is in fact a shoemaker, come to take Cécile's foot-measurements:

'Madame, a-t-il dit à ma mère, en me saluant, voilà une charmante demoiselle, et je sens mieux que jamais le prix de vos bontés.' A ce propos si positif, il m'a pris un tremblement tel que je ne pouvais me soutenir; j'ai trouvé un fauteuil, et je m'y suis assise, bien rouge et bien déconcertée. J'y étais à peine, que voilà cet homme à mes genoux. Ta pauvre Cécile alors a perdu la tête; j'étais, comme a dit Maman, tout effarouchée. Je me suis levée en jetant un cri perçant . . . ; tiens, comme ce jour du tonnerre. Maman est partie d'un éclat de rire, en me disant: 'Eh bien! qu'avez-vous? Asseyez-vous et donnez votre pied à Monsieur.' En effet, ma chère amie, le monsieur était un cordonnier.[27]

In some cases, this reverence for feet has the aspect of fetishism: it recurrently presents itself as such in the novels of Restif, whose obsession with these extremities carried over into his prescriptions for his illustrator Binet's plates—fantastically shortened projections from excessively elongated legs, normally perched on heels of a height

[25] p. 67.   [26] p. 68.   [27] pp. 12–13.

few but the sadistically inclined would dream up. (Alphonse Leroy's *Recherches sur les habillements des femmes* of 1772 contains a diatribe against high heels on women's shoes, which according to him push the body unattractively forward, deform and weaken the muscles of the leg, and give their wearers an unappealingly hopping gait. Country women, he writes, are not subjected to such perverse fashions, and have finer legs than town-dwellers.)[28] The petite foot admired and on occasion revered in the eighteenth century was sometimes criticized when it was the product of physical constraint. Maupertuis's *Vénus physique* attacks the Chinese habit of foot-binding, but observes that the French girl who pours scorn on this custom does so only because she finds it ridiculous to sacrifice ease of walking to the demands of doll-like daintiness. In general, however, 'elle ne trouve pas que ce soit payer trop cher quelque charme que de l'acquérir par la torture et la douleur.' She herself, Maupertuis continues, accepts the need to have her body imprisoned in a whalebone corset from childhood, or bent by an iron clamp more restrictive than any Chinese orthopaedic ligature.[29]

Seeing certain parts of the body as beautiful may, of course, depend on the amount of leisure which their possessor has to display them to best advantage, as well as the beholder's predisposition to find them attractive. The Edmond of *Le Paysan perverti* writes that in the city female charms are multiplied far beyond what is understood in the country. In his native village he had never heard a woman's hand praised, but in Paris 'une belle main a son prix.'[30] (The reader may be reminded of Climal's preliminary advance to Marianne, which takes the form of buying the bashful girl several pairs of gloves, on the grounds that 'cela conserve les mains, et quand on les a belles, il faut y prendre garde.'[31]

Some of the resources of the town simply depend on showing to advantage what country folk never think of in erotic terms: thus Parisiennes, freed from the coarse sabot or shoe of rustics, omit no opportunity to display an elegant foot, and with it a pretty leg.[32] Edmond's sister Ursule is actually convinced that her feet have shrunk since her arrival in the capital: she now wears shoes and slippers

---

[28] *Recherches sur les habillements des femmes et des enfants* (Paris, 1772), pp. 174–7.

[29] *Vénus physique, Œuvres* (Dresden, 1752), pp. 258–9.

[30] p. 39. [31] p. 31. [32] *Le Paysan parvenu*, p. 39.

that would never have fitted her before.[33] Other changes are a more explicable and natural product of living in a different environment. Ursule, like Marivaux's Jacob before her, remarks on the loss of ruddiness consequent on her move to the city (both characters seem to welcome their new pallor); and Edmond comments on the 'blancheur éblouissante' of Parisian women, 'qui ne se trouve presque jamais à la campagne'.[34] Jacob too is favourably struck by Mlle Habert's white skin, while noting that it is probably more flushed when its owner is not indisposed.[35] Diderot has the other nuns refer to Suzanne's white skin in *La Religieuse*, and provides expert commentary on female flesh-tints in the *Salon* of 1763. His notice on Vanloo's *Les Grâces* includes the accusation that the artist has failed to differentiate between the skin of a brunette, which is firm and white, but without transparency or glow, and that of a blonde, whose delicate flesh sometimes has a bluish tint from the veins underneath.[36]

Theorists of beauty, perhaps irked by the variable canons according to which it was judged, sometimes attempted to redefine the very concept, arguing that it should be discerned, not aesthetically, but in terms of utility. Rather than consisting in a collocation of planes and contours, beauty resided in the visible indication of an ability to perform pragmatic tasks (including that of procreation). In *Des femmes et de leur éducation* Laclos proposes that men should demand of a female's appearance only that it reveals her ability to be a satisfactory sexual partner: she may be deemed beautiful if she is strong-looking, large, and fresh. But he then concedes that different cultures will inevitably imprint their own moral desiderata on the judgement of beauty. The Swiss and English regard women who look gentle and modest as beautiful; the French, a gayer race, seek evidence of a capacity for liveliness and pleasure.[37]

Another resort for those faced with describing the phenomenon of human beauty was to evoke, not the beautiful object itself, but the observer's physical reaction to it. This was a natural enough consequence of the emphasis laid by Dubos and like-minded theorists on the importance (and universality) of the sensory response to emotion-arousing phenomena. In the eighteenth-century novel the detailing of physiological changes experienced by the person

[33] *La Paysanne pervertie*, p. 64.   [34] *Le Paysan perverti*, i. 39.
[35] *Le Paysan parvenu*, p. 54.   [36] *Salons*, i. 196.   [37] p. 428.

encountering beauty often does duty for a determined effort to isolate the qualities of the object that provoked the response. Beauty, in such cases, becomes virtually identifiable with the capacity to stimulate certain kinds of sensation.

In Crébillon's *Les Egarements du cœur et de l'esprit*, reference to such physical response in the beholder is sometimes of a very general kind: there is much on the 'touching' looks of Mme de Lursay, and on how the hero is 'taken aback' or 'stopped short' by her unexpected charms. But the effect on him of first encountering Hortense at the Opéra is described far more precisely. A 'mouvement singulier' suddenly shakes him; he is 'struck' by her beauty, utterly overwhelmed, surprised into a veritable 'transport'; he feels in his heart a disorder which spreads over all his senses. Meilcour is not the only member of the audience to be physically affected by the sight of such beauty: he notes the general surprise of onlookers, and concludes from it that Hortense must be appearing in public for the first time. This thought causes in him a 'movement' of joy; he notices two nobly attired women with her, but the surprise he feels at not knowing them either troubles him little. He is 'entraîné' by the charm of contemplating Hortense alone.[38]

The physical reference of all the words describing Meilcour's state is significant. Emotion here has its literal sense of 'movement'. If Meilcour's inner feelings find expression in relatively few external tokens, there is no missing the emphasis on physiological upset occasioned by the spectacle of beauty. A comparable process is described in the novels of Crébillon's contemporary Marivaux. The 'mouvements' occasioned by emotion are bodily: we read in *La Vie de Marianne* of twitches in the heroine's face,[39] shedding of tears,[40] and even paralysis.[41] Sometimes they are a product of the confusion caused by love—the 'mouvements inconnus' which announce themselves in a girl first acquainted with this emotion,[42] or gestures like the hand-kissing which is Valville's spontaneous response to the 'mouvement' of love he feels for Marianne.[43] Her appearance at the church service when she first encounters Valville, and at which she had coquettishly intended to cause a stir, has the desired effect of arresting all male glances;[44] the jealousy of the other women

---

[38] pp. 33–4.   [39] p. 408.
[40] p. 562.   [41] p. 568.
[42] p. 66.   [43] p. 74.   [44] p. 60.

present, too, is tellingly manifested in the 'distraction' and 'inquiétude' of their gaze when she enters the church.[45]

The transport occasioned by des Grieux's *coup de foudre* is described by him in physiological terms as a 'douce chaleur' coursing through his veins.[46] Subsequently, the power of Manon's beauty is made apparent through the physical effect she has on others. When she comes to Saint-Sulpice to visit des Grieux, his emotional condition at renewed sight of her charms is manifested in his downcast eyes and trembling;[47] he is overwhelmed by the 'mouvements tumultueux' of passion, shudders, feels himself 'transporté dans un nouvel ordre de choses', and is seized by an 'horreur secret'.[48] Even when Manon is being taken in the convoy of prostitutes to Le Havre, the force of her charms remains considerable. At Pacy, where the convoy stops at an inn, the sight of her is described as heartrending:[49] she moves not only an old woman of the village and Renoncour himself, but also the captain of guards conducting the convoy to the port. As the circumstances detailed here are harrowing ones (although Manon's beauty remains remarkably untouched by her adverse experiences), the physical response to her looks differs from what it might be under more fortunate conditions. The reaction which the latter cause in the onlooker is imagined by Burke in his treatise *A Philosophical Inquiry into the Origin of our Ideas on the Sublime and Beautiful* (1757). The encounter with beauty, he writes, arouses pleasure and therefore leads to the beholder's physical relaxation: the head reclines, the eyelids droop, the eyes roll gently towards the beautiful object, the mouth opens, breath is drawn in slowly, there is an occasional sigh, and the hands fall idly to the sides. The observation of beauty, Burke concludes, characteristically leads to a state of beatitude.[50]

The reaction to beauty of jealous observers—Marianne's female rivals in church, or Marianne herself when she realizes Valville is attracted to Mlle Varthon—will obviously be far from this easeful state. Crébillon provides further illustrations in *Les Egarements*, which describes a society much given to passing value-judgements on the basis of outward appearances. In this self-regarding world the sincere expression of admiration for beauty is often seen as a mark of ingenuousness. Meilcour, who is struck by Mme de Lursay's

---

[45] p. 61. [46] p. 21. [47] p. 44. [48] p. 45.
[49] p. 11. [50] *Works*, 12 vols (London, 1887), i. 232.

majestic looks, is encouraged by both Versac and Mme de Senanges to see her appearance in less exalted terms, in Mme de Senanges's case because she wants Meilcour for herself. When she walks with the young man in the Tuileries, accompanied by Mme de Mongennes, the two women pass the time belittling each other's looks for the benefit of Meilcour (who is revolted by their vulgarity).[51] In *Le Paysan parvenu* Marivaux presents a somewhat better-tempered instance of the same procedure in the discussion of Mlle Habert's real age: other interested females hazard more or less ungenerous guesses, but grudgingly concede that she is well-preserved for a woman of her years.

We have already noted Rousseau's qualification in *La Nouvelle Héloïse* of the idea that beauty is in the eyes of the beholder, and the above examples strongly suggest that its perception may be governed by mental attitude as much as ocular response. Senses other than that of sight may be involved, of course, but according to a story Rousseau recounts in the *Confessions* their different messages should not be undiscriminatingly combined. The evidence of the eyes, he writes, is quite distinct from that of the ears. On hearing the angelic tones of girls singing in one of the Venetian *scuole*, the music-loving Rousseau decided that the possessors of such celestial voices must themselves be sublimely beautiful. Imagine his chagrin, then, at finding on introduction to them that nearly all were physically deformed in some way:

M. Le Blond [his guide] me présenta l'une après l'autre ces chanteuses célèbres, dont la voix et le nom étaient tout ce qui m'était connu. Venez, Sophie . . . elle était horrible. Venez, Cattina . . . elle était borgne. Venez, Bettina . . . la petite vérole l'avait défigurée. Presque pas une n'était sans quelque notable défaut.[52]

But when he had withdrawn from their presence and begun to listen again to their singing, he was able to forget the distressing sight of the girls and believe once more that they were as perfect in appearance as in voice.

Rousseau was perhaps unusually impressionable and emotionally combustible. But many will share the opinion of *Emile* that illusion and love somehow belong together, whether the illusion be that conjured by the sense of sight or related to the more mysterious and intangible workings of the imagination. Yet shortly after the episode in the *Confessions* just recounted, and despite his assertion in *Emile*, Rousseau reveals how his own imagination proved fatally deficient in an amorous encounter. In the company of a charming courtesan, Zulietta, he was initially made impotent by the troubling thought that a girl so beautiful as herself, who should have been the paramour of a prince, must have some undiscovered blemish that explained why she was nothing but a common prostitute. On the point of success a second time, he noticed a small irregularity in her breasts. Imagination, and with it potency, failed him again; and the man who saw the angelic Julie in the plain Sophie d'Houdetot rejected the beautiful Zulietta because of a minor asymmetry in her physical form.[53]

In *Les Egarements du cœur et de l'esprit* Crébillon too illustrates the power and the limitations of imagination in affairs of the heart. The mounting excitement Meilcour feels in Mme de Lursay's presence at the end of the novel, it is suggested, is an instance of deception by the senses. Through a form of transference familiar to lovers, Meilcour manages to intensify his perception of her charms (despite the fact that he is haunted by the image of Hortense) so that he sees her as more beautiful than she actually is.[54] Yet the perfections he projects on to her are insufficient to guarantee him immunity from the intrusive memory of Hortense. Two images war with one another— the heightened one of a woman tangibly present, and the painful recollection of an untouchable beloved. Although during his night with Mme de Lursay, Meilcour claims, his imagination alone was stimulated by her presence,[55] his experience in fact seems to betoken a more complex blend of sense-perception, rational apprehension, and fantasy. Appearances are not definitively transformed by the consummation of love; the governing element of judgement retains its force over Meilcour, pronounces on the evidence of his senses, and finds it insufficient justification for his 'égarement'.

Meilcour's near-contemporary, Marivaux's Jacob, is less troubled by the inconvenient incursion of rationality into sensuous enjoyment. He is enough master of his will to govern its image-forming faculty, and rational enough to be able to interpret reality pragmatically. In him imagination is an aid to perception, not something that distorts it. He can rejoice in his possession of a woman thirty years

---

[53] pp. 320–2.   [54] p. 185.   [55] p. 186.

his senior by concentrating his attention on the physical endowments that make her still attractive despite her age, while retaining an intellectual awareness of the material advantages he may derive from consorting with women older than himself. It is rarely the case, as it is with Meilcour, that his eyes open to reveal an error in his earlier perception of females. He is too clear-sighted to be often a victim of disillusionment.

But describing female beauty in detail, whether the witness to it is a rational man or a star-struck lover, is much less frequently attempted in the eighteenth-century novel than is the description of a woman's apparel. The discussion of clothing is often bound up with matters of morality—what a female's garb may tell the observer about her taste for ostentatious display, modest concealment, or whatever. Rousseau, who in *Emile* cites Spartan women as models of their sex in both physical and moral development, states that they did not, for the sake of idle fashion, lock their bodies inside 'ces entraves gothiques . . . ces multitudes de ligatures qui tiennent de toutes parts nos membres en presse'. The excessive use of whale-bone corsets, which according to Rousseau has reached epidemic proportions in England, will surely culminate in a degeneration of the species.[56] It is disagreeable, he writes, to see a woman cut in two like a wasp, and this is because everything which constrains nature is in bad taste. Life, health, reason, and general well-being should be cultivated above all else. Grace belongs with physical ease; languor is not equivalent to delicacy, and the sickly are displeasing. People who suffer excite pity, not desire, and the latter demands a state of health and freshness.[57]

The same association of beauty with naturalness leads other eighteenth-century writers to abhor the artificial constraint of the corset, identified with the sophistication of 'civilized' life. There was good reason in their criticisms. Corsets displaced the breathing activity of the lungs in an upward direction, so that the respiratory movement drew attention to the bosom quite apart from the physical prominence they naturally gave to that part of the body.[58] As we have seen, Zimmermann believed the wearing of corsets to cause a variety of ailments in women, and issued a graphic warning against

---

[56] p. 458.   [57] Ibid.
[58] See James Laver, *Modesty in Dress* (London, 1969), p. 117, quoting Havelock Ellis.

the practice on account of the damage it did to the female 'machine', conferring on it the excessive delicacy so amply recorded by the 'sensible' novel of the eighteenth century.

At the same time, it seems unlikely that some of the more celebrated fainting-fits described in the literature of the age were attributable to sartorial constraint. Mlle Habert was probably too comfortable about the plumpness Jacob admires in her, as well, perhaps, as too elderly, for her swooning on the Pont-Neuf to have been caused by such constriction. Both she and her sister ascribe it to her having eaten insufficient breakfast for a frame which, as Jacob sees from her face, 'avait l'air d'être succulemment nourrie'.[59] And the seizure which overtakes the Présidente de Tourvel at a critical point in her reunion with Valmont[60] is more likely to have been produced by extreme emotion than by fashionably tight clothing, for although she wears a whalebone corset 'qui remonte au menton'[61]—a mark of the decent bourgeoise rather than of the fashionably décolletée aristocrat—she is doubtless too mistrustful of worldly vanity to be guilty of this particular form of self-mortification. Mme de Merteuil confirms the Présidente's unworldliness in this respect, remarking to Valmont (in another token of female jealousy?) that his new love is 'toujours mise à faire rire'. With her 'paquets de fichus sur la gorge' she would presumably derive no benefit from the upward projection of the bosom which the corset engineered.[62]

For Saint-Preux, the costume of Parisian women is an index of their self-preoccupation, and therefore reprehensible. He tells Julie that they dress bizarrely, but admits that they do so with extreme good taste, and that they understand the difficult art of adapting to their own advantage whatever fashion decrees—an art, he reflects, which their provincial sisters know nothing of. He notes that Parisiennes do not flaunt their wealth by dressing expensively, as rich women in other countries do: all classes wear the same materials, and '[il] n'y a point de peuple, excepté le nôtre, où les femmes surtout portent moins la dorure.'[63] But, he cautions, let not his reader suppose that this argues an essential modesty in the Frenchwoman's character. Not only are all the women of Paris addicted to tightly laced stays, but they take every care to display their bosoms to best advantage.[64] At least, to what they regard as best advantage; for

[59] p. 54.   [60] p. 295.   [61] p. 18.   [62] Ibid.   [63] p. 267.
[64] See also David Kunzle, 'The Corset as Erotic Alchemy: From Rococo Galanterie

they have no better understanding than other civilized races of the way beauty is enhanced by the imagination. Far from the expanse of pouter-pigeon breast provoking the onlooker's admiration, it curbs the flight of fancy by its very obviousness. Beauty is more likely to be perceived, Saint-Preux informs Julie, in what is hinted at rather than brutally uncovered. He quotes Montaigne's observation that 'la faim entière est bien plus âpre que celle qu'on a déjà rassasiée, au moins par un sens.'[65]

For one so recently arrived in the capital of fashion, Saint-Preux shows a remarkably acute awareness of the psychology of dress. But he does not draw the further conclusion implied by his observation about sartorial revelation and concealment, namely that the institutionalizing of immodesty causes it to lose its power of shocking and therewith its original purpose. The effects of dress and undress, in other words, are entirely relative to one another. Modesty is a social value with no absolute force, whose canons vary from one environment and one age to another. Most clothing is a token of an attitude rather than a response to a physical need: its utilitarian function of warming and protecting the body is far outstripped by its employment as the signal of abstract social values. And as it is judged relative to such values, it cannot contribute to changeless concepts of beauty.

Where beauty is seen as residing in a sense of modesty, apparel which enhances what is taken to be physical propriety will be highly prized; but propriety is itself a cultural variable, as memorably illustrated in *Paul et Virginie*. Whether or not Saint-Pierre meant Virginie's behaviour on the deck of the ship to exemplify cross-cultural confusion,[66] to the modern reader Virginie's coyness can only seem absurd. Perhaps the intervening two centuries have brought us closer to primitive ideals of beauty and modesty than Saint-Pierre, despite his two-year spell on Mauritius, felt prepared to countenance.

The morality of costume emerges in a rather different sense from *La Vie de Marianne*. Marianne is well aware that her God-given looks are enhanced by fine clothes, and suffers from the straitened

---

to Montaut's Physiologies', in *Woman as Sex Object*, ed. Thomas B. Hess and Linda Nochlin (London, 1973). [65] p. 266.
[66] *Paul et Virginie*, p. 206, note 1. The historical origin of her unwillingness to disrobe was apparently the refusal of the real-life captain of the *Saint-Géran* to do so, because his professional dignity prohibited him from reaching the shore naked, and also because he had important documents in his pocket.

circumstances we see her pitched into at the start of the novel: they do her vanity an affront as well as imposing material deprivation on her. The promise of a new dress by M. de Climal fills her with a delight that is immediately given physical expression:

mon aisance me donnait des grâces qu'il [Climal] ne me connaissait pas encore; il s'arrêtait de temps en temps à me considérer avec une tendresse dont je remarquais toujours l'excès, sans y entendre plus de finesse.[67]

Despite Marianne's growing suspicions as to Climal's motives in paying her such attentions, it is not until he offers to buy her some underwear that her doubts crystallize. Rather than purchasing the linen from Marianne's landlady, Mme Dutour, he offers to take her to an expensive supplier. But this proposal alerts Marianne to the real nature of his interest in her, for 'la charité n'est pas galante dans ses présents'[68] (the gallantry residing, of course, in Climal's intention to clothe the private parts of her body). Yet she deliberately complicates the situation in her mind in order to defer its resolution, postponing the morally necessary rupture with Climal so that she can keep what he has given her. Even after Climal's suggestion that he set her up in a love-nest, which Marianne meets with outrage (the more so for having in the meantime fallen in love with Valville), she feels a lingering regret at the need to dispose of his presents.

Literature of the eighteenth century comments as frequently as that of other ages on the connection between costume and seduction. The act of dressing up in order to attract a member of the opposite sex scarcely needs detailing. Of more interest, perhaps, particularly in the light of Climal's attempted seduction of Marianne, is the offering of apparel as a gift to the loved one. In Diderot's *Jacques le fataliste* the hero offers his beloved Denise, as a token of his passion, a pair of garters. This not only allows Diderot to focus on the leg as an erotic object, and thus lead to the ambiguous scene of Denise's massaging Jacques's wounded leg from the knee upward until the point is reached by her hand when he 'la baisa',[69] but also illustrates the shift in erogenous zones which the history of fashion documents, and which the wearing of particular clothes and ornaments of dress highlights. (There is a similarly erotic scene in Crébillon's *L'Ecumoire*, where Tanzaï fumblingly helps readjust his beloved Néardané's garter.) Diderot's novel, which contains many references

[67] p. 35.   [68] p. 39.   [69] p. 779.

to costume,[70] provides a further example of male wooing through the gift of clothes, although it shows not how the attempt to seduce in this way is frustrated by the scruples of the recipient (as in *La Vie de Marianne*), but how the imperious command of a third party prevents the seduction from bearing fruit. Mme de la Pommeraye, who is using the mother and daughter d'Aisnon as tools to aid her own revenge on the Marquis des Arcis, allows the younger woman to keep none of the finery he presents her with,[71] presumably in order to emphasize her deceptive virtue (for Mlle d'Aisnon is in reality a prostitute, and Mme de la Pommeraye's revenge will consist in having the Marquis des Arcis marry her in ignorance of this fact).

The garb of the 'dévote', which both mother and daughter have been obliged to adopt by Mme de la Pommeraye as part of her scheme, has a perverse charm for the former roué des Arcis. Austere dress often carries such an appeal. The costume of the convent is described in *La Religieuse* as conferring a singular beauty on the already attractive Suzanne Simonin: when she takes the veil the other nuns remark on how the stark blackness of her wimple enhances the whiteness of her skin, her coif rounds her face, and the habit offsets her waist and arms.[72] Simplicity in apparel is often found pleasing in societies where relief is sought from the artificialities of social life. The attraction may be moral or aesthetic, or a combination of the two. In Louvet's *La Fin des amours de Faublas* the idyllic retreat from the world which the Comtesse de Lignolle plans for herself and her lover is to be marked by simplicity of clothing as well as simplicity of other kinds: she will wear light taffeta in summer and calico in winter, and Faublas will make do with similar stuffs.[73]

Paul et Virginie, in comparable fashion, describes how the two families want no clothing save what can be made of cotton, which Mme de la Tour and Marguerite spin day and night;[74] the custom of the island is to leave infants like Paul and Virginie completely naked. When Virginie arrives in France, by contrast, her aunt presents her with new costumes for every season, in which she is dressed by maids who are themselves as splendidly attired as court ladies,[75] and who squabble over the possession of Virginie's garments before she has even cast them off. Bernardin means all this

---

[70] See Jeannette Geffriand Rosso, *'Jacques le fataliste': L'Amour et son image* (Pisa, 1981).
[71] p. 639.
[72] p. 239. [73] p. 1111. [74] p. 87. [75] p. 162.

to seem like corruption, although the attitude in which death freezes Virginie—one hand placed on her heart, and the other on her European costume—is apparently intended to be edifying. Paul, longing for her to return from Europe, cannot see that she needs any more adornment than a red handkerchief or flowers in her hair.[76]

It is no surprise to find Rousseau equating plainness of dress with moral virtue in *La Nouvelle Héloïse*. His description of the engraved plates illustrating the novel points to the fact. Julie, like Emile's Sophie, is clothed naturally and without affectation: she is elegant in the simplicity of her costume, which has the touch of negligence that flatters more than 'un air plus arrangé'.[77] In keeping with this, she wears little in the way of ornament, but her restraint bears witness to true taste and refinement. We are informed that neither she nor Claire ever wears panniers, though not whether Rousseau shares the view of one writer who deprecated them on the grounds that they permitted women of loose morals to conceal illegitimate pregnancies by drawing their skirts away from their bellies.[78] Rousseau's *Confessions* contains a statement of his preference in female looks which matches what is here suggested. He is attracted by neatness of dress, taste, and a moderate degree of refinement in the material of women's costumes. The least pretty girl who observed these conditions would always seem to him preferable to a naturally better-endowed one who did not.[79]

A noteworthy illustration of the connection between clothing and sexuality is given by Rousseau in the first part of *La Nouvelle Héloïse*. Waiting for his long-desired rendezvous with Julie in her 'cabinet', Saint-Preux loses himself in erotic raptures amidst the clothes she has taken off. All the 'vestiges' of his mistress contribute to his intoxication: the net she wears on her hair, 'cet heureux fichu contre lequel une fois au moins je n'aurai point à murmurer', the elegantly simple dishabille which so accurately indicates the taste of its wearer, her dainty slippers, the corset which touches and embraces a part of her body he dare not name, and which bears the imprint of her breasts . . . Overcome by surging desire, Saint-Preux is unable to complete the catalogue.[80]

---

[76] p. 184.   [77] p. 762.
[78] [Duguet], *Cas de conscience décidé par l'auteur de la 'Prière publique'*. On demande s'il est permis de suivre les modes, et en particulier si l'usage des paniers peut être souffert? avec les réponses aux objections (n.p., 1728), pp. 8–9.
[79] p. 134.   [80] p. 147.

The concept of the 'négligé', which Rousseau mentions in his description of Julie's appearance, has more than one connotation in eighteenth-century literature. It may betoken a genuine lack of interest in dressing carefully that serves to guarantee the wearer's moral character, her concern with higher (often more spiritual) things than worldly vanity. But it can equally signal a conscious desire to impress the onlooker with her unstudied grace, and give him the erroneous idea that she is either above the distractions of fashion or careless of human finery. Conversely, it can hint at a disposition quite opposite to the first-mentioned, namely a readiness to undress. To be incompletely or carelessly clothed is often more suggestive, and more erotically arousing to the observer, than total nakedness. In his *Salon* of 1765 Diderot notes apropos of Carle Vanloo's painting of the three Graces that it is the 'uncovered' ('découverte') woman who is indecent, not the fully nude one. To stick small rags of drapery on the buttocks of one of the Graces, and on the thighs of another, is to arouse a flood of improper thoughts. An indecent woman is one who would wear a 'cornette' on her head, stockings on her legs, slippers on her feet, and nothing more. This reminds Diderot of the way Mme Hocquet made a modest Venus the most immodest creature imaginable by adding, in plaster, a cloth between the hand concealing her pudenda and the private parts themselves. The ludicrous effect of this scruple was to make the goddess look as though she was drying herself.[81]

In *Emile* we learn that Sophie's apparent disdain for finery is really a kind of vanity: her taste for simplicity disguises an instinctive knowledge of what constitutes true elegance. She looks unaffected in her day-to-day wear, and has no idea what colours are fashionable, but is in fact exactly aware of what suits her. Although her appearance is modest, she is as coquettish as any other woman: she is conscious of the appeal made by the understated, preferring it to what is emphatically presented on the grounds that the imagination is more agreeably provoked by the former than the latter.[82] Similarly Mme de Tourvel, according to Valmont, senses that 'toute parure lui nuit; tout ce qui la cache la dépare.'[83] It is when she is 'en négligé' that her charms exercise their most powerful appeal. In the intense heat of summer she abandons more formal costume and dresses in a simple dishabille, covering her bosom with a piece of muslin that allows Valmont to guess agreeably at its curvaceousness.

In the Présidente's case it seems unlikely that such négligé is adopted in order to be sexually provocative. The casual dress of Mlle Varthon, by contrast, strikes Marivaux's Marianne as deliberately calculated to increase her rival's charms in comparison with her own appearance. Not that Marianne's apparel can itself be called anything but unstudied; but there is a subtle distinction between the appearance of the two girls. Mlle Varthon's simplicity of dress, ostensibly adopted for the sake of not competing with Marianne, is in fact 'un négligé . . . fort bien entendu'. Although her plain costume seems to exclude any intention of pleasing, and appears to give Marianne no reasonable grounds for complaint, the consideration it supposedly implies for Marianne is bogus. In such instances, Marianne observes, a betrayed and jealous woman sees more clearly than a loved one: she immediately senses the lack of good faith in her rival's action. 'La petite personne avait bien voulu se priver de magnificence, mais non pas s'épargner les grâces', Marianne comments, whereas she herself had risen from her sickbed and literally thought only to put on the first available item of clothing, a 'mauvaise robe', and show herself in public with all the disadvantages that recent illness does to the looks. She is left with the bitter reflection that female pride prohibits the idea of a lover abandoning his beloved on the grounds of her deficient charms. Far better that he should do so for reasons of mere inconstancy.[84]

The strategic point of being 'négligé' is fully worked out in *Les Egarements du cœur et de l'esprit*, where the kind of self-consciousness Marianne suspects in Mlle Varthon is directly acknowledged—at least by the male arbiter of fashion, Versac. In high society, he tells Meilcour, seeming casualness about physical appearance, which in women comes close to indecency and in men goes beyond the straightforward desire for ease and comfort, belongs to what is called 'bon ton'.[85] This aesthetic derives from the very artificiality of the world in which Versac and Meilcour live, and whose studiedness has already shocked the young man. In Mme de Lursay, of whom the word is recurrently used, négligé is in part a coquettish ploy intended to allure Meilcour and make her desire to seduce him and be seduced by him apparent. Her dishabille (which Meilcour is perceptive enough to call 'galant'), her 'coiffure négligée', and her restrained use of cosmetics, show her in a light that is less dazzling than touching; and although Meilcour calls her appearance noble, her beauty in this aspect is profoundly moving.

When he is shown into the boudoir of Mme de Senanges, however, and finds her despite the lateness of the hour still at her toilet, her unrepaired looks are disagreeable to him. Although her intention of seducing him is more directly expressed than Mme de Lursay's, the ravages of age have too thoroughly done their work for the kind of négligé that is an adornment of youth to be at all appealing.[87] Not so in the case of the Marquise de Merteuil. When the chevalier Belleroche arrives at her love-nest, he finds her in a 'déshabillé le plus galant', a delicious costume of her own invention which shows nothing but gives away everything.[88] Maliciously, she promises to give Valmont a version of it for his Présidente, once he has made her worthy of wearing it. As we should expect with an operator as calculating as the Marquise, every detail of the négligé has been worked out with the care she devotes to all aspects of seduction: although the hapless Belleroche is to be struck by the naturalness of her appearance, it is merely an example of art concealing itself.

The same is true of the episode describing the discomfiture by Mme de Merteuil of the much more sophisticated Prévan. As she triumphantly informs Valmont, she succeeds in out-manœuvring this experienced seducer by acting the part of an 'ingénue'. Once her other guests for the evening have left the house, she undresses and dons a 'toilette légère'; but since her plan depends for its success on Prévan's remaining dressed (and thus appearing to have forced himself on her uninvited), she does so very quickly. So Prévan enjoys her favours in his full evening wear, and is just making to take off his clothes (cursing the costume 'qui . . . l'éloignait de moi') when she rings for her servants and has him caught *in flagrante delicto*.[89] In this case the male's complete attire during the act of love-making is an essential part of the female's real purpose. But the illustrations to other erotic novels of the period suggest that it was far commoner for the male to preserve such appearances on these occasions than the female. The contemporary engravings to Mirabeau's *Le Rideau levé*, as well as to his *Ma Conversion*, more

[86] p. 56.   [87] p. 123.   [88] p. 30.   [89] p. 192.

frequently show women than men completely naked in the act of intercourse.[90]

If male négligé was less usual in eighteenth-century high society than female, despite what Versac tells Meilcour, that circumstance may perhaps be ascribed to the fact that it is for woman to invite amorous attack and for man to launch it. On the evidence of the contemporary novel, a male soliciting a female's attention did so by other than sartorial means. Indeed, while a man's interest in clothes was not necessarily found reprehensible, it may still have been thought to indicate effeminacy. Rousseau feels it necessary to mention in his remarks on the engravings to *La Nouvelle Héloïse* that Saint-Preux is very simply dressed—but in that he scarcely differs from the main female characters—and that Edouard's costume is like a typical English lord's, without showiness. The hints about men's fascination with dressing up which fiction of the period occasionally furnishes do not really argue a lack of masculinity. Indeed, the novel which pays the closest attention to this proclivity, Louvet's *Faublas*, shows the practice of transvestism starting as a game, and only continuing in force because of the mistaken beliefs which Faublas's initial act of disguise has put into the minds of some characters. So far is Faublas from effeminacy that he contracts two sexual relationships with women—three if we count his passing fancy for the maid Justine, and four if we include the night he spends 'instructing' the future wife of Rosambert—besides that with his adored wife Sophie. In any case, as is generally agreed, transvestism is not normally a sign of homosexuality, even if it is primarily a sexual obsession.[91] (La Bruyère thought it a far lesser vice than the use of make-up.)[92] Although as late as 1760 transvestites could be burnt to death in France, their habit had long been institutionalized in the upper reaches of society in the form of masked balls and masquerades like those held from 1715 at the Opéra. Casanova enjoyed swapping clothes with the mistress of the moment at such gatherings.

Although *Faublas* contains a certain amount of detail on the dressing of the youth's hair as part of his impersonation of women, and although *Manon Lescaut* contains a famous scene in which the

---

[90] See the plates in the edition cited.
[91] See Peter Ackroyd, *Dressing Up. Transvestism and Drag: The History of an Obsession* (London, 1979).
[92] *Les Caractères*, ed. Georges Mongrédien (Paris, 1954), p. 104.

hero's hair is dressed by his mistress as a token of her affection for
him, such concerns are more often shown as female ones in the
eighteenth-century novel. In Montesquieu's *Lettres persanes* the
Persian Rica recalls the time when the immense height of coiffures
put a woman's face in the middle of her body (and when the height
of heels did the same to her feet).[93] After detailing her preference
for high heels over low, and noting how 'les pieds et la tête sont le
plus important de la parure', Restif's Ursule writes in a letter to
Gaudet that her masterpiece of taste is the coiffure she daily assumes,
and which she alters even within a single day by the successive
donning of different articles of headgear:

La coiffure en Bacchante annonce une Cléopâtre; celle en folle, une badine,
qui leurre et couronne tour à tour; celle en naïve, une vierge, qui se défend
avec maladresse; celle en effrontée, que je veux prévenir, et faire un Encolpe
de mon amant.[94]

The moral point of the contrast, implied or explicit, between such
edifices and the natural look of those who wear their own hair
unadorned cannot be missed. Julie, needless to say, wears a modest
coiffure in the plates accompanying *La Nouvelle Héloïse*. Mme de
Morgennes alludes maliciously in *Les Egarements du cœur et de
l'esprit* to Mme de Senanges's indecent attempt to appear younger
than she is by favouring the hairstyle of a twenty-two-year-old
woman, although her critic's disclaimer of concern for the appearance
of her own has the ring of untruth.

Marianne, on the other hand, looking back to her youth in the
previous century, derives satisfaction from remembering the natural
beauty of her chestnut hair, thinking of how she wore it proudly
in an age when 'on se coiffait en cheveux' (that is, without wigs).
Clearly, the statutory blondeness of the seventeenth-century heroine
allowed for some variation; and by the eighteenth, as Mme de
Merteuil's contemptuous reaction to the Comte de Gercourt's
trichological theories suggests, the association of particular hair-
colours with particular moral states was far from automatic. (The
Marquise derides her former lover's belief that Cécile's blonde
hair is a guarantee of sexual restraint, proof positive that she will
not make a cuckold out of him, and declares her opinion that 'il

[93] *Lettres persanes*, ed. Paul Vernière (Paris, 1960), p. 206.
[94] *La Paysanne pervertie*, p. 384.

n'aurait jamais fait ce mariage si elle eût été brune.'[95] Julie's blondeness, mentioned by Rousseau apropos of Gravelot's illustrations,[96]
nevertheless seems intended to match the demure goodness of
her character, and Claire's brown hair somehow to reflect her
mischievous ebullience.) Faublas, meeting Sophie for the second time,
is enraptured by her long black hair, which contrasts with her
dazzlingly white complexion;[97] later on he is to be struck by the
beauty of the Marquise's black tresses, which similarly offset her
alabaster skin.[98]

One reason for this flexibility may have been the facility for
changing tints which women theoretically enjoyed; but this alone
would not explain why the eighteenth-century novel saw a shift from
the formerly rigorous apportioning of virtue to the blonde-haired,
for dyes had been known since antiquity. It would, of course, have
been unthinkable for a virtuous heroine, in the eighteenth century
or earlier, to tint her locks artificially. It is much likelier that the more
catholic view of hair colour evinced in novels of the Enlightenment
was part of the general favouring of non-uniformity.

Whether 'improved' or not through the use of hair dye, the
application of cosmetics, or the wearing of flattering clothes, beauty
itself was often seen by Christian moralists as dangerous both to its
possessor and to those who might be lured by its charms into immoral
acts. In that part-sensuous, part-ascetic novel *Paul et Virginie* the
heroine's beauty is described by the old man to her grieving lover
as an enticement which might have brought misfortune on herself
and those close to her by attracting the unwanted attentions of other
men. (From this it is a short step to the traditional Christian image
of woman as a temptress and an evil influence on man, familiar from
St Paul, Tertullian, and others.) By dying young, besides, Virginie
has preserved her beauty intact, like the Manon whom des Grieux
buries in the American desert.

The face was seen as the focus of female and male beauty, and
abundant reference was made to the changes it underwent under the
influence of emotion. Laclos makes few allusions to the appearance
of characters in *Les Liaisons dangereuses*, but Cécile mentions in
passing the fact that society women, unlike herself, seem not to blush
when they meet men—unless, she speculates, it is simply that the

---

[95] p. 14.   [96] p. 762.
[97] p. 424.   [98] p. 681.

rouge they wear prevents their blushes from showing.[99] (Mme de Merteuil's remark about the 'rosebud' Cécile, in the second letter of the book, may be intended to draw attention to the pinkness of her complexion, or merely to her extreme youth and freshness in comparison with the females Valmont habitually consorts with.) It is no doubt significant that the Présidente de Tourvel's face easily colours, as Valmont notes in a letter to the Marquise detailing his first experience of holding his quarry in his arms.[100]

The fact that the blush is so frequently, if not invariably, a sign of timid embarrassment or social discomfort means that it is much described by writers in connection with female modesty. In a letter to Mlle La Bussière, a woman whom he declaredly hoped never to know intimately, Rousseau wrote of her essential charm for him: it resided not in her eyes, however gentle and yet lively, nor her fresh complexion, nor her slim waist, nor her elegant shape and graceful contours, but in

cette rougeur aimable, fille de la pudeur et de l'ingénuité, dont j'aperçus votre front se couvrir dès que je m'offris à votre vue . . . Dieux, que vous étiez belle, tant il est vrai que la vertu est le fond le plus séduisant de la beauté![101]

In *La Fin des amours de Faublas* it is the pallor of the Comtesse de Lignolle's brow that signals her emotion—in this case, rage,[102] and on a later occasion, despair,[103] although in the earlier volume *Une année de la vie de Faublas* the flush on Sophie's forehead betokens her modesty as Faublas makes to kiss her.[104]

It may perhaps surprise the modern reader to learn that the forehead blushes, although eighteenth-century novelists clearly regard that part of the physiognomy as highly expressive. In his *Caractères des passions* Cureau de la Chambre had attempted a mechanistic explanation of the blushing brow, which he explicitly associated with love (whereas the flush of shame, according to him, is located in the cheeks, and that of anger in redness of the eyes). In the passion of love, he writes, joy causes the vital spirits to overflow into the parts of the superficies nearest the brain—in the event, the brow—or, alternatively, the stimulation of the brain by the imagination results in this organ being warmed up by the

---

[99] p. 15.   [100] p. 22.
[101] Quoted in *Chefs d'œuvre de l'amour sensuel*, ed. Jacques Sternberg, Maurice Toesca, and Alex Grall (Paris, 1966), p. 100.
[102] p. 920.   [103] p. 970.   [104] pp. 921, 478.

continual agitation of the spirits, which is visibly communicated to the forehead.[105] Mention of this part of the visage in connection with blushing may, however, owe less to direct observation than to literary tradition. Pliny the Elder, for instance, wrote that 'Frons homini laetitiae et hilaritatis, serenitatis et tristitiae [est]',[106] while Juvenal advised readers that there was 'Fronti nulla fides.'[107] In both these cases, *frons* designates the whole of the face. Besides, the flush of 'pudeur' was well known to spread to other parts of the body. Zimmermann refers to Haller's report of a girl whose breasts blushed in certain circumstances,[108] and remarks on how passionate movement made the actress Mlle Dumesnil flush at the bosom, although her cheeks were too whited-over for any rush of blood to show there.[109]

Most writers suggest, if only by implication, that the blush is an ungovernable phenomenon in a creature subject to the impulses of 'pudeur'. Greater *savoir-vivre*, greater familiarity with the world of emotion and amorous affairs, makes females less prone to this involuntary expression of feeling. (Eighteenth-century novels rarely describe male blushes.) When woman has advanced beyond such token revelations, she may have arrived at the state which many writers describe as definitive of the female nature, that of coquetry. The coquette is not necessarily without modesty—the very act of refusal with which sexual teases signal the limits of their availability is a debased form of such reticence—[110] but she delights in using her feminine attractions to lead admirers, at least in imagination, beyond the province of strict sexual propriety. To this end she employs the weapons which nature, art, and society put at her disposal; for if some of the female's alluringness is God-given, some is defined only by the (variable) code of the world she inhabits.

The eighteenth-century novel provides a detailed picture of what, in the erotic game where appearance is of paramount importance, was regarded as a stimulus to sexual interest—the physical tokens and gestures that indicated a person's openness to sexual approach. At the same time, it reveals how deceptive appearances may be when

---

[105] p. 104.
[106] ii. 37. 'A man's brow has the look of gladness and cheerfulness, of serenity and sadness.'
[107] *Satires*, 2.8. 'No trusting a brow.'   [108] p. 254.   [109] p. 255.
[110] See Georg Simmel, 'Flirtation', *On Women, Sexuality, and Love*, trans. Guy Oakes (New Haven and London, 1984), pp. 134 ff.

controlled by the supervening force of intellect. Unsurprisingly, the Marquise de Merteuil provides the best illustration of the misleading signals on which coquetry often relies. Aware that the physical consummation of a relationship is liable to be betrayed by a woman's changed tone in the presence of her lover, she takes care to consort in public only with men she intends to flirt with, but never grant the ultimate favours to; that privilege is reserved for males with whom she does not publicly appear. 'Je gagne encore par là d'ôter les vraisemblances, sur lesquelles seules on peut nous juger';[111] for society at large has every reason to suppose her habitually severe with the presumed suitor of the moment. Public appearance is taken to match private conduct.

The celebrated eighty-first letter of *Les Liaisons dangereuses* details a process of facial adaptation akin to what Marivaux describes in a practised coquette, of whom he writes in the *Mercure* of 1728. This woman complacently enumerates the ways she can modify her physiognomy at will, assuming an air of languor and indolence, or of vivacity, hinting at experience and cultivation, and sometimes adding a variation in the look of her eyes—changing her expression from anger to sweetness, or disingenuous incomprehension and tenderness.[112] The education of the self about which Mme de Merteuil writes in Letter LXXXI involves, in part, practising and perfecting facial transformations of this kind in the interests of something more devious than the average coquette intends, namely, forming a physiognomy which only its owner can reliably interpret. For the casual observer, the Marquise notes, her look seems to signify a temperament or a state of mind quite other than the true one. From her earliest years, she informs Valmont, observation and reflectiveness permitted her so to master her expression that she could deceive at will those whose company she frequented. Later, dissimulation became a habit as she perceived the need to hide her activities and intentions from others; and so she progressed from being able to govern her eye-movements (and so to assume 'ce regard distrait que vous m'avez loué si souvent')[113] to having physical control of her whole face. Thus, perversely, a feeling of grief would be translated into an expression of serenity or even joy; and just as she trained herself to appear overcome by pleasure when actually

[111] p. 175.   [112] *Journaux et œuvres diverses*, p. 86.
[113] p. 171.

suffering from self-inflicted pain, so real transports of delight would be converted into an appearance suggesting their opposite. It is, of course, a pleasant irony that at the end of the novel Mme de Merteuil's face, disfigured by smallpox, acquires a look she can do nothing about.

Clearly, then, scepticism about the necessary correspondence between inward state and external expression is in order. In the age of sensibility it was apparently all too easy to fall in love with a face for the abstract qualities—of kindness, compassion, tenderness, or whatever—it appeared to guarantee in its possessor. But if the success of such ploys as Mme de Merteuil and, to a lesser degree, the coquettish Marianne have recourse to is an adequate guide, this trust might have disastrous and disillusioning consequences. Many readers will still agree with Marianne, however, about one significant correspondence between inner and outer self, expressed by her in terms of the waning of beauty and of wit. She describes the misfortune of a society lady, formerly known for her charming conversational style, who became a 'babillarde incommode' when smallpox destroyed her looks: 'Voyez combien auparavant elle avait emprunté d'esprit de son visage!'[114] Although the change to which Marianne refers is in the reception of the woman's words by others before and after her illness, it seems likely that the change of mental attitude within the sufferer herself, attendant on an awareness of her altered looks, could actually have affected the quality of mind socially expressed in conversation. La Rochefoucauld's maxim to the effect that no woman's merit survives her beauty,[115] although a cynically extreme formulation, draws attention to a different link between external appearance and inward being, but perhaps one which also emphasizes the response of others as a governing influence over the latter. For as long as her looks last they will implicitly or explicitly compel her to act according to the expectation that virtue and beauty are coexistent; but when she loses that external beauty, and merit is no longer seen in her countenance, the shift in expectation prompts a shift in her moral behaviour.

The *Encyclopédie* article 'Physionomie', probably by Vauvenargues, sounds a cautionary note concerning the equation of fairness of countenance and goodness (or ugliness and vice). Accidents like

---

[114] p. 8.
[115] *Maximes*, ed. Jacques Truchet (Paris, 1967), p. 107 (number 474).

smallpox, the author writes, may prevent a beautiful soul from appearing; but are we to assume that the disease has actually expunged it? There are enough examples in the eighteenth-century novel of the 'belle âme' triumphing physically as well as morally over such adversity (one thinks particularly of Richardson's Pamela and Rousseau's Julie) for it to appear that writers of fiction, at least, still liked to match the inner and outward selves of their characters.

Yet many eighteenth-century thinkers seem to have believed that beauty might reside in an individual's non-physical being, and therefore be detectable otherwise than via the senses (although some, following Hutcheson, postulated the existence in humans of a special 'sixth sense' directed exclusively at the perception of beauty). Such a belief could more easily accommodate the non-rationalist bent of contemporary thought than the drily mechanistic explanations of philosophers were able to do. For the non-rationalist, beauty could be described in moral and spiritual terms rather than material ones, and might even—as in *La Nouvelle Héloïse*—be elevated above the merely physical. Such beliefs had an obvious affinity with the Christian emphasis on the vanity of worldly concerns, especially the concern with beauty of the face and body. They encouraged moralists to claim as a matter of principle that literature should address people's minds rather than their sensory appetites, and held a strong appeal for the many sententious novelists of the age.

We know that what Saint-Preux loves in Julie is not physical beauty of an outstanding kind: his response to the miniature portrait she sends him is sufficient proof of that. Rather, he reveres in her a depth of soul of which her imperfect visage is, in some complex and unexplained fashion, both the mask and the guarantee. However confusedly Rousseau expresses this idea, and however inconsistent Saint-Preux seems to be in his adversions to his beloved's appearance, there is still a kind of clarity in his perception. And in cases such as his own there is evidently no need of any caution against looking with the eyes of love. To do the latter, in conventional terms, is to be tricked by the process of 'crystallization'. To the lover who can see beneath the surface appearance, the fact of physical beauty represents no danger. For such a person, human aspect addresses the heart rather than the senses.

In her essay on friendship Mme Condorcet claimed that it was a feature of women's affections in particular to prefer a person's moral qualities to his or her possession of beauty.[116] She implies thereby that the former are not necessarily translated into the latter. For Marivaux it was anyway apparent that women with inferior looks find compensation for the fact through a subtle play of 'amour-propre'. In the *Spectateur français* he describes watching a crowd, which included several women who might with good reason have been heartily dissatisfied with their want of beauty, leaving a Paris theatre. Female coquettishness, he observes, prevented them from being so. When brought face-to-face with a visage more handsome than their own, their self-consoling expedient was to ignore it:

une fière sécurité revenait sur [leur] mine; il s'y peignait un air de distraction dédaigneuse, qui punissait le visage altier de la vanité de son étalage, mais qui l'en punissait habilement, et qui disait à la rivale qu'on n'avait pas seulement pris garde à elle.

When an ugly woman looks at herself in the mirror, he continues, she allows her eyes to slide over her individual bad features and focus on the whole composition of her face. This done, she is able to find the assemblage of comparative imperfections more piquant than the insipid regularity of a conventionally beautiful countenance. 'Et c'est avec ce visage de la composition de sa vanité qu'une femme laide ose lutter avec un beau visage de la composition de la nature. Et qui le croirait? quelquefois, cela lui réussit.'[117]

But even in Marivaux, so given in his novels to the painting of 'portraits', words convey little of the detail of a person's looks. It is noteworthy that the materialist thought of the age should have had such slight influence on modes of literary description. Aside from the attention paid by writers like Crébillon, Prévost, Diderot, and Laclos to the physiology of emotional arousal, there is little evidence that novelists translated contemporary discoveries about the physical constitution of man into fiction. Nor did the vogue for physiognomical studies in the last three decades of the century find much reflection in the imaginative literature of the period. Louvet's *Faublas* contains a passing reference to the fashion for reading a person's character in his face, but it is a thoroughly ironical reference: the Baron de ***, despite constantly claiming that 'je me connais

---

[116] *Lettres à C**** [*Cabanis*] *sur la théorie des sentiments moraux, ou Lettres sur la sympathie*, in Adam Smith, *Théorie des sentiments moraux*, trans. Mme Condorcet, 2 vols (Paris, 1798), II, 399.
[117] p. 125.

en physionomie', is unable to see that the young woman who consorts with his wife and occasionally shares her bed is really a male *en travesti*, Faublas himself. Montaigne's contention that 'c'est une faible garantie que la mine' was no doubt accepted by some and rejected by others in the age of Rousseau and Diderot. But novelists of the day, whatever their physiognomical persuasion, rarely attended to the human face as though they believed it to say something about character. Whether affirming the correspondence between soul and visage, cautioning as to all human appearances, or adopting a purely aesthetic attitude to mortal beauty, they seldom give the detailed account of looks which writers like Balzac were later to furnish.

6

The Language of Emotion

As the preceding chapters have suggested, the desire which constitutes love is not always easy to express. The lover's reticence, his inexperience in the art of love, or the social milieu he inhabits, may render the declaration of feeling difficult; uncertainty about the likely reaction of the beloved to what he says, or about his own emotions, may make him wary of translating them into words. In such cases he may choose to employ the distanced and indirect mode of expression which writing offers: this is the resort of Rousseau's Saint-Preux and Laclos's Valmont, both when they are denied access to the loved one and when they are close to her. Alternatively, the lover may avail himself of the ambiguities permitted by the spoken word, procuring for himself the politic kind of uncommittedness noted by the narrator in *Les Egarements du cœur et de l'esprit*:

Avec un homme expérimenté, un mot dont le sens même peut se détourner, un regard, un geste, moins encore, le met au fait s'il veut être aimé; et supposé qu'il se soit arrangé différemment de ce qu'on souhaiterait, on n'a hasardé que des choses si équivoques, et de si peu de conséquence, qu'elles se désavouent sur-le-champ.[1]

The rhetoric of love is not always so subtle, however; the persuasion which is its object may dictate open attack. If reticence about love-affairs characterizes some of the individuals and social groups presented in the eighteenth-century novel, others proclaim their intentions and successes with proud abandon. But the demands of modesty, as the Marquise de Merteuil knows, mean that comparative secrecy is enjoined on the female sex, while the male can declare his desires and their fulfilment to an admiring world as well as to the beloved.

To make love, in eighteenth-century parlance, was primarily to profess it through language. The Marquise de Merteuil complains to Valmont that Danceny is an innocent 'qui perdra tout son temps

---

[1] p. 19.

à faire l'amour, et ne finira rien'[2]—that is, waste time talking about love rather than consummating it physically. And the language of devotion is commonly one of remoteness and respect, a profession in which words have a distance additional to that of any linguistic code, where the relation of verbal sign to object or thought is purely conventional.

The sense of what can and cannot be said in addressing one's beloved is one that practice develops, and much of the training in love which neophytes like Meilcour receive is concerned with it. There is the 'jargon' Mme de Lursay complains of, which marks a person out as belonging to a peer group, and of which Meilcour's mentor Versac is a principal exponent. There is the preciosity of a Présidente de Tourvel, a distant and elevated language through which she attempts to persuade Valmont of her emotional remoteness from him. There is the gallant discourse of the unpractised young (when Danceny sends the Marquise de Merteuil a love-letter, she criticizes his recherché expression). There is the hyperbole of passion, familiar in des Grieux and in the various characters of *La Nouvelle Héloïse*. There is the lyrical epithalium, part-pagan and part-Christian, which Paul and Virginie offer each other, a latter-day Song of Songs. There is the confused silence of the young, the artless, the timid, and the inexperienced; there is the reticence of *bienséance*, which prohibits direct reference to sexual acts. And, partly in answer to this, there is the appeal to other modes of communication. There are the looks of shy or uncertain lovers who cannot or dare not commit themselves to greater explicitness, the eyes through whose window the soul is glimpsed in its tumultuous passion or serene repose[3]—which lights in love, although Meilcour is too unknowing to interpret its change,[4] and permits the wordless exchange 'je vous aime' between Marianne and Valville.[5] There are the sighs which signal to the loved one that a woman is deeply moved by him. There is the absent play with extraneous objects, a favourite diversion of the flirt; the part-caress, which may or may not be charged with meaning; and the involuntary blush, that uncontrollable sign of emotion.

The epistolary form put a variety of expressive means at the disposal of the eighteenth-century novelist. The correspondence of

---

[2] p. 20. [3] See Buffon, op. cit. 281–2.
[4] *Les Egarements du cœur et de l'esprit*, p. 70; also Cureau de la Chambre, op. cit. 38, 93 ff., and Diderot, *Eléments de physiologie*, p. 224. [5] p. 67.

Abélard and Héloïse provided Rousseau with a model for the literary translation of physical passion in *La Nouvelle Héloïse*, although his eighteenth-century correspondents write with a lyrical floridness foreign to their twelfth-century forebears. But other novelists influenced by the latter's exchange of letters, where Héloïse movingly reveals the persistence of erotic desire in herself after the castration of her lover, show the limitations of a merely verbal discourse on love. The words and other tokens to which Mme de Beauharnais's 'Abailard supposé' is reduced are, perhaps appropriately, ultimately frustrating for him and his beloved, while the preachiness of Restif's new Abélard is as emotionally unfulfilling as Julie's sermonizing in *La Nouvelle Héloïse*.

Among the obvious constraints involved in talking about love is the fact that love-words bear only an indirect relation to reality. This limitation may, it is true, be turned to advantage in literature, where the property Burke ascribed to it of presenting impressions rather than concrete actuality is rarely so marked as in its reference to sexual activity. Euphemism and *double-entendre* are the stock-in-trade of the *libertin* writer, providing an additional indirectness that may compound the reader's pleasure—the pleasure of part-concealment, which has an equivalent in the different idiom of dress and undress. But the fact that words require interpretation for their meaning to be understood can result in misunderstanding, more or less crucial according to circumstances. Much of *Les Egarements du cœur et de l'esprit* concerns the misconstrual of the verbal (and other) tokens of emotion; learning the hidden significance of these tokens constitutes the initiate's sentimental education. Failure to grasp the register of particular words may lead to social discomfiture, an outcome Valmont anticipates in teaching Cécile the 'catéchisme de débauche' for the girl to rehearse to her husband on her wedding-night. A more tragic instance of such incomprehension occurs in Mme de Grafigny's *Lettres d'une Péruvienne*, a novel about a naïve Peruvian girl who is transported to France, and her experiences there. Imitating the sounds uttered by Déterville, a young Frenchman who has fallen in love with her and decided to teach her his language, she unknowingly tells him that she loves him in return:

Il commença par me faire prononcer distinctement les mots de sa langue . . . dès que j'ai répété après lui, 'oui, je vous aime', ou bien, 'je vous promets d'être à vous', la joie se répand sur son visage, il me baise les mains avec transport.[6]

But for those who grasp its meaning and can use it as they wish, language has the ability to move its listener or reader according to the speaker's or writer's will. The plot of *Les Liaisons dangereuses* rests entirely on this premiss, in contrast to that of *La Nouvelle Héloïse*. Rousseau argues in the second preface to his novel that its epistolary rhetoric is entirely innocent. The looseness of the correspondents' style guarantees their moral purity, the fact that they have no underhand designs on their reader. Not so in Laclos's work, where style betokens calculation. One must learn, the Marquise de Merteuil tells Cécile,[7] to tailor one's expressions to the character of one's correspondent, to write as 'a reflection of the other person';[8] and if the two *meneurs de jeu* never entirely succeed in teaching their pupil to follow this instruction, their own activities as letter-writers amply demonstrate the practical realization of the theory.

The merit of such adaptability, in the case of these two characters as of other, less criminal beings, is that it enables the writer to do as he will with the target of his eloquence. It makes the latter subject to the former's designs, the object of an 'ars movendi' potentially as powerful as that of the classical orator. But the Marquise and the Vicomte of *Les Liaisons dangereuses* lack the quality which, in the rhetor of antiquity as in the naïve correspondents of *La Nouvelle Héloïsè*, ensured that such power would not be turned to immoral ends. They are without the probity which Aristotle described as *êthos*, and Cicero as *mores*, the goodness which validates oratorical art and prevents it from becoming a tool of vice.[9] The rhetor's ability to move his audience, according to ancient precept, must be subordinate to the controlling quality of uprightness, and the art that shaped his discourse should be free from the suspect taint of the 'ornatus'.[10] Rousseau's characters seemed to their creator to comply

---

[6] *Lettres d'une Péruvienne*, in *Lettres portugaises, Lettres d'une Péruvienne et autres romans d'amour par lettres*, ed. Bernard Bray and Isabelle Lany-Houillon (Paris, 1983), p. 284. [7] pp. 242–3.

[8] See a letter from Virginia Woolf to Gerald Brenan, 4 Oct. 1929, in Virginia Woolf, *Letters*, ed. Nigel Nicolson and Joanne Trautmann, 6 vols (London, 1975–80), iv.

[9] See, for example, Alain Michel, *Rhétorique et philosophie chez Cicéron* (Paris, 1960), pp. 15–16.

[10] See Marc Fumaroli, *L'Age de l'éloquence* (Geneva, 1980), pp. 55 ff.

with this requirement; Laclos's protagonists manifestly do not, at least for the greater part of the book. Valmont and the Marquise do not necessarily write in an ornate or florid fashion, but as the motive behind their use of language is always persuasion they select and combine words with singular artfulness.

It is undeniable, however, that the rhetoric of Rousseau's 'belles âmes' often displays a more straightforward kind of ornateness. In its lyrical expansiveness the author's prose seeks to effect the 'persuasion passionnelle'[11] that later ages were to see as typical of the eighteenth century's irrationalism, its emphasis on the feeling expression of emotion rather than the systematic exposition of Descartes's 'clear and distinct ideas'. For an age impregnated with Lockeian sensationalism, yet still imbued with the rhetorical precepts of classical antiquity, the power of language to persuade its listener or reader appeared a function of its ability to stir the passions. The old notion that rhetoric was primarily concerned with the arousal of emotion (frequently in the speaker as well as his audience) was stressed anew in a century which accorded to the passions a dignity and value the previous one had often denied them, and which, in ·Diderot's words, saw how they could be the source of man's pleasures as well as his misfortunes.[12] Although, as he also declared, the feeling man who gave way to his sensibility might never attain the greatness in public life of the controlled rationalist,[13] his proneness to be moved could also constitute moral greatness. Diderot's drama *Le Fils naturel* tries to show that to be open to the passional discourse of one's fellows is to be accessible to the benign promptings of humanitarianism: the quality of fellow-feeling depends on being susceptible to the tender emotions of love and sympathy. (In *Le Fils naturel* this quality, and its extension into a condition of fullblown passion, are expressed in the broken and exclamatory tones of the so-called 'style haletant', a panting mode of speech intended to convey states of intense feeling.)[14]

---

[11] See B. Munteano, 'Survivances antiques: l'abbé Du Bos esthéticien de la persuasion passionnelle', *RLC*, 30 (1956).

[12] *Pensées philosophiques*, *Œuvres philosophiques*, p. 9.

[13] See the fragment 'Sur le génie', *Œuvres esthétiques*, pp. 19–20; also *Paradoxe sur le comédien*, *passim*, and *Le Rêve de d'Alembert*, p. 357.

[14] See A. Brun, 'Aux origines de la prose dramatique: le style haletant', in *Mélanges de linguistique française offerts à M. Charles Bruneau* (Geneva, 1954), pp. 41–7.

Eighteenth-century treatises on rhetoric, following the spirit of the times in their dual attention to the language of passion and the enlightened discourse of reason, often discuss the forms of eloquence governed by heart and mind respectively. In his *Essais sur divers sujets de littérature et de morale* (1735) Trublet refers to the degree of conviction carried by 'bookish' rhetoric and that which seems to grow spontaneously from a situation. True eloquence of thought and expression, he writes, will never inform a speech composed at leisure as it does one which is 'fait sur-le-champ par un homme naturellement éloquent'.[15] This statement is reflected in some remarks made by Mme de Merteuil in *Les Liaisons dangereuses*. Valmont's thought-out letters to the Présidente de Tourvel, she tells him, betray their origin in his rational mind rather than impassioned heart by the orderliness of their composition.[16] This is a general fault in novels, she continues, where emotion is always manufactured, and announces its own artificiality: 'l'auteur se bat les flancs pour s'échauffer, et le lecteur reste froid.' *La Nouvelle Héloïse* is the only novel she knows where this fault is not apparent, and for that reason it has always persuaded her that its origin lay in real life.

Her following remarks echo an observation Trublet makes in the same essay on rhetoric, that the written word can never match the spoken one in forcefulness, partly because it lacks the accompaniment of the speaker's bodily action (the *actio* which Cicero and Quintilian called an essential part of the orator's eloquence).[17] Valmont's insistence on seducing Mme de Tourvel slowly and indirectly, by letter, means that his efforts carry far less weight than they would if he engaged in face-to-face confrontation:

espérez-vous prouver à cette femme qu'elle doit se rendre? Il me semble que ce ne peut être là qu'une vérité de sentiment, et non de démonstration, et que pour la faire recevoir, il s'agit d'attendrir et non de raisonner.[18]

But his endeavour must be further frustrated by the fact that he cannot be in the presence of his beloved to drive home his advantage. Even if his written rhetoric produces the intoxication of love in the Présidente, she will soon allow reasoned reflection to curb the impulses of passion. Were he to be actually with her, on the other hand, he

---

[15] [Abbé Trublet], *Essais sur divers sujets de littérature et de morale* (Paris, 1735), p. 66.   [16] p. 68
[17] See my *'Actio' and Persuasion: Dramatic Performance in Eighteenth-Century France* (Oxford, 1986), *passim*.   [18] p. 67.

would find his actorly powers a persuasive force in the battle against her virtue:

L'habitude de travailler son organe y donne de la sensibilité; la facilité des larmes y ajoute encore: l'expression du désir se confond dans les yeux avec celle de la tendresse; enfin le discours moins suivi amène plus aisément cet air de trouble et de désordre qui est la véritable éloquence de l'amour; et surtout la présence de l'objet aimé empêche la réflexion et nous fait désirer d'être vaincues.[19]

Words may initiate the process of desire, as we see on reading the Présidente's piqued reaction to Père Anselme's report that Valmont has returned to the ways of God; but physical presence will convert desire into concession. Finding Mme de Tourvel in this vulnerable state, bereft at his apparent desertion of her, Valmont is able to profit from the moment in the way he had earlier refused to do (a refusal to which Mme de Meurteuil refers in Letter XXXIII). The Présidente's relief at finding the scene she so bitterly imagines in Letter CXXIV belied by reality can only be expressed in the act of giving herself to Valmont, as he had calculated it would. Although, as he gleefully points out to the Marquise, it is his intellect that triumphs, Valmont's letter to her after his 'victory' over Mme de Tourvel fails to acknowledge that the power of presence won out against his earlier attempts at epistolary persuasion.

Since Valmont's letter to Mme de Meurteuil is a self-conscious one, written by a man desirous of impressing a woman to whom all sentimental attachments seem a form of weakness, we cannot tell whether the detached rationality he professes corresponds to the state he was really in during the crucial encounter with the Présidente. In his dealings with the Marquise he must always emphasize the superiority of mind to feeling; in the encounter itself a different priority may have been asserted. The contrast between the two orders of reason and emotion is one which eighteenth-century rhetorics discuss as readily as do novels of the period. In Trublet's *Panégyriques des saints, suivis de Réflexions sur l'éloquence en général, et sur celle de la Chaire en particulier*, a distinction is drawn between the fine speaker and the true speaker which matches that proposed in Crébillon's *Egarements* and Rousseau's *La Nouvelle Héloïse*. The fine speaker or 'disert' is

[19] p. 68.

'un homme d'esprit, et d'un esprit cultivé', whose persuasiveness is the product of art; the true speaker or 'éloquent' is 'un homme de sentiment, et même un homme de génie, s'il est très éloquent', whose persuasiveness comes from nature.[20]

When Meilcour enters the world of high society, he finds it peopled with fine speakers, whose language is the brittle and affected jargon of 'mondanité'. Mme de Lursay detests the studied rhetoric of Versac, who professes a social doctrine of unnaturalness and insincerity which matches his mode of speech. Meilcour is repelled by the verbiage of the 'haut monde', which reduces true feeling to a subject for detached discussion:

J'ignorais entre beaucoup d'autres choses que le sentiment ne fût dans le monde qu'un sujet de conversation, et j'entendais les femmes en parler avec un air si vrai, elles en faisaient des distinctions si délicates, méprisaient avec tant de hauteur celles qui s'en écartaient, que je ne pouvais m'imaginer qu'en les connaissant si bien elles en fissent si peu d'usage.[21]

Mme de Lursay, we are given to understand, is excepted from this general condemnation; for if she 'parlait d'une façon d'aimer . . . singulière',[22] she nevertheless also spoke with great sincerity, but without being a 'précieuse'. Later on, we remember, she teaches Meilcour of the fine gradations governing sentiment and its expression, and so contributes to his moral understanding of the world.

Saint-Preux attacks the artificial idiom of Parisian society in a celebrated letter of *La Nouvelle Héloïse*, and one which reflects the theories about honest and dishonest rhetoric set out in Rousseau's second preface. In Letter XVII of Part Two Julie's banished lover describes the grand discussion in the 'haut monde' of what its members call sentiment, but which bears no relation to the 'épanchement affectueux dans le sein de l'amour et de l'amitié' which Julie's circle knows.[23] Parisians are *par excellence* the 'diserts' of the abbé Trublet: they expend all their sentiment in wit, and define feeling so exhaustively that there is none left over for the purposes of ordinary living. Despite Saint-Preux's fulminations against this perversion of language, however, Julie has already detected the taint of Parisian jargon in his own style, and accuses him of having

---

[20] Abbé Trublet, *Panégyriques des saints, suivis de réflexions sur l'éloquence en général, et sur celle de la chaire en particulier*, 2nd ed., 2 vols (Paris, 1764), i. 252.
[21] p. 18.   [22] p. 19.
[23] p. 249; see also *Emile et Sophie*, *Œuvres complètes*, iv. 886.

descended into the pit of 'bel esprit' himself.[24] Saint-Preux appears already to have forgotten that wit, according to their compatriot Muralt, is 'la manie des Français', and is allowing its corrupting influence to show through his own discourse.[25]

Julie does not prohibit the lively, animated turn of phrase that betokens force of feeling; what she criticizes is 'cette gentillesse de style qui, n'étant point naturelle, ne vient d'elle-même à personne, et marque la prétention de celui qui s'en sert'.[26] It was the true voice of sentiment that pervaded their own passionate conversations, and it necessarily prohibited the intrusive self-consciousness of 'esprit'. How much more abhorrent, then, is the latter's presence in letters, where absence always adds a measure of bitterness to what is written, and where heart addresses heart with a corresponding 'attendrissement'. This is not to say, she goes on, that love should be declared sadly; but its gaiety should not be the false garb of ornament and art. The love-language of letters, in short, should shine with its native grace, not with artificial brilliance.

The experienced old courtier Sélim of Diderot's *Les Bijoux indiscrets* draws comparisons resembling Muralt's between the French and English nations. He tells the Sultan's favourite that both profess a concern for decency in sexual matters which their conduct belies. English women speak of modesty as readily as their Gallic counter-parts, but both offend against its dictates: 'celles-ci aiment le jargon des sentiments; celles-là préfèrent l'expression du plaisir.'[27] A few years later the author of *L'Ami des femmes* settles the blame for such denaturing of language on men as well as women, declaring in the spirit of Crébillon that the former distort the notion of sentiment by their excessive protestations. These grand professions, together with the simpering response of women, are nothing but a game in which each sex gives the other lessons in imposture.[28]

None of this really means that only the plainest expression is acceptable in the honest lover's discourse, or that eloquence necessarily signifies calculation and therefore untrustworthiness. There is a difference between what Julie calls the 'jargon fleuri de la galanterie'[29] and the lyrical expansiveness of passion. Saint-Preux takes his beloved sharply to task for her criticism of his letter-writing style since his

---

[24] p. 237.   [25] p. 238.   [26] Ibid.
[27] p. 180.
[28] *L'Ami des femmes* (Paris, 1759), p. 112.
[29] p. 238.

arrival in Paris. A degree of rhetorical finesse, he tells her, is wholly natural in the language of love, because the bare bones of prose suggest nothing of the exaltation a lover feels and the heightened quality of his perception in this privileged state of emotion. The time-honoured figures of eloquence occur naturally in the expression of exceptional feeling, because it is normal for the lover to want his sentiments to stand out as unusual. His appeal to linguistic procedures that amplify statements, intensify them through exclamation, extend their reference by comparison, or analyse the quality they try to capture by enumeration, is therefore simply an implicit acknowledgement of the extraordinary nature of passion. It is far from betokening the dominance of mind over heart:

Pour peu qu'on ait de chaleur dans l'esprit, on a besoin de métaphores et d'expressions figurées pour se faire entendre . . . je soutiens qu'il n'y a qu'un géomètre et un sot qui paraissent parler sans figures. En effet, un même jugement n'est-il pas susceptible de cent degrés de force? Et comment déterminer celui de ces degrés qu'il doit avoir, sinon par le ton qu'on lui donne? Mes propres phrases me font rire, je l'avoue, et je les trouve absurdes, grâce au soin que vous avez pris de les isoler; mais laissez-les où je les ai mises, vous les trouverez claires, et même énergiques.[30]

The fact that rhetorical theory had reduced to rule the most efficient means of arousing emotion or conveying its consequences in no way detracted from the honesty of employing these procedures in the declaration of true love. To remark, as Batteux did in the section on 'figures touchantes' of his *Principes de littérature*, that using such figures as deprecation, interrogation, amplification, imprecation, and commination (threat) helped the speaker to touch his listener's heart was not necessarily to suggest that they were contrived or unspontaneous.[31] After all, there can be no knowing whether rhetorical theory merely gave systematic form to natural procedures or, conversely, created them *ex nihilo*; but common sense suggests that the former hypothesis is at least partly true. Besides, the history of literature shows that there is considerable divergence over the ages, and across social groups, in opinions about what constitutes natural discourse. We might now regard the amplitude of passionate expression in the eighteenth-century novel as betokening a suspect

---

[30] p. 241; see also Jean-Louis Bellenot, 'Les Formes de l'amour dans *La Nouvelle Héloïse*', *Annales de la Société Jean-Jacques Rousseau*, 33 (1953–5), pp. 153 ff.

[31] *Principes de la littérature*, 6 vols (Lyon, 1802), iv. 93 ff.

self-consciousness or desire to impress in the speaker, but to do so would perhaps be to misinterpret its author's intentions. That period was simply more inclined to the grandiose declaration of feeling than is our own. The eighteenth-century lover is not shown to be false just because he protests his love fulsomely, although he may indeed be insincere in his affections. Eloquence, as Saint-Preux observes, is a very variable thing, and naturalness a far from absolute quality.

The innocence or otherwise of love-rhetoric, then, can rarely be determined by the appraisal of isolated speeches or passages of writing. What Rousseau does not invite his reader to consider, and what the antagonist N in the second preface to *La Nouvelle Héloïse* rather surprisingly omits to mention, is that disorder or looseness in a writer's prose style may be willed, something conjured up for the requirements of the moment. The protagonists of *La Nouvelle Héloïse* may deliberately write badly (as Rousseau, if his protestations in the preface are sincere, must be assumed to have done) in order to make a particular impression on their correspondent or the reading public. If taken too far, of course, this effect would be self-defeating. (Diderot must have realized this in the case of Suzanne Simonin, whom he permits for aesthetic reasons to write more stylishly than is strictly plausible in a girl of her professed naïvety and her age.) There is a limit to our willing acceptance of bad prose, however appropriate it may be as a rhetorical ploy or as a guarantee of verisimilitude. In *Les Liaisons dangereuses* the judicious mixing of different prose-styles lends a conviction to the narrative which is lacking from Rousseau's novel, where, as N remarks, those inhabiting Julie's little world all speak with the same voice. But Laclos showed that linguistic variation might be less an indication of truth than a demonstration of intellectual control by the two protagonists.

Throughout *Les Liaisons dangereuses* Mme de Merteuil and Valmont give practical demonstration of the precept the Marquise impresses on Cécile: when writing to someone one must fit one's words to him, not suit oneself. The writer should say less what he thinks than what he judges to be pleasing to his correspondent. But Valmont and Mme de Merteuil carry this principle of dissimulation further: in the interests of intrigue both 'ghost' letters from another party to a fourth person implicated in their plot, so demonstrating the ease with which styles may be adopted or exchanged.

According to Rousseau, it is in the world of high society that such dissimulation comes most naturally to people. It inevitably entails the practising of rhetorical skills, particularly with respect to the declaration of feeling. The 'haut monde' encourages individuals to lay claim to sentiments they do not possess simply in order to impress others; and the habit of constantly professing what one does not believe means that one seeks a persuasive formulation in the absence of inner conviction.[32] Somewhat against the argument of Saint-Preux in the novel proper, Rousseau denies in the second preface that truly impassioned speakers employ a highly coloured and intense mode of discourse:

Non; la passion, pleine d'elle-même, s'exprime avec plus d'abondance que de force; elle ne songe même pas à persuader; elle ne soupçonne pas qu'on puisse douter d'elle. Quand elle dit ce qu'elle pense, c'est moins pour l'exposer aux autres que pour se soulager. On peint plus vivement dans les grandes villes; l'y sent-on mieux que dans les hameaux?[33]

A love-letter penned in the study will, so to speak, burn the very paper on which it is written; the reader will be moved by it, but with a merely passing agitation which leaves only the memory of words behind. A letter dictated by true love, on the other hand, will be diffuse and untidily constructed, full of longueurs and repetitions, wholly lacking in striking turns of phrase and arresting formulations; yet its reader will be moved in a manner he cannot account for rationally, but feels strongly.

Si la force du sentiment ne nous frappe pas, sa vérité nous touche; et c'est ainsi que le cœur sait parler au cœur. Mais ceux qui ne sentent rien, ceux qui n'ont que le jargon paré des passions, ne connaissent point ces sortes de beautés, et les méprisent.[34]

Valmont, who likes 'les méthodes nouvelles et difficiles', wants to demonstrate to the sceptical Marquise that epistolary rhetoric may seduce as effectively as the persuasive force of physical presence. To this end he mimics the written styles which, in the above quotation, Rousseau describes as those of true feeling. Valmont informs Mme de Merteuil that he has even imitated the mode of expression typical of women, filling his letters to the Présidente with the rambling illogicalities characteristic of her sex:

j'ai mis beaucoup de soin à ma lettre [Letter LXVIII], et j'ai tâché d'y répandre ce désordre qui peut seul peindre le sentiment. J'ai enfin déraisonné

---

[32] *La Nouvelle Héloïse*, p. 14.   [33] pp. 14–15.   [34] p. 15.

le plus qu'il m'a été possible: car sans déraisonnement, point de tendresse; et c'est, je crois, par cette raison que les femmes nous sont si supérieures dans les lettres d'amour.[35]

Letter LXVIII, in which Valmont replies to Mme de Tourvel's offer of friendship against the love he has declared for her, is a compendium of rhetorical commonplaces, containing interrogations, exclamations, anaphora, parallelism, personification, amplification, antithesis, and chiasmus, and forming a whole which is rather less marked by 'déraisonnement' than he suggests to the Marquise.

Danceny grants readily enough to Mme de Merteuil, now his lover, that conversation by letter is superfluous when they are able to see each other freely.[36] Knowing her fear of being compromised by the existence of tell-tale love-letters, he accedes to her desire for secrecy despite feeling, as Saint-Preux surely did before him, that epistolary communication can be reassuring and enriching even when correspondents are virtually in each other's presence. But making love by letter, according to Danceny, inevitably means most when lovers are apart. Even the sight of a missive from the other consoles at such times, before its contents have been read, just as pleasure may be derived from touching the beloved's portrait in the darkness of night.[37] A letter, he continues, is a picture of the writer's soul. But a letter has a temporal dimension, for it accompanies its recipient's changing moods; '[elle] n'a pas, comme une froide image, cette stagnance si éloignée de l'amour; elle se prête à tous nos mouvements; tour à tour elle s'anime, elle jouit, elle se repose.'[38] It translates his joy or sadness, his successive feelings of tranquillity or oppression. However wayward or fluctuating the spirit that dictates them, his emotions can be captured and stabilized on paper.

The successive quality of language, which makes it a suitable vehicle for conveying passing moments, also means that it can be adapted to the listener's or reader's own life in time. Words, which acquire meaning in conjunction with other words, do not make their effect immediately and in isolation, but do so concurrently with the interpreter's own existence, made up of a concatenation of instants. For Rousseau, this property gave verbal discourse a richness and an evocativeness lacking in visual art. In his *Essai sur l'origine des langues*, where the comparative natures of the seen and the heard

---

[35] p. 139.   [36] p. 344.   [37] p. 345.   [38] Ibid.

are discussed, Rousseau notes that gestural language affords lovers a less effective means of capturing feeling and moving another's heart than does the language of (spoken) words. Although in general we address the eyes much more clearly than the ears, because visual communication is less dependent than verbal on convention, the successive impressions afforded by discourse strike harder and more lastingly than does the single one our eyes receive. Mime without speech, he reasons (against thinkers like Diderot who emphasized the effectiveness of mute action divorced from the word), will leave its beholder almost unmoved, whereas discourse unaccompanied by gesture will wring tears of emotion from the listener. While visible signs may give a more exact imitation than verbal, the latter arouse much greater interest.[39] For Valmont, with certain qualifications to which I shall return, this opinion remains a valid one, although he chooses to stress the power of the written rather than the spoken word. Mme de Merteuil, in contrast, remains wedded to the view that the conjunction of spoken word and physical action carries greater persuasive power than the word in isolation.

Julie de Lespinasse believed that words fit to convey the feelings of lovers must always fall short of what the heart desires. Merely thinking creatures find language an adequate instrument for communication, but finer sensibilities are aware of its deficiencies: 'les expressions sont faibles pour rendre ce que l'on sent fortement, l'esprit trouve des mots, mais l'âme aurait besoin d'une langue nouvelle.'[40] For Mme de Merteuil, on the other hand, it is the ability of a person supposedly without emotion to find the words conveying a simulated state of passion that is in question. Yet she is also aware that each feeling has its own language, which must be learnt by those who aspire to the rhetorical manipulation of emotion. She tells Danceny—whom she significantly addresses as 'mon trop jeune ami'—that his impatient letter of longing is couched in the jargon which is permissible only between lovers, and inadmissible for all others.[41] What he must learn is that 'chaque sentiment a son langage qui lui convient; et se servir d'un autre, c'est déguiser la pensée qu'on exprime.' His letter lacks the tone of frankness and simplicity that characterizes hers; Danceny translates the plain statement she might make to him as a friend into a lover's hyperbole.

---

[39] *Essai sur l'origine des langues*, ed. Angèle Kremer-Marietti (Paris, 1974), pp. 88–94.   [40] p. 97   [41] p. 278

The whole of the young man's letter is phrased in the second-hand rhetoric of romantic love. Starting with the acceptably familiar assertion that for lovers time passes at a different speed from that known by ordinary humans ('Si j'en crois mon almanach, il n'y a . . . que deux jours que vous êtes absente; mais si j'en crois mon cœur, il y a deux siècles'),[42] he immediately slips into a gallant word-play on the fact that his new friend ('mon adorable amie') has left Paris to attend to a legal matter: 'Comment voulez-vous que je m'intéresse à votre procès si, perte ou gain, j'en dois également payer les frais par l'ennui de votre absence?' In Danceny's defence it might be observed that other characters adopt legal terminology in this novel, where a rake tries to seduce the wife of a Président. Thus Valmont describes his intention to win the Mme de Tourvel who is 'loin de penser qu'*en plaidant*, pour parler comme elle, *pour les infortunées que j'ai perdues*, elle parle d'avance dans sa propre cause';[43] he mentions his attendance at her court on the 'jour préfix et donné par l'ingrate',[44] and tells the Marquise that he is about to confront the Présidente to 'faire signer mon pardon'.[45] Mme de Merteuil, too, remarks of her former lover that '[le procès] de Belleroche est fini: hors de Cour, dépens compensés.'[46]

In his letter of accusation to the Marquise, Danceny succumbs to the temptation of cliché, telling her that she has committed a 'noire trahison' in allowing him to languish far from her while she attends to her legal affairs. There is a self-conscious archness in his further reference to the reason for her absence: 'vous m'avez tant dit que c'était par raison que vous faisiez ce voyage que vous m'avez tout à fait brouillé avec elle. Je ne veux plus du tout l'entendre; pas même quand elle me dit de vous oublier. Cette raison-là est pourtant bien raisonnable.' He employs a familiar range of rhetorical devices, including paradox, anaphora, and the tricolon: 'on rêve, on fait des châteaux en Espagne, on se crée sa chimère.' And where he attempts a more imaginative use of language, he provides an analysis of love whose daintiness outdoes the subtleties of a Marivaux: 'Les secrets de l'amour . . . sont si délicats qu'on ne peut les laisser aller ainsi sur leur bonne foi. Si quelquefois on leur permet de sortir, il ne faut pas au moins les perdre de vue; il faut en quelque sorte les voir entrer dans leur nouvel asile.' Mme de Merteuil's

---

[42] p. 273.   [43] p. 23.   [44] p. 289.
[45] p. 322.   [46] p. 314.

reaction to all this is predictable enough, if unfair in its blanket condemnation:

Mon ami, quand vous m'écrivez, que ce soit pour me dire votre façon de penser et de sentir, et non pour m'envoyer des phrases que je trouverai, sans vous, plus ou moins bien dites dans le premier roman du jour.[47]

To mention Marivaudage is to be reminded how frequently Laclos's characters, like Marivaux's, reflect on the essence of the feelings aroused in them by love. This is by no means invariably the case in the eighteenth-century French novel, where language as often serves to describe action as to permit meditation on states of being. In *Manon Lescaut*, for example, it is fairly unusual to encounter the musing about the psychological implications of being in love we recurrently meet with in *La Vie de Marianne*; as unusual, indeed, as to come across descriptions of physical appearance. In the memorable scene of Manon's visit to des Grieux after his triumphant debating exercise at the Sorbonne, we learn little about his innermost feelings during the reunion beyond the fact that he is taken aback and unwilling to face her directly.[48] There is no speculation about her motives—tender recollection? curiosity? boredom?—in seeking her former lover out, although Manon herself declares that it is impossible for her to live without his love.[49] And the references to Manon's transport of joy in his presence are, if conventionally hyperbolic in tone, curiously perfunctory: 'Elle m'accabla de mille caresses passionnées. Elle m'appela par tous les noms que l'amour invente pour exprimer ses plus vives tendresses.' The rhetoric of des Grieux's love is more explicit, but he still stops short of analysing his emotions in depth. What we are presented with is a tableau-like scene in which physical disposition and bodily reaction are detailed, and where we read of Manon's approaching des Grieux to kiss him, his trembling in the grip of 'mouvements tumultueux', and their sitting down opposite each other in silence.

This bears a certain resemblance to episodes in *Les Egarements du cœur et de l'esprit*, where a similar sense of manneredness and embarrassment combines with an effort to convey the obtrusiveness of the physical. One thinks especially of the scene following Mme de Lursay's reception, when Meilcour lingers after the departure of the other guests and is paralysed by his confused emotions. The tempo of

---

[47] p. 278.   [48] p. 44.   [49] p. 45.

the two passages, however, is markedly different. While Prévost hastens the action onwards, passing rapidly from des Grieux's report of his new *coup de foudre* at sight of his beloved to her recounting the manner in which she lived as B . . .'s kept woman, Crébillon lingers over the details of Meilcour's discomfiture, and so underlines the sensation of unease felt in this claustrophobic little world. There is the same attention to physical attitude as in Prévost—here, Meilcour's standing in mute disarray while Mme de Lursay reclines on a sofa, absently knotting—but a far more extended report of the uneven exchange between the characters. Meilcour's awkwardness is expressed in his breathtakingly gauche 'Vous faites donc des nœuds, Madame?',[50] and Mme de Lursay's experience in her lengthy analysis of his emotional state and the way it is betrayed by his changed physiognomy.

The literary effect of this long-drawn-out scene resides in Crébillon's masterly portrayal of mutual reserve and of Mme de Lursay's gradual shifting into a position of strength, revealed in the transition from a silent rhetoric of look and gesture into one of language. Throughout, the reader senses her skilled deployment of the theory of 'gradations', her control of the passage from physical to verbal expression, and the latter's intensity. This enables her to be mistress of Meilcour's emotions in the absence of his own ability to declare their nature with due restraint. A practised woman, she knows that the refusal to grant certain favours increases the desire of the lover who seeks them; hence her determination to impose a language of moderation on the young man, and her (unconvincing) affirmation, as she tries to hurry his departure, that they have 'said all' to each other. Meilcour can only reply, as he is meant to, that their conversation has been far from complete: 'si je garde quelquefois le silence auprès de vous, c'est bien moins parce que je n'ai rien à vous dire que par la difficulté que je trouve à vous exprimer tout ce que je pense.'[51]

It is not Meilcour's verbal reticence that serves him ill in this encounter, however, but his inability to convert word into action. Mme de Lursay's gestures and observations, discreet though they are, have yet made her attraction to Meilcour explicit; or, rather would have made it explicit were he more experienced than he is in interpreting the muted language of female concession. As it is, her

[50] p. 65.   [51] p. 69

offering of herself becomes clear to Meilcour only after he has left. Her delicacy, assumed or otherwise, has worked against her, for Meilcour is not sufficiently conversant with the ways of love to know how ambiguous are its formulations:

Savais-je, moi, que toute femme qui, en pareille occasion, parle de sa vertu s'en pare moins pour vous ôter l'espoir du triomphe que pour vous le faire paraître plus grand?[52]

The smokescreen of etiquette, insubstantial though it was, nevertheless appeared to the unversed youth too thick to be dissipated. Only with hindsight is it evident that Mme de Lursay's words about respect and virtue were wholly misplaced in an intimate *tête-à-tête*.

Words are often a kind of armour, which may become a weapon, for the inhabitants of the privileged world Crébillon describes. They form a carapace that insulates the man or woman of quality from the roughness of reality, and their coded nature makes comprehension impossible for the non-initiated. Mme de Lursay speaks contempt-uously of Versac's 'dazzling' jargon: it combines the frivolity of foppishness with the decisive tone of pedantry.[53] Meilcour regards it as extraordinary and artificial, yet paradoxically natural-seeming.[54] 'Bon ton', Meilcour is informed by his mentor, consists in avoiding the reflective in one's speech, however naturally expressed, for it seems affected, signifying the user's desire to stand out from the 'gens de bonne compagnie' (another paradox). To fit in, one must cultivate finesse in one's turns of speech and puerility in one's ideas; nothing, according to Versac, betokens good breeding so much as the elegant mouthing of absurdities. Together with the rhetoric of look and gesture, this language takes time to interpret and to use, and its very indirectness performs an important function in the game of love whose rules Meilcour is intent on learning.

A different kind of reticence in the articulation of feelings marks many of Marivaux's characters, and has indeed been described as the essence of Marivaudage.[55] In Marivaux's dramas such reserve is the product of various circumstances: the earlier disappointment in love experienced by Hortense and Lélio in *La Surprise de l'amour*,

---

[52] p. 71.   [53] p. 81.   [54] p. 72.
[55] See Frédéric Deloffre, *Une Préciosité nouvelle: Marivaux et le marivaudage* (Paris, 1955).

the loyalty felt towards an unsuitable but long-standing partner in
La Double Inconstance, or the desire to test the foundations on which
a proposed union rests in *Le Jeu de l'amour et du hasard*. This kind
of verbal distancing (which produces the deferment Barthes has called
the 'coitus reservatus' of Marivaudage)[56] is perhaps a product of the
salon world familiar to Marivaux, an ambience in which concepts
like the essence of love and friendship were detachedly and theoretically discussed; a world in which women, for whom direct avowal
of love has traditionally been less easy than for men, played an
important part.[57]

It is in Marivaux's novels, however, that the most sustained
analysis of feeling takes place—unsurprisingly, given the greater
extent of that literary genre and the facility it offers for reporting
unspoken thought rather than or as well as articulated speech.
Appropriately, too, it is the female Marianne rather than the male
Jacob who indulges in the most prolonged reflections on emotion.
Fascinated though she is by her own sentiments, Marianne also
considers those displayed, and on occasion concealed, by others. It
is true that she affects for some time not to understand the meaning
of Climal's advances to her; but she is also able to dissect their nature
dispassionately. What characterizes his assault on her virtue, she
decides, is its unsavoury origin in the physical. Since desire rather
than respect is its essence, it cannot easily take rebuttal:

Les passions de l'espèce de celle de M. de Climal sont naturellement lâches;
quand on les désespère, elles ne se piquent pas de faire une retraite bien
honorable, et c'est un vilain amant qu'un homme qui vous désire plus qu'il
ne vous aime.[58]

This is not to say, she continues, that the most delicate of lovers
does not also feel real desire, in his way; but in him the feelings of
the heart mingle with sensuous urges, producing a love that is tender
rather than corrupt. Love performs the miracle of transforming
sensuality into sentiment through the intervention of the soul, so that
in love's name 'on fait très délicatement des choses fort grossières.'[59]

The imperious call of love, acknowledged so instantaneously by
Marianne that at first sight of Valville in the church she forgets the

---

[56] See Roland Barthes, *Fragments d'un discours amoureux* (Paris, 1977), p. 26.
[57] See Ruth Kirby Jamieson, *Marivaux: A Study in Sensibility* (New York, 1941),
pp. 17 ff.   [58] p. 40.   [59] p. 41.

coquetry she has been practising on the other men present, is analysed by her with a subtlety and profundity quite foreign to Prévost. By its very nature, the *coup de foudre* is hard to describe. The kind of sentiment which rests on shared experience, community of interest, similarity of background, and the like can be rationally grasped and articulated; the passion which suddenly engulfs a person, rendering him helpless before its advance, is much less easy to understand and hence express. Unlike des Grieux, however, Marianne tries to do so. She succeeds, if not in explaining its origin, at least in establishing its complex composite effect. Fear is an important element, at any rate for the person who has no idea where emotion will lead, and who has no previous experience of this sudden mystification: 'ce sont des mouvements inconnus qui l'enveloppent, qui disposent d'elle, qu'elle ne possède point, qui la possèdent; et la nouveauté de cet état l'alarme.'[60] Even pleasure, she learns in this new state, acquires a troubling aspect it has never had before because of the threat it seems to pose to native modesty. The initiate both sees and does not see what must follow on the revelation of love. Individual will is put in abeyance and the inevitability of yielding to passion acknowledged,

Car, en vérité, l'amour ne nous trompe point: dès qu'il se montre, il nous dit ce qu'il est, et de quoi il sera question; l'âme, avec lui, sent la présence d'un maître qui la flatte, mais avec une autorité déclarée qui ne la consulte pas, et qui lui laisse hardiment les soupçons de son esclavage futur.[61]

Love puts Laclos's Cécile in a comparable position, half-knowing and half-unknowing, but one to which she can give more profound expression as she matures emotionally. In her early letters to Sophie Carnay Cécile declares that she cannot put her feelings for Danceny into words, and attempts to convey her emotion merely by weak repetition ('[Danceny] était devenu si triste, mais si triste, si triste, que ça me faisait de la peine'),[62] or to describe her and her lover's state with the most inexpressive of pronouns ('ça n'a pas empêché qu'il n'ait eu la complaisance de chanter avec moi comme à l'ordinaire, mais toutes les fois qu'il me regardait, cela me serrait le cœur'). Later on, while never approaching the eloquence of most other characters in the novel, Cécile becomes able to dissect feelings with a degree of perspicuity. In Letter LV, also to Sophie, she is still preoccupied with the idea of a lover's sadness, part of the paradoxical

---

[60] p. 66. [61] Ibid. [62] p. 38

pleasure of the tender emotion: 'Il t'est bien aisé de dire comme il faut faire, rien ne t'en empêche; mais si tu avais éprouvé combien le chagrin de quelqu'un qu'on aime nous fait mal.'[63] In Cécile's reflection on the difficulty of saying no when one wants to say yes there is a vestige of the sentiment Mme de Lursay tried to make Meilcour understand, although Cécile's incomprehension of why this difficulty should be sets her apart from the older woman. The fashionable emphasis (of a kind also encountered in Mme de Tourvel's letters) in 'moi-même qui l'ai [the difficulty] senti, bien vivement senti' reveals Cécile's growing familiarity with the expressive style of her wordly milieu, although the old lapses into inarticulacy also occur: 'je suis bien sûre que c'est comme ça.' The happiness brought by love is now described in the classic tricolon, after being introduced by rhetorical exclamation: 'Mais l'amour, ah! l'amour! . . . un mot, un regard, seulement de le savoir là, eh bien! c'est le bonheur.' The emotional plenitude procured by love is conveyed in an elegant antithesis: 'Quand je vois Danceny, je ne désire plus rien; quand je ne le vois pas, je ne désire que lui.' The use of anaphora to intensify is revealed in the accumulative 'Quand il n'est pas avec moi, j'y songe; et quand je peux y songer tout à fait, sans distraction, quand je suis toute seule par exemple, je suis encore heureuse'—an effort of abstract thought which would certainly have been beyond the Cécile we meet at the start of the novel, newly emerged from the convent. Her reflections climax in an image combining vividness—the *enárgeia* of ancient rhetoric—with oxymoron: 'je sens un feu, une agitation . . . Je ne saurais tenir en place. C'est comme un tourment, et ce tourment-là fait un plaisir inexprimable.'

This love-language is more intense than anything Cécile uses to or about Valmont, who becomes her only lover in the physical sense, but as a rhetoric of praise it falls short of what other characters in the novel bestow on their beloved. We have already seen Valmont and the Présidente borrowing the language of religion in their references to each other, and noted the different interpretations to be put on their words. While for Valmont they mainly betoken an affectation of pride (albeit one which seems to be transformed into something else when he falls more deeply in love with the Présidente), for the latter such apparently blasphemous eulogy is simply the consequence of a transference. A human being is now seen as the

---

[63] p. 111.

legitimate object of the old divine worship, and a worthy recipient of the ultimate sacrifice, her womanly virtue. The admission of her fall is made by the Présidente to Mme de Rosemonde in the plainest possible words, contrasting with the studied eloquence of her past letters to Valmont:

Tout ce que je puis vous dire, c'est que placée par M. de Valmont entre sa mort et son bonheur, je me suis décidée pour ce dernier parti. Je ne m'en vante, ni ne m'en accuse: je dis simplement ce qui est.[64]

Whereas the terms of the Marquise's pact with Valmont demanded that the avowal of Mme de Tourvel's 'defeat' be made by her in a letter to him (written evidence being the only kind Mme de Merteuil is willing to accept), it is Mme de Rosemonde the Présidente chooses to make the admission to. The only epistolary testament Valmont ever receives is the tormented plaint of Letter CXXXVI, where his victim refers to the way she has sacrificed all to him, and 'perd[u] pour [lui] seul [ses] droits à l'estime des autres'.[65]

In other words, there is none of that direct and explicit verbal caressing which the Marquise insisted on seeing in writing. What Valmont heard from his 'belle prude' in private he does not reveal. The evidence provided by other 'dévotes' of the eighteenth-century novel suggests, however, that it may have been religious in its intensity. Valmont anticipated this eventuality early on in his campaign: 'le premier pas franchi, ces prudes austères savent-elles s'arrêter? leur amour est une véritable explosion; la résistance y donne peu de force.'[66] At the start of his enterprise, we remember, Valmont imagined the Marquise herself venerating him in his hour of success with an almost mystical devotion: 'Vous-même, ma belle amie, vous serez saisie d'un saint respect, et vous direz avec enthousiasme: 'Voilà l'homme selon mon cœur.''[67]

This kind of reversal is anticipated by Rousseau in the second preface to *La Nouvelle Héloïse*, where 'enthousiasme'—the 'divine madness' of the ancients—is described as the ultimate degree of passion, and closely associated with the rhetoric of love:

comme l'enthousiasme de la dévotion emprunte le langage de l'amour, l'enthousiasme de l'amour emprunte aussi le langage de la dévotion. Il ne voit plus que le paradis, les anges, les vertus des saints, les délices du séjour céleste.[68]

---

[64] pp. 299–300.   [65] p. 317.   [66] p. 224.

[67] p. 17.   [68] pp. 15–16.

According to Faublas, the erotico-religious ecstasies of the 'dévote' Mme Desglins found this type of verbal expression: 'Divins transports! bonheur des élus, joies du paradis!'[69] Rousseau observes that it is unthinkable for the lover in such circumstances as these to employ base language. He will necessarily elevate his speech, conferring on it the nobility and dignity of his emotions. In writing to the loved one, he aspires to the divine register: 'ce ne sont plus des lettres que l'on écrit, ce sont des hymnes.'[70]

Diderot presents a different aspect of the religious enthusiasm for earthly things in *La Religieuse*. There is no certainty that the praise lavished on Suzanne by the nuns at her first convent is the excessive outpouring of a love denied ordinary outlets, however, as with the 'dévotes' Valmont mentions. None the less, their admiring chants have a religious cadence:

'Mais voyez donc, ma sœur, comme elle est belle! comme ce voile noir relève la blancheur de son teint! comme ce bandeau lui sied! comme il lui arrondit le visage! comme il étend ses joues! comme cet habit fait valoir sa taille et ses bras!'[71]

Not so the eulogies uttered by the lesbian Mother Superior of Sainte-Eutrope. In her, Diderot implies, claustral life has shifted passion from its conventional channels so that it is focused on the fairest objects of her sex. If the woman's panegyric bears a superficial resemblance to Mme de Merteuil's praise of Cécile in *Les Liaisons dangereuses*, the similarity should not mislead. In her encomia of the young girl the Marquise is apparently concerned primarily to arouse the jealousy of Valmont. Thus in Letter LXIII she paints a provocative portrait of Cécile's charms in her grief-stricken state at Mme de Volanges's discovery of her liaison with Danceny:

Vous ne sauriez croire combien la douleur l'embellit! . . . Frappée de ce nouvel agrément que je ne lui connaissais pas, et que j'étais bien aise d'observer, je ne lui donnai d'abord que de ces consolations gauches qui augmentent plus les peines qu'elles ne les soulagent . . . Je lui conseillai de se coucher, ce qu'elle accepta; je lui servis de femme de chambre: elle n'avait point fait de toilette, et bientôt ses cheveux épars tombèrent sur ses épaules et sur sa gorge entièrement découverte; je l'embrassai; elle se laissa aller dans mes bras, et ses larmes recommencèrent à couler sans effort. Dieu! qu'elle était belle! Ah! si Magdeleine était ainsi, elle dut être bien plus dangereuse pénitente que pécheresse.[72]

[69] p. 772.   [70] p. 16.   [71] p. 239.   [72] p. 125.

The lesbian Superior's advances to Suzanne are both more explicit and more threatening than this. The first indications of her sexual nature occur in Suzanne's recounting of the way the woman 'punishes' another nun's misdemeanour, when physical chastisement is succeeded by a litany of praise: 'qu'elle a la peau blanche et douce! le bel embonpoint! le beau cou! le beau chignon! . . . la belle gorge! qu'elle est ferme!'[73] The panegyric is resumed a little later on as the Superior, after applauding Suzanne's spinet-playing, turns to admiration of her hands, figure, eyes, mouth, cheeks, and complexion—attentions which embarrass Suzanne and awaken the jealousy of the woman's former favourite, Sœur Thérèse.[74] The climax is reached in the Superior's plaint for Suzanne's suffering at the hands of other nuns, which is described as a veritable passion. The tone, again, is that of a litany:

'Les cruelles! serrer ces bras avec des cordes! . . . ' Et elle me prenait les bras, et elle les baisait. 'Noyer de larmes ces yeux! . . . ' Et elle les baisait. 'Arracher la plainte et le gémissement de cette bouche! . . . ' Et elle la baisait. 'Condamner ce visage charmant et serein à se couvrir sans cesse des nuages de la tristesse! . . . ' Et elle le baisait . . . [75]

— and so on. Her advances end soon after with the conducting of a love-catechism in which Suzanne has to answer questions about her conduct alone in bed that are intended to elicit information about her indulgence or otherwise in solitary pleasures.[76]

Given the convent environment, it is unsurprising that such wooing as this should be couched in the language of religious devotion. The great preachers of the seventeenth and eighteenth centuries standardly used the contemporary vocabulary of love in their references to the 'amour divin' of Christ. Bossuet was bold enough to reverse the process, describing creatures visited by divine love as prey to as many forms of suffering as those inflicted on any Racinian heroine.[77] The pervasive influence of books of piety led to the general adoption in literature and polite society of images and rhythms familiar in religious prose. As Duclos noted in *Les Confessions du comte de . . .* (1741), 'Une dévote emploie pour son amant tous les termes tendres et

---

[73] p. 329.   [74] p. 335.
[75] p. 348; see also Leo Spitzer, 'The Style of Diderot', *Linguistics and Literary History* (Princeton, 1948), p. 149.   [76] p. 352.
[77] See Thérèse Goyet, *L'Humanisme de Bossuet*, 2 vols (Paris, 1965), i. 204, note 245.

onctueux de l'Ecriture, et tous ceux du dictionnaire le plus affectueux et le plus vif '[78] – a process we have already observed in Mlle Habert and Mme de Tourvel.

In this climate it seemed natural for Saint-Pierre to insert into *Paul et Virginie* an epithalium recalling the Song of Songs. When Paul sings the praises of his beloved: 'tu me parais au milieu de nos vergers comme un bouton de rose . . . Lorsque je t'approche, tu ravis tous mes sens. L'azur du Ciel est moins beau que le bleu de tes yeux . . . ',[79] or when Virginie celebrates the beauty of her companion by comparing it with the brilliance of morning sunshine, and likens their love to the sibling-love of all creatures brought up together in nature,[80] they seem to echo the biblical singer in his hymning of the comely sister-spouse, and the beloved whose fairness is as the ravishing glory of the natural world.

In the *haut monde* of the eighteenth century, lyrical wooing of this kind would have appeared an absurdity. It has its place in the sentimental novel—above all in *La Nouvelle Héloïse*—but is excluded as an anachronism, a relic of past *courtoisie*, from the society Crébillon describes in *Les Egarements du cœur et de l'esprit*, or which Saint-Preux encounters during his stay in Paris. Meilcour quickly discovers how superfluous are elaborate protestations of love in this rarefied atmosphere, and how readily women give themselves to a man who shows interest in them.[81] Like Saint-Preux after him, he believes (at least until Mme de Lursay and Hortense, in their different ways, effectively show him otherwise) that sensual enjoyment is all that either sex seeks in this ostensibly refined milieu. In his implied regret at such a state of affairs Crébillon echoes Marivaux, who wrote for the *Spectateur français* of 1723 an elegy on the passing of the old love-rhetoric. In his youth, he declares, people professed respectful sentiments for the object of their affections:

j'entends que j'ai eu de ces sentiments qui aboutissent à faire dire des choses bien tendres, de cela qu'on appellerait en ces temps-ci élégie ou églogue; enfin de cet amour qui n'est qu'un soupir perpétuel, et qui vise bien respectueusement à surprendre une belle main, qu'on baise avec un ragoût si ravissant qu'une femme en est toute honteuse, à cause du plaisir qu'elle vous y voit prendre.[82]

[78] See Roger Mercier, *La Réhabilitation de la nature humaine (1700–1750)* (Villemonble, 1960), p. 324.   [79] p. 130–1.   [80] pp. 131–2.
[81] p. 15.   [82] *Journaux et œuvres diverses*, p. 206.

Such a love is satisfied with what, in the society Marivaux now knows, would be dismissed as the merest token of *politesse*, but which was able to 'piquer l'âme' of those conversant with the past's rhetoric of word and gesture.[83] Finding erotic fulfilment, in other words, is a matter of attitude. One type of language or one part of the body is not inherently better at arousing and assuaging desire than another; all depends on a prior state of mind. But at present, Marivaux reflects, love is taken as a purely physical quantity with which the spirit has nothing to do, and as a consequence the rhetoric of respect had died.

Crébillon concurs with this verdict, writing at the start of *Les Egarements du cœur et de l'esprit* that the women of olden days were more eager to inspire respect than arouse desire. And perhaps, he continues, they were made happier thereby than are their modern counterparts; for although men spoke less readily of love to them, the love they excited was more satisfying and more durable. It is his disposition to respect her that Mme de Lursay praises in Meilcour and contrasts with the quality of mere timidity which she would quickly have tried to dispel in him. People err, she tells him, in imagining that it is a proof of love to abandon consideration.[84] On the contrary, it is 'la plus mauvaise façon de penser qu'il y ait au monde'. Although men are perfectly reasonable in expecting a return for their attentions, they must set about obtaining it with circumspection. Respect lies in refraining from pressing home an advantage one has gained. 'Par exemple, nous sommes seuls, vous me dites que vous m'aimez, je vous réponds que je vous aime, rien ne nous gêne'; but the greater the freedom she seems to grant his desires, the greater is his worth in not abusing it. She tells Meilcour that he is possibly the only man she knows capable of such conduct,[85] and she therefore feels it perfectly proper to proceed with him as openly as she does.

The inconsequential nature of respectful love-talk, however, is also underlined by Crébillon's contemporary Marivaux. In *Le Cabinet du philosophe* he discusses the veiling of sexual desire through language, a method of palliating brutally direct attack. Women ascribe grossness, he writes, to men who proceed without the preliminaries of *politesse*, and who tell their beloved without further ado, 'Madame, je vous désire, vous me feriez grand plaisir de m'accorder vos faveurs.' But to say to the same woman 'Je vous aime,

---

[83] pp. 206–7.   [84] p. 67.   [85] p. 69.

madame, vous avez mille charmes à mes yeux' will give only pleasure
to her, and lead to the granting of the same favours as will infallibly
be denied in the first case. All females, Marivaux continues, are the
same, and know that they are simply the dupes of language. 'Le vrai
sens de ce discours-là est impur; mais les expressions en sont honnêtes,
et la pudeur vous passe le sens en faveur des paroles.'[86] When vice
speaks, it does so with a crudeness that repels; but 'qu'il paraît
aimable quand la galanterie traduit ce qu'il veut dire!' All such
translations spare the ears alone, for a woman's soul is not deceived
by them. Only a libertine, however, would insist for that reason on
denying her the harmless pleasure of feeling her modesty respected.
Other men see that nothing is lost in flattering her sense of virtue,
provided that the object of their longing is eventually yielded up
to them.[87]

For the young Meilcour it appears insufficient even to couch
professions of love in gallant and respectful terms. All declarations,
he believes, must offend a woman equally—a misapprehension which
explains why Crébillon is able to devote most of his novel to one such
avowal. Meilcour's lack of experience, Mme de Lursay says, should
have alerted her to the fact that the declaration needed to come from
her rather than him: 'Que je suis fâchée de n'avoir pas su plus tôt
que vous vouliez qu'on vous prévînt!'[88] It mattered little to her, she
tells him, that he himself should inform her of his own feelings,
although there was always the possibility he might trick her into
believing erroneously that he loved her. This is the way of the world:

Souvent on le dit à une femme parce que sans cela on ne saurait que lui
dire, qu'on est bien aise d'essayer son cœur, que l'on croit flatter son orgueil,
ou que l'on veut soi-même s'accoutumer à ce langage et essayer à quel point
et comment l'on peut plaire.[89]

To do likewise is simply to follow convention, and would not render
the young man blameworthy.

The difficulty of professing love is a topic which has already been
raised in a conversation between Meilcour and Mme de Lursay
apropos of a play both have just seen. Meilcour is persuaded of the
real complexity of declarations such as the one made in the drama,
Mme de Lursay disagrees: 'dire qu'on aime est une chose qu'on fait

---

[86] *Journaux et œuvres diverses*, pp. 337–8.   [87] p. 338.
[88] p 31.   [89] p. 42.

tous les jours, et fort aisément', although the ease is apparently all on the side of the male.

'Je suis persuadée, dit-elle, que cet aveu coûte à une femme; mille raisons, que l'amour ne peut absolument détruire, doivent le lui rendre pénible; car vous n'imaginez pas qu'un homme risque quelque chose à le faire?'[90]

This is precisely what Meilcour does imagine, at least in cases where he is uncertain of being loved in return. But, Mme de Lursay retorts, it is for the man to risk the first move, since only his avowal can authorize a woman to make her own. Few men could feel any regard for a female who offered them such an easy conquest as Meilcour seems to consider appropriate. Women, conversely, will always listen to a man who has such a declaration to make. Meilcour remarks that the phrasing of the profession is far from easy, to which she is reassuring:

Les déclarations les plus élégantes ne sont pas toujours . . . les mieux reçues. On s'amuse de l'esprit d'un amant, mais ce n'est pas lui qui persuade; son trouble, la difficulté qu'il trouve à s'exprimer, le désordre de son discours, voilà ce qui le rend à craindre.[91]

Despite her encouragement, Mme de Lursay observes to Meilcour that the rhetoric of love can be as deceptive as other forms of intended persuasion. Men are creatures of artifice, wily enough to feign confusion and a state of emotion when they are barely touched by desire; and in such cases they are justly disbelieved. Finally, however, for a person in Meilcour's position

il est plus avantageux, même plus raisonnable, de parler que de s'obstiner à se taire. Vous risquez de perdre par le silence le plaisir de vous savoir aimé; et si l'on ne peut vous répondre comme vous le voudriez, vous vous guérissez d'une passion inutile qui ne fera jamais que votre malheur.[92]

But Meilcour himself, she adds helpfully, is speaking like a man who has just such a difficult declaration to make.

Where the female is, or appears to be, in a stronger position than the male, the burden of wooing may happily be assumed by her. In *Les Egarements du cœur et de l'esprit* Mme de Senanges arrogates this role to herself, but the directness of her methods—praising Meilcour's looks to his face, rather than allowing him to pay such

---

[90] p. 21. [91] p. 22. [92] p. 23.

compliments as the circumstances of her age and decrepitude permit—
sickens him in its want of delicacy. Mme de Lursay proceeds
differently. In keeping with the tastes she revealed when discussing
the play with Meilcour (praising the extreme delicacy with which
the avowal of love was made),[93] she aims at persuading the man
whose affections she wants to secure that he rather than she has made
the running. This is not very different from the flattering of female
vanity which Marivaux described in *Le Cabinet du philosophe*. When
Mme de Lursay makes it appear that he, not she, conceived the plan
of having his coach arrive late to pick him up, Meilcour knows that
appearances alone have been saved by her fabrication: 'je sortis . . .
riant en moi-même de ce qu'elle me faisait honneur du stratagème,
pendant qu'elle aurait pu à si juste titre s'en attribuer l'invention.'[94]

Given Rousseau's opinions about the relative status of men and
women, and the due proportion of attack and defence in their
respective approach to physical union, it is perhaps surprising that
his novels should show women wooing as decisively as they do. *Emile*
instructs its reader that since man is active and strong, woman should
be passive and weak, making herself agreeable to her partner rather
than provoking him. This does not mean that Rousseau regards
woman as being without desires, and consequently unlikely to make
advances to man. But he takes the traditional view that nature has
qualified the female's immoderate appetites by giving her modesty,
a principle of regulation. Her true destiny, Rousseau proclaims, is
to be useful to man, caring for him and making his life pleasant.

Why, then, does Rousseau show in *La Nouvelle Héloïse* a
domineering female, and in *Emile* a girl who makes a proposal of
marriage to her man? The second case is rather easier to resolve than
the first. The true suitor in *Emile* is the boy's tutor: he finds Sophie
and leads Emile to her at the opportune time. Therafter Emile pays
his suit according to his mentor's instructions. Sophie's effective
offering of herself occurs in part-penance for the impatience she
showed at his arriving late for a rendezvous, a tardiness which was
caused by his performing a charitable act on the way to see her. And
the language of her proposal is in any case markedly restrained:
'Emile, prends cette main: elle est à toi. Sois, quand tu voudras, mon
époux et mon maître; je tâcherai de mériter cet honneur.'[95] Even
here, in other words, the male is to be given the impression that he

---

[93] p. 21.   [94] p. 63.   [95] p. 813.

remains in control, that the crucial decision has been his. The female knows differently, but also knows how to qualify the terms in which her offer is made in order to comply with general expectation.

But *La Nouvelle Héloïse* presents a different state of affairs. Here there is no question but that the female, contrary to the doctrine professed in *Emile*, is the strong and active partner. Saint-Preux's subjection to her, unvirile but as chivalric as his name, is too plain to need much illustration. From the early letter in which he depicts Julie as a 'belle dame sans merci', his masochistic desire to be punished by her and incur her disapproval is apparent:

Cent fois le jour je suis tenté de me jeter à vos pieds, de les arroser de mes pleurs, d'y obtenir la mort ou mon pardon. Toujours un effroi mortel glace mon courage; mes genoux tremblent et n'osent fléchir; la parole expire sur mes lèvres, et mon âme ne trouve aucune assurance contre la frayeur de vous irriter.[96]

His later confession of the drunken visit to a brothel, or of indulging in masturbation, seems calculated to draw forth censure from his beloved. It is not unexpected, then, to find her preaching to Saint-Preux on the very subject on which she may suppose him most vulnerable, whether he is fit to be a husband. Julie's doubts about him on this score, he learns, stem from the very intensity of his passion: she informs him that people in love make bad partners in wedlock because they think only of themselves. Greater community spirit and more selflessness befit spouses. Love is the great creator of illusion, namely the conviction that it will last for ever. Julie's sermon, it later becomes evident, is itself based on a false premiss, as she is obliged to admit at the end of the novel. But at this stage of the book nothing saps her absolute certainty of being right.

As Aristotle realized several centuries before Julie, there is nothing illogical in the attempt to argue a person out of an emotion. Reason and feeling, so often presented as each other's antagonist, are in fact bedfellows. Since it depends on beliefs as well as on bodily states, emotion can be altered by argument because beliefs can be so altered.[97] (This, incidentally, is why Mme de Merteuil is mistaken, twenty years after Julie, in chiding Valmont for trying to *prove* to the Présidente that she should yield to him. The whole project on

---

[96] pp. 35–6
[97] See W. W. Fortenbaugh, *Aristotle on Emotion* (London, 1975); also Ronald de Sousa in Rorty (ed.), *Explaining Emotions*, p. 127.

which the Marquise and Valmont embarked was based on the premiss that people's feelings can be manipulated by the rationalist, and to an impressive extent they are proved right in their assumption.) Julie's efforts at persuasion ultimately fail because her passionate love for Saint-Preux is too strong to be uprooted.

The rhetorical skills of her twelfth-century forebear were also turned to the matter of marriage, if from a different direction, and on their own terms they achieved a brilliant success.[98] In her first letter to Abélard, Héloïse explains, with the skill of a dialectician far more formidable than Julie, why she is unwilling to wed him; she dwells on the notion of dignity, a topic which Abélard discusses in relation to women in a subsequent letter.[99] While Abélard's argument is based on the special favour shown to women by Christ, an implicit indication of their worth, Héloïse is concerned with the glory of preserving fidelity to a partner outside the bonds of marriage. It says much for her female subordination, and nothing against her powers of reasoning, that Abélard persuaded her nevertheless to become his wife.

In *Les Liaisons dangereuses*, as we know, two languages of sexual attack are used. One is the language of military tactics,[100] employed by Valmont both to convey the complexity and precision of the manœuvres in which he engages, and to suggest his emotional detachment from the liaison with Mme de Tourvel. The other language, which belies that detachment, is the language of sentiment. At times, of course, the apparent declarations of sentiment have really been dictated by tactical considerations; but at others it is less clear, as the Marquise observes to him, that the emotion he professes is feigned. The fact, revealed at an early stage of the novel,[101] that Valmont makes drafts of his letters to the Présidente proves nothing about the inauthenticity or otherwise of the protestations contained therein. Whatever the circumstances, and despite the prefatorial claim by the 'editor' that all the correspondents write faulty French, Valmont rarely pens anything unreflectingly.

---

[98] See Régine Pernoud, *Héloïse et Abélard* (Paris, 1970), pp. 75 ff.

[99] See Mary Martin McLoughlin, 'Abelard and the Dignity of Women', in *Pierre Abélard-Pierre le Vénérable: Les Courants philosophiques, littéraires et artistiques en occident au milieu du XIIᵉ siècle (Colloque international du C. N. R. S., Abbaye de Cluny, 2 au 9 juillet 1972)* (Paris, 1975).

[100] See Denis de Rougemont, *L'Amour et l'occident*, revised ed. (Paris, 1939), p. 229, on the extent to which, in the 12th and 13th centuries, the language of love was enriched by borrowings from that of war.   [101] p. 71.

Valmont's wooing is, to borrow from Batteux's list of 'figures touchantes', a compound of confession, deprecation, exclamation, threat and imprecation. Confession is present in the admission of past faults to obtain the Présidente's pardon: Letter LII, where Valmont describes how he was early set on the path of dissipation by being thrown into the 'tourbillon' of the world, is a typical example. Deprecation, which Batteux defines as the recourse to prayers and tears when the speaker despairs of gaining his end by other means, is well illustrated by Letter XXIII, where Valmont tells Mme de Merteuil how he sank to his knees as he begged Mme de Tourvel, in an elegant ternary construction, to take pity on him in his lovelorn state: 'O vous que j'adore! écoutez-moi, plaignez-moi, secourez-moi!'[102] Exclamation is too common a feature of the sentimental style adopted by various characters in the novel to need specific reference, although it is worth noting that Valmont's cry in Letter XXIII is echoed by Danceny much later on, when he addresses Mme de Merteuil as 'ô vous que j'aime! ô toi que j'adore! ô vous qui avez commencé mon bonheur! ô toi qui l'as comblé!'[103] Threats feature in those letters which report Valmont's allusion to his approaching death from unhappy love: Mme de Rosemonde relays to the Présidente her nephew's counselling her to 'ne trouble[r], par aucun regret, l'éternelle tranquillité dont il espère jouir bientôt';[104] and in the letter relating his triumph over Mme de Tourvel's virtue Valmont notes how he had expressed his desire to 'terminer, avec quelque tranquillité, des jours auxquels je n'attache plus de prix, depuis que vous avez refusé de les embellir'.[105] Finally, imprecation (glossed by Batteux as the expression of fury and despair) features in Valmont's communication with the Marquise after his quarry has given him the slip, if not in his letters to the Présidente herself:

Mon amie, je suis joué, trahi, perdu; je suis au désespoir: Mme de Tourvel est partie. Elle est partie, et je ne l'ai pas su! et je n'étais pas là pour m'opposer à son départ, pour lui reprocher son indigne trahison! . . . Ô femmes, femmes! plaignez-vous donc si l'on vous trompe! Mais oui, toute perfidie qu'on emploie est un vol qu'on vous fait.[106]

As this last quotation suggests in its portrayal of the roué's self-disgust, the rhetoric of love may readily be transformed from a

---

[102] p. 51.   [103] p. 339.   [104] p. 281.
[105] p. 290   [106] p. 225.

rhetoric of persuasion into one of dissuasion—often, as here, directed by the writer or speaker at himself. Like Valmont, but for different reasons, Mme de Tourvel tries to argue herself out of a passionate state with a show of deliberative eloquence; but as we saw in Letter LVI, her efforts convince neither herself nor the man she attempts to discourage. The array of hypotheses and rhetorical questions: 'Supposé que vous m'aimiez véritablement', 'les obstacles qui nous séparent en seraient-ils moins insurmontables', 'aurais-je autre chose à faire qu'à souhaiter que vous pussiez bientôt vaincre cet amour', 'S'il existe des plaisirs plus doux [than the tranquil ones of marriage], je ne veux point les connaître', 'comment affronter ces tempêtes [of an affair with Valmont]', and so on, merely lessen the force of the message she tries to impress on Valmont.

But the more sustained rhetoric of dissuasion in *Les Liaisons dangereuses* is revealed in one character's attempt to argue another out of infatuation. Mme de Volanges practises this type on the Présidente, warning her against any kind of dealings with Valmont. To this end she describes his libertine's reputation and his utter disregard for the fate of those women who become involved with him. Like the Présidente, Mme de Volanges shows some attention to style in her effort to convince:

Encore plus faux et dangèreux qu'il n'est aimable et séduisant, jamais, depuis sa plus grande jeunesse, il n'a fait un pas ou dit une parole sans avoir un projet, et jamais il n'eut un projet qui ne fût malhonnête ou criminel.[107]

There is a studied rhetorical balance in this sentence, beginning with the antithesis of the binary phrases, continuing through the parallelism of 'fait un pas ou dit une parole', and culminating in the anaphora and amplification of the last three clauses. Despite the force of her argument, which Mme de Volanges resumes in Letter XXXII, she fails to sway the younger woman, who remains of the opinion that the country air has effected a remarkable transformation on the erstwhile rake.

More determined still is the attempt made by the Marquise in many letters to Valmont to draw him away from Mme de Tourvel. She starts with a note of contempt: how could the all-conquering Vicomte conceive the plan of seducing a prude, and a prude with insipid looks and no style?[108] She then riles Valmont with the assertion that,

[107] p. 26.   [108] p. 18.

whether he realizes it or not, he is in love.[109] At the same time she
attacks his methods for their inefficiency, which implies that he is
frightened of succeeding: 'Mon ami, quand on veut arriver, des
chevaux de poste et la grande route!' In order to make him hungry
for more substantial fare than he is likely to enjoy with the Présidente,
she appends a detailed description of her own affair with Belleroche,
thinking that Valmont may yet be won over from sentiment to
sensuality. She then details the various charms of Cécile, hoping to
tempt him back to the real business of the day, namely their vengeance
on Gercourt. Before the end of the novel she does, of course, persuade
Valmont to end his relationship with Mme de Tourvel, taunting him
with a letter which he copies out and forwards to the Présidente,
and which deals with the involuntary nature of the emotions: 'On
s'ennuie de tout, mon ange, c'est une loi de la nature; ce n'est pas
ma faute.'[110] Her argument that it is indecent for a man of
Valmont's stature to be tied to a woman who diminishes his libertine's
stock in the world has finally struck home.

On the evidence of other eighteenth-century novels, it is not
uncommon for the sexually experienced to employ their persuasive
arts in this way, seeking to detach a man from a woman seen as
unworthy of him. Although Meilcour and Valmont have little in
common, not least because of the former's ignorance of areas the
latter knows intimately, they are both subjected to the rhetoric of
dissuasion by one practised in sexual matters. The mentor-figure in
Les Egarements du cœur et de l'esprit is Versac, who tries to argue
Meilcour out of his attraction to Mme de Lursay and into a liaison
with Mme de Senanges. Versac's motives are no purer than Mme
de Merteuil's, admittedly; and although he eventually comes to take
a protector's interest in Meilcour, wanting his schooling to be
entrusted to the most suitable woman for the purpose (in this case
Mme de Senanges), his initial attempt to disaffect the young man
with Mme de Lursay seems purely malicious in origin. He enjoys
calumny for its own sake, and is an incorrigible gossip.[111] But at
a later stage, when he delivers his long lecture to Meilcour about
the nature of the society they both inhabit and the ways of advancing
in it, there is more seriousness in his advice; and, rightly or wrongly,
Meilcour immediately afterwards finds his attitude to Mme de Lursay
crystallized into one of contempt: 'Versac me m'a pas trompé,

---

[109] p. 28.   [110] p. 328.   [111] pp. 80–1, 90 ff.

me disais-je, et je ne sais pas comment on ne donne que le nom de coquette à une femme de cette espèce.'[112] This conviction is not, however, proof against Mme de Lursay's own eloquence. In the course of their meeting soon after Meilcour's session with Versac, she succeeds in talking the former back into a state of respect for her. That very day, with a mixture of tenderness and regret, Meilcour loses his virginity under Mme de Lursay's tutelage.

It is scarcely surprising that religious arguments against indulging in carnal pleasure should find no place in this book's rhetoric of dissuasion. If the *libertin* novel tries to warn against love, it does so for the sake of a greater sensual fulfilment than one particular liaison seems to offer. The threatening language we encounter in works which, like Mirabeau's *Le Rideau levé*, preach the desirability of moderate sexual enjoyment is intended to warn against the physical, not the spiritual or moral, dangers of excessive sexual activity. In *Les Liaisons dangereuses* the Présidente has little success in weaning Valmont from rampant sensuality into a condition of greater moderation: the scheme by which he finally overcomes her virtue involves merely the *pretence* that he has abjured the flesh for the sake of eternal salvation.

Télémaque, a pedagogical novel by a man of the cloth, shows the limited power of language to turn a lover from a path of dissoluteness back to virtue and self-denial. Fénelon's Mentor perceives the danger of Télémaque's absorption by the nymphs of Calypso's isle, and heaps reproaches on his charge. Propriety, the young man is told, prescribes his return to his homeland and prohibits his yielding to mad passion.[113] Love is a shameful tyrant which tries to lull Télémaque into enjoying 'une vie molle et sans honneur au milieu des femmes'. Télémaque feels the sting in these words, but remains undecided; and the stern moralist Fénelon deploys a rhetoric of terror to full effect, describing the sorry state to which the youth is reduced:

Il était devenu maigre; ses yeux creux étaient pleins d'un feu dévorant . . . Sa beauté, son enjouement, sa noble fierté s'enfuyaient loin de lui. Il périssait, tel qu'une fleur qui, étant épanouie le matin, répandait ses doux parfums dans la campagne et se flétrit peu à peu vers le soir: ses vives couleurs s'effacent; elle languit, elle se dessèche et sa belle tête se penche, ne pouvant plus se soutenir; ainsi le fils d'Ulysse était aux portes de la mort.[114]

[112] p. 167.   [113] p. 174.   [114] p. 175.

Apart from its lyricism, this is barely different from the accounts
furnished by eighteenth-century doctors of the physical decrepitude
ensuing on unwise indulgence in sexual activity, whether solitary or
otherwise. For moral reasons or on narrowly medical grounds, the
individual must be discouraged from giving free rein to his libidinal
urges. Godly intervention, fortunately, saves the day in *Télémaque*.

No such supernatural support is offered to another champion of
chastity in the eighteenth-century novel, Prévost's Tiberge. Sadly,
the eloquence of des Grieux's friend is less than divine in its
persuasiveness, and when he tries to argue his companion out of
earthly love he meets with greater articulacy than his own. When
Tiberge visits him at Saint-Lazare, des Grieux expects a homily on
dishonesty and its punishment: the circumstances of his imprisonment
make this natural enough. But he quickly informs his pious friend
that misfortune has not cured him of his passion. The chastisement
of heaven has not enlightened him, nor is his 'un cœur dégagé de
l'amour et revenu des charmes de Manon'. Rather, he is still the
victim of 'cette fatale tendresse dans laquelle je ne me lasse point
de chercher mon bonheur'.[115] Tiberge is unable to comprehend how
a man may recognize the unworthiness of his fleshly attachments
and yet refuse to try breaking them. His expression of this puzzlement
is the signal for des Grieux to launch into a piece of deliberative
rhetoric in which he draws his friend's attention to an equivalent
paradox in his own reasoning. For Tiberge's opinion takes no
account of the strength of the case against virtue, namely that it gives
man a hard bed to lie on. How dare Tiberge contend that what
mortifies the body is bliss for the soul, des Grieux challenges, and
gets into his argumentative stride:

Or si la force de l'imagination fait trouver du plaisir dans ces maux mêmes,
parce qu'ils peuvent conduire à un terme heureux qu'on espère, pourquoi
traitez-vous de contradictoire et d'insensé dans ma conduite, une disposition
toute semblable?[116]

If the path des Grieux follows is stony, 'je tends à travers mille
douleurs à vivre heureux et tranquille auprès d'elle [Manon].'
Moreover, the happy goal at which he aims is close, while Tiberge's is
far away; the former is known, 'c'est à dire sensible au corps', whereas
the latter is unknown, intangible, and glimpsed only through faith.

[115] p. 90.   [116] p. 91.

Predictably enough, this reasoning shocks Tiberge. On its own materialist terms, however, it is unimpeachable. The argument from the known will always have a certain advantage over that from the merely surmised. As his listener recoils in horror, des Grieux warms to his theme. The debating skills he was taught at school and practised at Saint-Sulpice—the very skills which had won him applause on the day Manon chose to rediscover him—are easily adapted to the purposes of secular persuasion. His choice of phrases reveals his consciousness of engaging in a contest of wills which he can win by superior verbal presentation: 'ce n'est pas sur [cette comparaison] que porte mon raisonnement', 'j'ai eu dessein d'expliquer', 'je crois avoir fort bien prouvé', 'j'ai traité les choses d'égales, et je soutiens encore qu'elles le sont.' The force of his plea for yielding to erotic love becomes irresistible as he anticipates his less nimble adversary's conventional objections:

Répondrez-vous que le terme de la vertu est infiniment supérieur à celui de l'amour? Qui refuse d'en convenir? Mais est-ce de quoi il est question? Ne s'agit-il pas de la force qu'ils ont, l'un et l'autre, pour faire supporter les peines?[117]

The number of deserters from severe virtue is infinitely greater than the number from love.If Tiberge wishes to object that virtue is not always painful to practise, that there are no longer tyrants or martyr's crosses to bear, des Grieux can rejoin that love, too, may similarly be peaceful and blessed;

et, ce qui fait une différence qui m'est extrêmement avantageuse, j'ajouterai que l'amour, quoiqu'il trompe assez souvent, ne promet du moins que des satisfactions et des joies, au lieu que la religion veut qu'on s'attende à une pratique triste et mortifiante.[118]

Seeing his friend's distress, des Grieux makes the kind of concession that marks a debater's consciousness of his superiority to his opponent. Tiberge's vulnerability, his friend assures him, lies in the weakness of his method rather than the wrongness of his thesis. Preachers may convincingly demonstrate that virtue is necessary in some absolute sense, but they should not disguise its difficulty and rebarbativeness. Arguing *sub specie aeternitatis*, they may establish that the delights of love are transitory and will be followed by an eternity of damnation, and even that, the more seductive they are, the more magnificent

[117] pp. 91–2.   [118] p. 92.

will be heaven's reward for him who abjures them; but they should not deny that love constitutes the greatest of earth's felicities. This conclusion restores most of Tiberge's equanimity, and he grants that there is some reason in des Grieux's contentions. There is little else that he can do, so comprehensively has he been outmanœuvred.

The moral victory won, des Grieux now turns his eloquence to the more pressing matter of enlisting Tiberge's sympathy, a vital preliminary to persuading him to deliver a letter that will set in train des Grieux's escape from gaol. This he attempts with the argument his friend calls Jansenist, and which is based on the alleged fatality of his passion: 'c'est mon devoir d'agir comme je raisonne! mais l'action est-elle en mon pouvoir? De quels secours n'aurais-je pas besoin pour oublier les charmes de Manon?' Persuaded that des Grieux is weak rather than bad, Tiberge obediently does as he is bidden.

At this stage, in contrast to what we see both earlier and later in the novel, des Grieux's love is merely checked by outside agency. There is no cause for him actually to bewail the departure of his loved one, with whom he expects soon to be reunited. But at the time of her first abandonment of him, and again at the end of the book when she perishes in the American desert, circumstances prompt that time-honoured lover's song, the lament. There are many such laments in the eighteenth-century novel, often based on classical models and like them employing set rhetorical patterns. As we have seen, after Manon has deserted him for M. B . . . des Grieux, a prisoner in his father's house, writes a 'commentaire amoureux' on the fourth book of the *Aeneid*, reflecting that Dido deserved a lover as faithful as him instead of the inconstant Aeneas. The latter's desertion of his queen became the subject of one of Ovid's *Heroides*, a series of poems in which women bemoaned the loss of their lover, and which gave rise in the eighteenth century to the genre of the heroid.[119] Although the Ovidian model prescribed a female's lament for her beloved, the eighteenth century was ready to accommodate male complaint as well.[120]

Des Grieux's vituperation against women when he discovers Manon has left him again, this time to be the mistress of G . . . M . . ., is perhaps in too aggressive a vein to be associated with this genre.

---

[119] See Heinrich Dörrie, *Der heroische Brief* (Berlin, 1968).

[120] See Susan Lee Carrell, *Le Soliloque de la passion féminine ou le dialogue illusoire* (Tübingen and Paris, 1982).

It fits, rather, into the type of revenge-rhetoric also familiar in the literature of love, and of which Valmont's imprecations against the Présidente after her desertion of him are a further example. 'Ah, tu es une femme,' des Grieux tells the girl who delivers Manon's valedictory note to him, 'tu es d'un sexe que je déteste et que je ne puis plus souffrir.'[121] This sentiment echoes the one he gave voice to after Manon's first desertion: 'je ne mettais plus de distinction entre les femmes, et . . . après le malheur qui venait de m'arriver, je les détestais toutes également.'[122] By the time of Manon's death des Grieux's feelings have radically altered. His distraught state, however, prohibits appeal to the lover's eloquence, and his listener Renoncour is left to imagine the depths of an inexpressible grief.[123]

The desire for death which des Grieux declares at the end of this sad scene, after he has buried 'dans le sein de la terre ce qu'elle avait porté de plus parfait et de plus aimable',[124] is a standard appeal in the rhetoric of love. Saint-Preux professes it at a particularly unhappy stage of his affair with Julie, darkly threatening suicide in one of his letters to her,[125] and trying to justify it in a later one to Milord Edouard.[126] Robust as ever, Julie chides him for his weak complaints, declaring that the lot of women is far harder than that of men. Having dealt with the general she turns to the particular:

Comment peux-tu donc ne sentir que tes peines? Comment ne sens-tu point celles de ton amie? Comment n'entends-tu point dans ton sein ses tendres gémissements? Combien ils sont plus douloureux que tes cris emportés! Combien, si tu partageais mes maux, ils te seraient plus cruels que les tiens mêmes![127]

Men are not required as women are, she continues, to preserve their sexual virtue; they thus have less to weep over in the gloom of an unhappy love-affair than the opposite sex.

Even without Saint-Preux's cause for lament, the eighteenth-century male is perhaps more vocal in bewailing a lost love than his modern counterpart would be. Valmont gives emphatic expression to his feelings for the Présidente both before and after her terminal decline. If his initial threat of death as the consequence of frustrated love is mere posing, it is at least possible that his actual death in the duel with Danceny is a kind of suicide, since he is unlikely to

---

[121] p. 136.   [122] p. 37.   [123] p. 199.   [124] p. 200.
[125] p. 38.   [126] pp. 377 ff.   [127] p. 212.

have lacked proficiency in swordsmanship. His final professions in the novel are ambiguous enough to raise doubts about his desire to continue living. In the conclusion to Letter CLV, where he offers Danceny the choice between pleasure and happiness in a relationship with Mme de Merteuil and Cécile respectively, Valmont informs the young man with apparent sincerity of his distress at the fate of the Présidente:

c'est que je regrette Mme de Tourvel; c'est que je suis au désespoir d'être séparé d'elle; c'est que je paierais de la moitié de ma vie le bonheur de lui consacrer l'autre. Ah! croyez-moi, on n'est heureux que par l'amour.[128]

But given that in the same letter Valmont tells Danceny how Cécile, the representative of love and contentment, has been 'entertaining' Valmont himself ('elle a trouvé le moyen de me faire aussi parvenir jusqu'à elle'), we may prefer to think that he is simply playing a libertine's literary game. Another letter of Valmont's, this time to Mme de Volanges, professes the same feelings of acute sorrow at his loss of the Présidente, but the fact that it remains an unpublished draft may indicate that Laclos wished the uncertainty about Valmont's true sentiments to remain. Although the former rake declares in the letter that 'j'ai outragé indignement une femme digne de toute mon adoration . . . mes torts affreux ont seuls causé tous les maux qu'elle ressent',[129] Mme de Volanges expresses some doubt about the genuineness of his assertions.[130]

What is beyond question, despite the 'publisher's' preliminary note stating that Mme de Tourvel's unlikely fate proves the in-authenticity of the entire correspondence, is the cause of her own demise. The Présidente dies of grief when her lover deserts her. But mental distress does not interfere with her eloquence. The letter she dictates in her delirium is as 'literary' as any she wrote under happier conditions. The exordium, with its imprecation, rhetorical questions, tricolon, and poetic circumlocution, shows that even in a state of overwhelming grief Mme de Tourvel retains a stylist's grip on language:

Etre cruel et malfaisant, ne te lasseras-tu point de me persécuter? Ne te suffit-il pas de m'avoir tourmentée, dégradée, avilie? Veux-tu me ravir jusqu'à la paix du tombeau? Quoi! dans ce séjour de ténèbres où l'ignominie m'a forcée de m'ensevelir, les peines sont-elles sans relâche, l'espérance est-elle méconnue?[131]

---

[128] p. 355.   [129] p. 1391.   [130] p. 352.   [131] p. 361.

The echo of *Phèdre* in Mme de Tourvel's reference to inhabiting a land of shadows after the shame love has brought upon her scarcely needs pointing out, nor the alexandrine of her later 'le cruel souvenir des biens que j'ai perdus'. From classic revenge-speech the letter moves, via the elegiac lament of 'Où sont les amis qui me chérissaient, où sont-ils?', the parallel constructions 'Je suis opprimée, et ils me laissent sans secours! Je meurs, et personne ne pleure sur moi!', the invocation of an absent husband whose cause heaven has adopted, 'qui a lié ma langue et retenu mes paroles', to the Rousseauist apostrophe of the penultimate paragraph: 'Mais quoi! c'est lui . . . Oh! mon aimable ami! reçois-moi dans tes bras, cache-moi dans ton sein.'[132] And finally, after the brief heroid of 'Pourquoi te refuser à mes tendres caresses? Tourne vers moi tes doux regards! Quels sont ces liens que tu cherches à rompre?', the Présidente descends into the depths inhabited by Virgil's maddened Dido—'quelle nouvelle fureur t'anime? . . . Tu redoubles mes tourments'—and becomes the new Clarissa Valmont once boasted he would make of her, who fears the deadly pursuit of the male and shuns his very name: 'pourquoi me persécutez-vous? que pouvez-vous encore avoir à me dire? ne m'avez-vous pas mise dans l'impossibilité de vous écouter comme de vous répondre?'

The note of classical tragedy and the tone of the heroid are heard in another, earlier lament of the Présidente's, this time one based on the mistaken belief that Valmont is about to abandon her. In the letter she writes Mme de Rosemonde after hearing of her lover's alleged conversion and consequent determination to renounce her, the Présidente imagines his leaving her, in words recalling Bérénice's: 'Enfin, je le verrai s'éloigner . . . s'éloigner pour jamais, et mes regards qui le suivront ne verront pas les siens se retourner sur moi!'[133] The alexandrine contained in the second sentence is wholly in keeping with the poetic and elegiac mood of the entire paragraph, with its metonymic transference and its *enárgeia*, a vividness so intense that the whole scene is presented as a theatrical tableau:

Je verrai ses regards se porter sur moi, sans émotion, tandis que la crainte de déceler la mienne me fera baisser les yeux. Ces mêmes lettres qu'il refusa si longtemps à mes demandes réitérées, je les recevrai de son indifférence . . . et mes mains tremblantes, en recevant ce dépôt honteux, sentiront qu'il leur est remis d'une main ferme et tranquille!

[132] p. 362.　　　　[133] p. 285.

The nobility associated with the heroid, and a measure of poetry, infuse even Cécile's letters to Danceny, the lover she is forbidden to see. Letter LXIX shows how far she has advanced from the schoolgirlish style characterizing her early scribblings to Sophie Carnay: 'Vous [Danceny] me demandez ce que je fais; je vous aime, et je pleure.' The alexandrine hemistich sets the tone for what follows, a lyrical evocation that would have been beyond the reach of the Cécile who, barely a month before, left the convent for the world: 'Mon Dieu! ne plus vous voir! . . . Ces mots tracés au crayon s'effaceront peut-être, mais jamais les sentiments gravés dans mon cœur.' It is a tone captured by Danceny in the letter he sends her less than a week later, in which we hear the lover's lament at his mistress's coldness:

A présent, que me reste-t-il? des regrets douloureux, des privations éternelles, et un léger espoir que . . . [votre silence] change en inquiétude . . . que sont devenus . . . vos sentiments si tendres, et qui vous rendaient si ingénieuse pour trouver les moyens de nous voir tous les jours?[134]

The pathos of this letter is matched, if not exceeded, by that of another hero's grieving words. In *Paul et Virginie* the degree of eloquence displayed by the young Paul as he mourns his beloved's imminent departure is in a sense implausible. But in another respect it appears natural, as Danceny's recherché expressions do not. If Danceny, with the rhetoric of the educated and literary man at his disposal, occasionally abuses the language of love (we may recall Mme de Merteuil's criticism of his letter on that score), Paul's is the simple eloquence of the child of nature. Imagining Virginie's seduction by the riches of France, he bewails her anticipated forgetfulness of their homeland:

pour être plus heureuse, où voulez-vous aller? Dans quelle terre aborderez-vous qui vous soit plus chère que celle où vous êtes née? Où formerez-vous une société plus aimable que celle qui vous aime?[135]

But it is in its evocation of his own abandonment that Paul's plaint is most moving. Simple in its dignity, it epitomizes the rhetoric of sensibility that won the hearts of Bernardin's readers:

Cruelle! je ne vous parle point de moi: mais que deviendrai-je moi-même quand le matin je ne vous verrai plus avec nous, et que la nuit viendra sans

[134] pp. 166–7.   [135] p. 151.

nous réunir; quand j'apercevrai ces deux palmiers plantés à notre naissance, et si longtemps témoins de notre amitié mutuelle?[136]

If Paul's utterance has the cadence of verse, the poetic ring does not seem out of place: that rhythm has been well established by earlier passages of description and dialogue, and so rendered natural on the novel's own terms. Saint-Pierre's 'espèce de pastorale' achieves the same eloquence as *La Nouvelle Héloïse*, and its rhetoric may be justified by appeal to the same aesthetic, part realist and part imaginative, as governs the earlier work.

Paul and Virginie are not savages, although they do not learn to read and write until Virginie's sojourn in Europe makes literacy imperative for the purposes of communication. They are not primitives, although they are much less well educated than the inhabitants of Julie's little world. But if they have the articulateness Bernardin associates with a life lived according to nature's rhythms, they know nothing of the perverted language which is generated, according to Saint-Preux, by the corrupting demands of an artificial society. Yet even the latter, on the evidence of the eighteenth-century novel, occasionally substituted a 'natural' (that is, non-conventional) mode of discourse for the rhetoric of words, and nowhere so frequently as when the expression of feeling was at issue.

Debased sentiments, Saint-Preux notes, readily find a vocabulary to translate them in the utterances of the unfeeling. For those whose emotions are truer, as well as for those whose passions need the cover of silence, other forms of eloquence are available. Speaking the language of love does not always necessitate having recourse to words. Zilia, the heroine of *Lettres d'une Péruvienne*, writes that the French nation into which she has been transported uses the body as an instrument of mute conversation.[137] In his *Essai sur l'origine des langues*, Rousseau too remarks how Europeans gesture as they speak, and how all the force of their language seems to be in their arms.[138] He goes on to observe that people speak to the eyes much more directly than to the ears—primarily, no doubt, because verbal language is a code (what Diderot calls a system of hieroglyphs),[139] whereas the visual has little need of translation to be understood.

But the non-verbal rhetoric of love varies in its comprehensibility. Just as the language of the fan (a mode of communication rarely

---

[136] Ibid.   [137] p. 282.   [138] p. 88.
[139] See his *Lettre sur les sourds et muets*.

mentioned in the eighteenth-century French novel) was seldom clear except to initiates, so other coquettish gestures might contain a significance which the inexperienced observer was unable to grasp immediately. It is a mark of Meilcour's naïvety that he sees no particular meaning in Mme de Lursay's nonchalant knotting as she attempts to encourage him to make an advance; and his fumbling comment on her activity leaves her incredulous at the extent of his gaucheness. Marivaux's Jacob sees a clearer intention behind Mme de Ferval's prolonged testing of different quills as she prepares to write a letter on his behalf. After her vain efforts to find and then sharpen a suitable instrument she passes the task over to Jacob, who is happy to prolong the process in order to extend their agreeable *tête-à-tête*.[140] Jacob discovers along with many other initiates that in learning to love one learns to interpret small signs as carrying a greater weight of meaning than at first sight appears.[141]

The signs of flirtation may repel when they are overtly stated. Tokens, particularly in a 'civilized' society, must be discreet: the sideways glance with head half-turned, the fleeting gesture, the hesitant motion. Open declaration never has the inviting character which moves such as these possess. To be plain would be straightforwardly to grant what the coquette prefers to be equivocal about, to concede in part, and in part to refuse. Flirtation often operates, as with Mme de Lursay and Mme de Senanges, by using extraneous objects—a handkerchief, a pack of cards, a flower—which serve to arrest the person whose attention the flirt wishes to attract.[142] By seeming to concentrate exclusively on the object she means to excite the jealousy of the interested observer, making him aware of the privilege enjoyed by that object, and which he himself covets. But she simultaneously wants to make him realize—as Jacob realizes and Meilcour does not—that she is playing a game for his benefit, and to provoke a counter-move from him.

Failing to join in the game can have unfortunate consequences: the man who does not respond may appear unchivalrous at best, hurtfully negligent at worst. But normally only the male's refusal has such an effect. For a woman to rebuff a man's advances, whether

[140] p. 132.
[141] See Niklas Luhmann, *Liebe als Passion: zur Codierung der Intimität* (Frankfurt am Main, 1982), p. 24.
[142] See Simmel, op. cit. 135.

or not they have been made with discretion, is entirely permissible, entirely within the rules of the game.[143] This is the female prerogative to which Versac alludes in general terms when he tells Meilcour that the only reliable path to social success is seeming to be at all times attentive to female wishes and guided by female opinions.[144] This alone permits a man to dominate the opposite sex.

There are other types of unspoken love-language which Meilcour comes more or less readily to understand. Even he immediately knows how to interpret a woman's dishabille, noble in Mme de Lursay's case, immodest and indecent in that of Mme de Senanges. Learning how to decode the silent squeeze of the hand takes a little longer. When Mme de Lursay grants him his first experience of this favour, Meilcour knows no better than to return it:

elle m'en remercia en redoublant d'une façon plus expressive; pour ne pas manquer à la politesse, je continuai sur le ton qu'elle avait pris. Elle me quitta en soupirant, et très persuadée que nous commencions enfin à nous entendre, quoiqu'au fond il n'y eût qu'elle qui se comprît.[145]

The effect on Meilcour, in any case, is less electric than the merest touch of Hortense's hand, which makes his whole body tremble so violently that he can barely remain upright.

The physician Doppet devotes a section of *Le Médecin de l'amour* to describing the eloquence of hands in conveying feeling. The lover who presses his beloved's fingers and palm tells her of his love as clearly as he would do with the most impassioned words.[146] When the two exchange a tender squeeze of the hands, the lover can be sure that his emotion is reciprocated; but woe betide him if he takes the liberty of extending manual exploration to other parts of the body:

Quand une main veut s'égarer sous une gaze importune, et peint par sa démarche la légèreté et l'audace du galant, la main de sa beauté offensée punit par un soufflet le peu respectueux personnage. Ce langage n'est-il pas encore des plus énergiques?[147]

Some such tokens, however, are as difficult to interpret as words, and possess the same purely conventional character. The deferential posture of kneeling, for example, may not be lover-like in intention. There is much bending of the 'pregnant knee' in *La Vie de Marianne*

---

[143] Ibid. 140.   [144] p. 153.   [145] p. 45.
[146] Amédée Doppet, *Le Médecin de l'amour* (Paris, 1787), p. 98.   [147] Ibid.

and *Les Liaisons dangereuses*, but some of it is either hypocritical or merely businesslike. We have already noted Valville's betrayal of his interest in Marianne by his imitating, as though involuntarily, the posture of the surgeon who bends over her injured foot, although he is later to reveal his faithless switch of affection to Mlle Varthon by kneeling at her side to offer her a reviving draught when she falls ill.[148] Although knees may give way as part of a general state of excitement or physical collapse, their flexibility can betoken a conscious policy—that of assuming the aspect of love rather than actually suffering its debilitating effects. Valmont appears the great deceiver here, or at least professes to be for the sake of convincing Mme de Merteuil that all his moves in the planned seduction of Mme de Tourvel are rationally controlled. The Présidente, to her ultimate misfortune, must take his periodic obeisances to her as indicative of genuine emotion. Her own pious kneeling, Valmont remarks in an early letter, will one day be before him, the new God of her universe.[149] (In this same letter, however, he describes how her devout posture was preceded by expedient kneeling on his own part, one attitude among many noted in the tableau of protestation and refusal he conjures up for the Marquise.) Much later, when Valmont finally overcomes Mme de Tourvel's virtue, there appears to be a blend of ham acting and true emotion in his adoption of the same posture. Before his 'victory' over her Valmont's kneeling is no more than a familiar expression in the language of love. But afterwards, he informs Mme de Merteuil, he rose from the Présidente's arms only to fall before her to swear eternal devotion; 'et, il faut tout avouer, je pensais ce que je disais.'[150]

Where such actions are involuntary, they may be the truest pointers to states of emotion—truer than the words that can compose a lie, or the looks that can be assumed (as the young Meilcour learns to assume them) in order to convey unfelt passion. If the 'air', 'trouble', and 'désordre' so frequently described in the emotional climaxes of eighteenth-century novels are the genuine eloquence of love, other forms of emotional rhetoric are created tokens—products of an art which, like the art of prose, may conceal its artistry, but can never completely belie it.

The theory of sympathy, as we have seen, rests on the belief that the 'chemistry' of love is not subject to human will. According to

---

[148] p. 351.   [149] p. 52.   [150] p. 295.

La Rochefoucauld, this is particularly apparent in the eye-language of lovers. Describing the encounter of sympathetic souls in *La Justification de l'amour*, he notes that the emanation of sympathetic vapours occurs above all around the eyes. Meeting with a person of like disposition, they are in turn received by him through these organs, forming what is called the love of correspondence. Having in his own eyes only the image of the beloved, a man in love betrays his submission to her by ardent looks. The spirits then penetrate to the heart and thence to the understanding, where the correspondence can be expressed in words. The initial language of love, however, is ocular.[151]

Whether or not it is formulated in La Rouchefoucauld's materialist terms, this notion pervades the eighteenth-century novel. Eye-language is the communicative mode *par excellence* of a love which dares not betray itself, or which is still gathering the 'wits' with which to do so.[152] If the modest heroine succeeds despite the constraints of propriety or adverse past experience in informing a man that he is loved, it is often through these windows of the soul that she does so. When she realizes how mistaken she has been in adjudging herself without sexual feeling, the Comtesse d'Olnange conveys her changed attitude to Rosebelle through mute looks, fixing on him eyes that are 'remplis d'une langueur qui disait tout et ne défendait rien'.[153]

Cécile tells Sophie Carnay of her embarrassment at meeting Danceny's look, which 'me décontenance, et me fait comme de la peine'; otherwise she would gaze at him uninterruptedly.[154] She shares with the Présidente de Tourvel the lover's timidity, contrasting with the brazen discountenancing practised by the immodest. For those who find it expedient to affect bashfulness, like Mme de Merteuil, the language of the eyes is of great assistance. We have already noted her proud description of the 'regard distrait' she can summon at will; it is followed in Letter LXXXV by her detailing of the lowered eyes through which, in conjunction with trembling hand and quickened breathing, she hopes to persuade Prévan of her susceptibility to his charms.

Laclos was well aware of the perfidious possibilities contained in

---

[151] La Rouchefoucauld, *La Justification de l'amour*, ed. J. D. Hubert (Paris, 1971), pp. 79–80.
[152] See Jean Rousset, *Leurs Yeux se rencontrèrent* (Paris, 1981).
[153] p. 194.   [154] p. 36.

the eye-language of lovers. He observed in *Des femmes et de leur éducation* that

On sait assez que les grands mouvements de l'âme et des sens se peignent dans les yeux en surmontant même les obstacles qu'on leur oppose. Tel est le droit de la nature; l'art a cherché à l'imiter, et y est parvenu: l'usage en est fréquent au théâtre, l'abus s'en est glissé dans la société et les regards sont devenus menteurs et perfides.[155]

But for the honest in heart the emotional giveaway threatened by those organs means that, in critical circumstances, the only resort is to lower the gaze. Thus the Présidente, anticipating the painful scene of Valmont's rupture with her, also foresees how she must meet his look: 'Je verrai ses regards se porter sur moi, sans émotion, tandis que la crainte de déceler la mienne me fera baisser les yeux.'[156] And Cécile's 'morning-after' air, from which Valmont promises himself such pleasure, consists of ashamedly downcast eyes as well as the difficulty in walking which Cleland's Fanny Hill also experiences after her sexual induction.

Sometimes love is communicated by way of tears, a moving token of feeling, if not invariably a reliable one. The 'quelques larmes' which Manon sheds during her reunion with des Grieux at Saint-Sulpice may or may not be indications of true emotion; Prévost declines to settle the matter, but the highly theatrical terms in which this scene is recounted certainly suggest the possibility of feint on her part. Although Marianne calls the language of tears instinctive, she also senses its expediency in her distraught state after being knocked down by Valville's carriage. The mortification she feels at having to reveal her lowly circumstances finds expression in a fit of sobbing which, she believes, illustrates her true nobility more effectively than could any words:

cet abattement et ces pleurs me donnaient, aux yeux de ce jeune homme, je ne sais quel air de dignité romanesque qui lui en imposa, qui corrigea d'avance la médiocrité de mon état . . . [les pleurs] font foi d'une fierté de cœur qui empêchera bien qu'il ne [me] dédaigne . . . Il y a certaines infortunes qui embellissent la beauté même, qui lui prêtent de la majesté. Vous avez alors, avec vos grâces, celles que votre histoire, faite comme un roman, vous donne encore.[157]

---

[155] p. 434.   [156] p. 285.   [157] pp. 80–1.

Shedding tears, as Marianne perceives, belongs to the distinctive mode of communication of the romance; it immediately fits its subject into a tradition whose central theme is and always has been love. It talks the language of the heart without needing the support of words. Moreover, in seeming to call for consolation, tears permit an intimacy which verbal language may take long to establish. Suggesting vulnerability, they elicit the observer's protectiveness or demonstrativeness. This is apparent in the scene just described by Marianne, where the sight of her weeping prompts Valville first to sink to his knees, then to take her hand and cover it with kisses in a manner which is legitimized by Marianne's lachrymose condition. As she says, 'dans ma consternation, je semblais lui abandonner [ma] main avec décence, et comme à un homme dont le bon cœur, et non pas l'amour, obtenait de moi cette nonchalance-là.'[158]

The convenient idiom of tears is underlined by Valmont in Letter XXIII of *Les Liaisons dangereuses*, but in terms which leave their origin uncertain. Uncertain, that is to say, with respect to him; in Mme de Tourvel they clearly betoken a state of desperation in the face of erotic pressure, as her exclamation 'Ah! malheureuse!' before the onset of sobbing reveals.[159] Immediately prior to their joint collapse the Présidente has been describing to Mme de Rosemonde and the village priest Valmont's philanthropic gesture towards the peasant family. Her 'enthousiasme', he writes, which suggested that she was delivering the panegyric of a saint, was accompanied by tell-tale signs that promised much to the eager lover—'son regard animé, son geste devenu plus libre, et surtout ce son de voix qui, par son altération déjà sensible, trahissait l'émotion de son âme'.[160] The game of look and gesture is continued throughout the novel, sometimes signalling the resistance of one party to another's advances, and sometimes revealing their complicity.

The inadvertent betraying of love may occur through or in spite of silence. Mme de Merteuil knows that the lover's requited state is conveyed by involuntary signs which are virtually impossible to conceal. Mme de Lursay acknowledges this truth too, and points out its dangerous implications to Meilcour:

Je sais que vous allez me promettre toute la circonspection possible; je suis même certaine que vous vous en croyez capable; mais moins vous êtes accoutumé à aimer, moins vous aimeriez d'une façon convenable. Jamais vous ne sauriez contraindre ni vos yeux, ni vos discours; ou par votre

---

[158] p. 81.   [159] p. 51.   [160] p. 50.

contrainte même, trop avant poussée, et jamais ménagée avec art, vous feriez connaître tout ce que vous voudriez cacher.[161]

Mme de Lursay's own emotional state is habitually expressed by a languorous look, at once touching and sweet, or by the sparkling fire of her eyes.

But it is also signalled by sighs, a mark of female emotion which is a veritable leitmotiv of *Les Egarements du cœur et de l'esprit*. Its connotations vary, however, according to the character of the woman in question. In Mme de Lursay it always at first indicates readiness to receive the male's advances, although its meaning is sometimes missed by the naïve young Meilcour. On such occasions it may be transmuted into the sigh of impatience or irritation. In Hortense its significance is less easily interpretable. As reported by the woman whom Meilcour hears addressing the girl in the Tuileries, it seems to imply a state of regret and frustration at being removed to the country away from the unspecified male object of her affections. After Hortense's meeting with the man, sighing was reportedly accompanied by those other tokens of womanly susceptibility, a 'langueur douce et tendre' and an unsettled air, together with the tell-tale blush that now suffuses her cheeks as her companion speaks.[162]

The association of female weakness and the womanly sigh is explored by Prévost in *Manon Lescaut*, but to initially cheapening effect. Announcing her departure to become the kept woman of the old voluptuary M. de G . . . M . . . , Manon writes to the stunned des Grieux that she has temporarily left the latter in order to augment their joint fortune. Tenderness, she tells him, is incompatible with hunger, and their present state threatens them with destitution. 'La faim me causerait quelque méprise fatale; je rendrais quelque jour le dernier soupir, en croyant en pousser un d'amour.' Des Grieux is fully justified in his reaction: 'Quelle grossièreté de sentiment! et que c'est répondre mal à ma délicatesse!'[163] At the end of the novel, however, the sigh regains all its dignity. Describing Manon's last moments in the American desert, her lover details her tokens of love, first in tending his wounds, and then in seeming careless of her own comfort. His sighs as he tries to warm the chill hands of the woman who is now 'cette amante incomparable', and then her own as she weakly intimates that her death is nigh, are proof both of their deep mutual love and of Manon's extreme sensibility.

Speaking to women may call for the same restraint as governs their own utterances. Everything depends on circumstances and the

[161] p. 26. [162] p. 50. [163] pp. 69–70.

prevailing conventions. Scientific and technical discourse, as the dialoguists of *Le Rêve de d'Alembert* realize, legitimizes the use of language which the conversation of lovers prohibits. 'Je crois que vous dites des ordures à Mlle de Lespinasse,' d'Alembert accuses Dr Bordeu as he briefly awakens. 'Quand on parle de science', is the rejoinder, 'il faut se servir de mots techniques.' D'Alembert concedes the justice of the argument. 'Vous avez raison; alors ils perdent le cortège d'idées accessoires qui les rendraient malhonnêtes.'[164] And in the subsequent discussion Mlle de Lespinasse declares herself ready to listen unflinchingly to the riskiest topics, asking only for 'de la gaze, un peu de gaze' to spare her womanly blushes.[165] Later in the century Mercier echoed the principle to which Bordeu appeals, remarking in *L'An 2440* that 'il n'y a point de mots réputés bas. Car si les mots ne sont autre chose que les signes représentatifs de nos idées, dès que les idées sont nécessaires, l'expression devient nécessaire.'[166]

A potential objection to verbal explicitness in discussing matters related to love, however, is that such declaration solidifies what may be quintessentially a non-material experience. Such is the view proposed by Mme de Genlis in her work *De l'influence des femmes sur la littérature française*. Writing by women, she contends, is characterized by delicacy and sentiment, because education and social ideas of propriety oblige women to contain, concentrate, and soften the expression of nearly all the feelings they can describe. But this, she argues, does not force them into dissimulation. The art of women lies, generally speaking, not in concealing what is felt, but in revealing it by non-explicit means: nothing must be said that constitutes a positive avowal. Mme de Genlis extends this notion to embrace the nature of love itself as well as its description by female writers:

l'amour surtout rend cette délicatesse ingénieuse; il donne alors aux femmes un langage touchant et mystérieux, qui a quelque chose de céleste, car il n'est fait que pour le cœur et l'imagination; les paroles articulées ne sont rien, le sens secret est tout, et ne peut être bien compris que par l'âme à laquelle il s'adresse.

The rhetoric of desire employed by libertine writers is to be prohibited because of its crass overstatement, which renders base an experience that is imaginative as well as physically actual:

---

[164] p. 329.   [165] p. 374.
[166] Quoted in Ferdinand Brunot, *Histoire de la langue française des origines à nos jours*, new ed., 13 vols (Paris, 1966–72), vi, part 2, 1, 1206.

quel mauvais goût il faut avoir pour dévoiler tout ce mystère, pour anéantir toutes ces grâces, en présenter dans un roman ou dans un ouvrage dramatique une héroïne sans pudeur, s'exprimant avec tout l'emportement de l'amant le plus impétueux.[167]

In seeking to give women energy, in short, male writers have divested them of dignity. The use of unduly referential language is inevitably cheapening.

But indecency, it might be argued, is more in the mind of the reader or listener than in the fabric of words themselves. Given the conventional nature of language, indeed, it could hardly be otherwise. Diderot asks in *Jacques le fataliste*,

Que vous a fait l'action génitale, si naturelle, si nécessaire, et si juste, pour en exclure le signe de vos entretiens, et pour imaginer que votre bouche, vos yeux et vos oreilles en seraient souillés?[168]

In a climate of hypocritical word-shame, he implies in *Les Bijoux indiscrets*, it is just that organs not usually connected with speech should pronounce what the mouth is afraid to utter; so the tattling jewels of women reveal their owners' most intimate secrets.

Women, Toussaint observes in *Les Mœurs*, 'ont l'intelligence aisée et l'oreille délicate . . . ce serait leur faire injure que de s'exprimer devant elles avec trop de clarté; leur imagination, dit un écrivain moderne, aime à se promener à l'ombre.'[169] Some may find such delicacy reprehensible. Dr Bordeu seems to draw that conclusion from Mlle de Lespinasse's fright at the mention of certain subjects in the third dialogue of *Le Rêve de d'Alembert*: having displayed masculine fortitude and intellectual fearlessness for several minutes, 'voilà que vous reprenez votre cornette et vos cotillons et que vous redevenez femme.'[170] None the less, even the Diderot who speaks out so strongly against word-shame in *Jacques le fataliste* shows conventional discretion at periodic intervals in the novel. Thus the sexual climax reached by the landlord's wife in one of the inns Jacques stays at is conveyed by modestly expressive 'points de suspension' and the vague reference to a 'je ne sais quoi';[171] the successive orgasms Jacques enjoys with Justine are indicated merely by a 'et tout alla fort bien.—Et puis très bien encore.—Et puis encore très

---

[167] *De l'influence des femmes sur la littérature française* (Paris, 1811), pp. xii–xiv.
[168] p. 715.   [169] *Les Mœurs* (Amsterdam, 1762), p. 165.
[170] p. 380.   [171] p. 512.

très bien?—C'est précisément comme si vous y aviez été';[172] and Dame Marguerite's pleasure is transmitted simply by her 'je rêve . . . je rêve . . . je rêve', accompanied by the detailing of her trembling, closing her eyes, and opening her mouth.[173]

This sparing of the female sensibility is necessarily bound up with the notion of woman's own reticence, whether socially imposed or inborn. Jacob, who saves Mlle Habert many blushes, knows how to interpret her modest intimations of desire. The new Mme de la Vallée, 'dans l'impatience de me voir à son aise' on their wedding-night, repeatedly takes her watch from her pocket and tells their guests the hour.[174] Her 'Couchons-nous, mon fils, il est tard' of course signifies, as Jacob helpfully tells us, 'Couche-toi, parce que je t'aime.' Her proposal to him—for such it effectively is—would have had the appearance of a business transaction but for Jacob's gallantry in covering over the directness of her advances. '[Je] pense qu'à présent tu vois bien de quoi il s'agit,' she tells him, and Jacob pretends that she has made her intentions far from obvious:

il me paraît que je vois quelque chose; mais l'apprehension de m'abuser me rend la vue trouble, et les choses que je vois me confondent à cause de mon petit mérite. Est-ce qu'il se pourrait, Dieu me pardonne, que ma personne ne serait pas déplaisante à la vôtre?[175]

In Jacob we also see how using the language of love results in the speaker as well as the listener being persuaded: 'Je fis si bien que j'en fus la dupe moi-même, et je n'eus plus qu'à me laisser aller sans m'embarrasser de rien ajouter à ce que je sentais.'[176] We may be reminded of Valmont's remark to Mme de Merteuil that 'femme qui consent à parler de l'amour finit bientôt par en prendre.'[177]

Periphrasis like Jacob's is common coin in the discourse of love. It is, to some readers' retrospective surprise, a marked feature of *Les Liaisons dangereuses*, the novel which shocked so many sensibilities. The world Laclos describes in 1782 is still, like the world Crébillon depicts in *Les Egarements du cœur et de l'esprit*, a world of muted statement. The 'empressements' Meilcour discerns in Mme de Lursay, the 'mouvements' he experiences in her presence, the very 'égarements' into which he slips at the end of the novel, are matched by the 'hommages' of *Les Liaisons dangereuses*, the Présidente's veiled

---

[172] pp. 696–7.   [173] p. 706.   [174] p. 175.
[175] p. 97.   [176] p. 96.   [177] p. 150.

admission to Mme de Rosemonde ('Pourquoi serait-il [Valmont] devenu plus tendre, plus empressé, depuis qu'il n'a plus rien à obtenir?')[178] and the imprecision of the 'bonheur' at which the roué aims. The language of *politesse*, at half a century's distance, prescribes the same guardedness, the same recourse to euphemism. There is no such detailing of sexual parts, their appearance, and their function as we find in Mirabeau's *Le Rideau levé* or Sade's *La Philosophie dans le boudoir*. When Mme de Merteuil sarcastically refers to Valmont's having 'had' women, not 'raped' them, the verbs 'avoir' and 'violer'[179] shock by their unwonted explicitness, as the Marquise intended them to. Far commoner in this artificial society is the resort to *double entendre*, especially by those who enjoy the intellectual exercise of verbal deception. The suggestion that personal relations are being pursued at a level above the physical may, of course, be thoroughly misleading; but it may, as we suspect in the case of Valmont and the Présidente, convey some truth that is not easily expressible in words.

In the end, the fact of referring to love indirectly perhaps has less to do with the requirements of modesty and propriety than with the idea that to materialize the emotions is, not indecent of shocking, but false. The point of vagueness may be its ability to reflect the nature of love more truly than the most precise language can do. It suggests something whose essence is to be elusive and insubstantial, and which has to do with the spirit as well as the empirical self. One of the oddities of eighteenth-century culture is that the age so often called an age of sensibility, with all that the word implies of externalized feeling and declamatory sentiment, was also one that recognized the inadequacy of language as an expressive instrument. The wordy Diderot of *Le Fils naturel* was also the man who, in the accompanying theoretical dialogue, emphasized the need to suppress speech on occasion in the interests of greater expressive clarity. Unlike performed drama, the novel has only words with which to work. But the evidence provided by its eighteenth-century practitioners suggests, at least with respect to the descriptions of love and the language of emotion, some scepticism as to the efficacy of verbal evocation. Its power to imply (rather than be suggestive), to convey imaginatively (rather than portray directly), perhaps epitomizes the distinctive quality of literature as an art-form.[180]

---

[178] p. 308.   [179] p. 28.

[180] See E. Allen McCormick, '*Poema pictura loquens*: Literary Pictorialism and the Psychology of Landscape', *Comparative Literature Studies*, 13 (1976); Gita May, 'Diderot and Burke: A Study in Aesthetic Affinity', *PMLA*, 75 (1960).

7

Love and Friendship

TO judge by the extreme tones in which it was commonly professed, eighteenth-century friendship was not far removed from the passion of love. The fact that the verb 'aimer' unites what the English language separates into two different concepts reflects this coalescence in the French tradition, although the latter distinguishes clearly enough between an 'ami' and an 'amant'.[1] On the other hand, it is well known that the sentimental tenor of the age encouraged an intensity of protestation which actual feeling may have belied: the man or woman of sensibility was not above simulating genuine affection with the rhetoric of 'amour-passion'. This should not mislead, however. Love and amity were seen as sharply distinct from each other by many writers of the time, whether they regarded that difference as principally one of degree or saw it as having mainly to do with the sex of the objects concerned. In literature we find 'amitié' used as a bargaining-counter by those who want to deflect 'amour', or, conversely, as a prerequisite of love (if not a quality transmutable into it). Friendship can appear a haven from sexual threat, a dangerous preliminary to it, or a happy state of subdued and possibly unacknowledged sexuality.

Eighteenth-century novels portray all kinds of friendship, and their characters discuss the topic exhaustively and sometimes exhaustingly. Conscious of their status as creatures of sentiment, they examine the moral connotations of amity, contrasting the temperate impulse of mere friendliness with the serious and practical virtue of benevolence and its close relative, philanthropy. They express the desire to do well by others for reasons of shared humanity, and debate the worth of friendship within and between the sexes (despite occasional doubts about the wisdom of its cultivation between man and woman). They introduce the concept of rationality into their deliberations, sometimes declaring friendship to be based on the reasoned apprehension of qualities in a particular person, and sometimes denying that mind

---

[1] But cf. 'petite amie'.

has any legitimate part to play in the matter of affection. They pursue the notion that regard for their own selves may be an indispensable preliminary to regard for others, or, conversely, consider the possibility that the ethical value of friendship may be compromised by its origin in self-concern rather than disinterestedness.

None of the foregoing should be taken to imply that the eighteenth century had anything very new to say about friendship, for most of the observations just noted echo discussions familiar from the ancient philosophers. The same potential ambiguity as inheres in the French 'aimer' is found in the Greek *phileîn*, translatable as either 'to love' or 'to like'.[2] In his *Rhetoric* Aristotle glosses *phileîn* as to want good things for others rather than for oneself,[3] and in the *Nicomachean Ethics* he calls the *phílos* a person who entertains such an altruistic desire. This person and this desire represent special cases of loving, for *phileîn* and *philía* may also be applied to affections like that of a mother for her child: they denote something stronger, in other words, than what is conventionally meant by friendship. In the *Nicomachean Ethics* Aristotle differentiates between self-interested *philía* and *philía* directed at the good of another, arguing that people whose affection for their fellows is based on selfish considerations of personal advantage are morally inferior to people whose principal concern is for their friend.[4] If the latter ceases to be useful to the *phílos*, he is no longer loved.[5] A prime example in the eighteenth-century French novel of a person who cultivates friendship according to the demands of expediency is Prévost's des Grieux, who appeals to Tiberge as a supplier of help when needed, and otherwise shows little consideration for him. Yet of Tiberge des Grieux tells Renoncour at the start of his narrative that he showed 'un zèle et une générosité en amitié qui surpass[aient] les plus célèbres examples de l'antiquité'.[6]

The notion Aristotle develops in the *Nicomachean Ethics*, that friendship in the complete sense is impossible with more than a handful of people, because such friendship needs experience and a degree of familiarity with them, is reflected in what *La Nouvelle Héloïse* later describes among 'civilized' and 'natural' people respectively.

---

[2] See Gregory Vlastos, 'Plato: The Individual as an Object of Love' in *Philosophy through its Past*, ed. Ted Honderich (Harmondsworth, 1984), p. 18.

[3] 1380^{b}35–1381^{a}1.   [4] 1156^{a}1.

[5] See also Seneca, *Epistulae morales*, ix. 9: fair-weather friends are those who are chosen for the sake of their utility, and are discarded when they cease to be needed.

[6] p. 18.

The former, epitomized by the Parisians Saint-Preux writes about to Julie, play with the language of 'amitié' without troubling to develop the relationship seriously; the latter, in contrast, cultivate it for a small number of other beings, consorting with them constantly and putting their friends' interests above their own. The idea that love, as opposed to friendship, can be felt only for a single individual is present in both Aristotle[7] and Rousseau. Just as in real life Rousseau experienced complete love for one person alone, Sophie d'Houdetot, so in *La Nouvelle Héloïse* Julie and Saint-Preux's mutual passion is unique, the consuming emotion of their lives. In both writers, too, we encounter the assertion that the good man may love himself without detriment to the moral worth of his love for others. This is the 'amour de soi' which Rousseau opposes in *Emile* to 'amour-propre'.[8] To learn the value of one's own being is to be enlightened about that of one's fellows, and necessarily precedes the latter state. It has nothing to do with the egotistical self-concern with Aristotle discerns in the wicked.[9]

For Cicero likewise, true friendship must be marked by goodwill towards others; it is an inferior variety that springs from personal weakness or a desire to secure some profit for oneself.[10] Expediency, he argues in *De amicitia*, is no real basis for amity. Not material gain, but a friend's affection, should constitute the delight of friendship.[11] This does not mean that friends should never need anything of each other, but the bestowing of benefits ought to attend friendship rather than be its *raison d'être*. If we seek tactical advantage in place of disinterested pleasure in the relationship, similarly, we inevitably demean it—a thought which Laclos surely intended to strike the readers of *Les Liaisons dangereuses* as they watched Mme de Merteuil and Valmont cultivate the affections of the young Cécile and Danceny for their own disreputable ends.

This particular novel may suggest to us as well that friendship can be admixed with erotic love. Something of the kind is hinted at in the relationship between the Marquise and Cécile. The primary sense of amity, however, appears to preclude sensual affection. When Rousseau wrote in the *Confessions* that he was born for friendship,[12] he was deliberately distinguishing it from the type of relationship in which tender and non-aggressive affection was subordinated to

---

[7] 1158ᵃ1.   [8] See Williams, op. cit., ch. 2.   [9] 1168ᵇ1–1169ᵃ1.
[10] *De amicitia*, v. 19–20.   [11] xiv. 51.   [12] p. 104.

developed physical sexuality. The prospect of losing his virginity, even to the Mme de Warens whose nature he knew to be sexually undemanding, filled him with alarm; the act itself was followed by a kind of sadness. He retained his 'pucelage'[13] until a comparatively late age, he notes in the same work, and was never able completely to love a woman with whom he had had intercourse. His own needs were gratified by less overtly sexual behaviour: exposing his bottom to female onlookers in his boyhood, sitting at the feet of the charming Italian Signora Basile and pressing her fingers against his lips, and, especially, being spanked by Mlle Lambercier. For Rousseau, contact of this kind constituted an emotional state which was superior to ordinary friendship; it was at once more voluptuous and more tender, and he doubted whether it could ever be felt between people of the same sex.[14] He needed an 'amie' instead of an 'ami', but a 'société intime' which could not be procured by 'la plus étroite union des corps'.[15]

It is hardly surprising, given such evidence, that Saint-Preux should make the exalted claims he does for a non-carnal relationship with Julie. Even in the days preceding her marriage, when he could plausibly have regarded his erotic feelings for her as legitimate, Saint-Preux was at pains to make it clear that Julie's physical self was not her principal attraction for him.[16] The alleged absence of a primarily sexual compulsion from their relationship made it an 'amitié amoureuse' rather than a mainly erotic 'amour'. Yet their mutual feelings were unquestionably more deeply rooted in the erotic than Saint-Preux and Julie were always ready to admit; and this is apparently why Wolmar's efforts to pasteurize their affection in the clean environment of Clarens could not succeed. But an earlier novel, Mme de Tencin's *Les Malheurs de l'amour*, had shown a lover successfully transforming himself into a friend, tasting with his former mistress (since become a nun) 'les charmes de la plus tendre et de la plus solide amitié'.[17]

Uncertainty—whether assumed or genuine—about the exact nature of their feelings for another is frequently professed by characters in the eighteenth-century novel. In the absence of settled

---

[13] Rousseau distinguishes in the *Confessions* between 'pucelage' (technical virginity) and 'virginité' (the state of one who has never desired carnal knowledge) (p. 108). [14] *Confessions*, p. 104. [15] p. 414. [16] p. 32. [17] See Paul Kluckhohn, *Die Auffassung der Liebe in der Literatur des 18. Jahrhunderts und in der deutschen Romantik* (Halle, 1922), p. 52.

divisions between the states of love and friendship, this is not unexpected. Language can be a useful means of disguising the true nature of a relationship, suggesting an innocence which actions seem to belie. The mention of 'amitié', in such cases, may be part of a desperate attempt to refute the damaging imputation of 'amour'. Jacob's Geneviève has recourse to this expedient early on in *Le Paysan parvenu*, faced with Jacob's well-founded suspicions about her relationship with her employer:

C'est mon maître, il a de l'amitié pour moi; car amitié ou amour, c'est la même chose, de la manière dont j'y réponds; il est riche: eh! pardi, c'est comme si ma maîtresse voulait me donner quelque chose, et que je ne voulusse pas.[18]

As befits one who declares himself repelled by such reasoning ('on ne refuse pas ce qu'une maîtresse vous donne, et dès que monsieur ressemble à une maîtresse, que son amour n'est que de l'amitié, voilà qui est bien'), Jacob is more circumspect in his own professions. While Mme de Ferval prolongs their first encounter by searching for a suitable quill, she interrogates him about his feelings for Mlle Habert. If the latter unquestionably 'aime d'amour', Jacob's own emotions may be more vaguely defined; for while it is perfectly natural for a spinster in her fifties to feel passion for a man over thirty years her junior, there is no necessity for reciprocation. Mme de Ferval's teasing 'mais a-t-elle quelques charmes à vos yeux, tout âgée qu'elle est?'[19] inevitably calls forth a degree of prevarication in him. However fixed his affections, nevertheless '[en] fait d'amour, tout engagé qu'on est déjà, la vanité de plaire ailleurs vous rend l'âme si infidèle, et vous donne en pareille occasion de si lâches complaisances!' So he does not declare to Mme de Ferval that he loves Mlle Habert with all his heart, because the former did not wish to hear it, and because he was flattered that she did not.

Jacob, a man of robust sexuality, would have had no truck with the Platonist beliefs professed in *Les Egarements du cœur et de l'esprit* by Mme de Lursay, for whom the ladder of love necessarily rested on a foundation of non-carnal friendship. According to her theory, to attempt to scale the ladder before testing the firmness of its support was both foolhardy and unseemly (the latter opinion, in the malicious

[18] p. 36.   [19] p. 131.

view of Versac, being a rather recent acquisition of hers). Crébillon's contemporary Marivaux, a supreme psychologist of the feminine, well understood the need felt by women to test love by affecting friendship. The female leads of his drama often display this compulsion, particularly the Silvias of *La Double Inconstance* and *Le Jeu de l'amour et du hasard* (although the Dorante of the latter play shows the same urge in the male). So, in a rather different fashion, does the heroine of *La Vie de Marianne*, whose attitude is more ambiguous than that of her dramatic sisters. In the face of Climal's advances Marianne reflects that he probably loves her as a lover does his mistress,[20] for she has seen lovers in her village, heard love spoken of, even read a few novels. All this, together with the lessons provided by nature, had made her dimly aware that a lover was very different from a friend. But Climal's status has been fixed in her eyes by his initial acts of charity towards her, and that should preclude his adoption of the lover's role:

je ne sache point de manière de connaître les gens qui éloigne tant de les aimer de ce qu'on appelle amour: il n'y a plus de sentiment tendre à demander à une personne qui n'a fait connaissance avec vous que dans ce goût-là.[21]

The humiliation suffered by the beneficiary closes his or her heart to tender feeling, and self-respect gains the upper hand. If a benefactor really knew how to oblige those to whom he did good, Marianne reflects, they would do anything for him in return, for there is nothing sweeter than a sense of gratitude when self-esteem is not compromised by it. But since humans are as they are, two virtues are required in beneficiaries, one to prevent them from being indignant at receiving charity, and the other to make them grateful for it. Marianne's musings here have an Aristotelian ring, for in the *Nicomachean Ethics* Aristotle observes that benefactors love their beneficiaries more than the latter love them, probably because the first are in the position of creditors, and the second, that of debtors.[22] Debtors would prefer their creditors not to exist, whereas people who do good want the object of their benefaction to exist in order to receive his thanks. In *La Vie de Marianne* Climal clearly hopes that the latter will come about. Had he been a better philosopher or psychologist, he would probably have

[20] p. 37.   [21] p. 38.   [22] 1167[b]1.

been less surprised than he is to find his calculation of reward mistaken.

As this and subsequent episodes in the book reveal, Marianne's sensibilities are rather cruder than she would wish them to appear. Like Climal himself, if for a different purpose, she plays the hypocrite's game to fullest advantage. In common with her benefactor, she professes innocence for the sake of material benefit, he hoping for her virtue and she for more of the gifts that flatter her womanly vanity. She toys with a male's attention, as Mme de Lursay does, for ends more concrete than she is prepared to admit.

Other heroines of the eighteenth-century novel are different, or seem to be. The Comtesse d'Olnange of *L'Abailard supposé*, for example, believes that the physical marks of a man's affections are precisely what she does not want; and she feels able to admit love rather than liking for Rosebelle only when persuaded that he cannot give the ultimate expression to his passion for her. His *soi-disant* castration means that, as she says, 'mon âme peut s'abandonner à une volupté pure, rassurée par la certitude que la vôtre n'appartient qu'au seul désir d'aimer.'[23]

If this sounds suspiciously like the Platonism of Mme de Lursay, its origins are quite other. While Mme de Lursay veils desire for reasons of propriety, Mme de Beauharnais's heroine abjures mature sexual relationships on account of her unhappy first marriage. Other women in eighteenth-century fiction seem readier to renounce the satisfactions offered by a male lover—husband or otherwise—for the sake of the greater fulfilment available elsewhere. For creatures like the Comtesse d'Olnange or the heroine of Mme de Tencin's *Les Malheurs de l'amour*, renunciation of the flesh is apparently complete, even though offset by other forms of gratification such as the widowed Claire derives from her friendship with Julie. But as we discover, the Comtesse is no more proof against the urges of sexual love than most of her novelistic contemporaries. It generally takes an altogether decisive event, like the true castration of the twelfth-century Abélard, to bring about total withdrawal from the world of the flesh; most women in the eighteenth-century novel, including those modelled on the historical Héloïse, show less fortitude. Rousseau's Julie is no exception, however loath she appears

[23] p. 107.

to admit the fact. Incapable of the direct admission that would destroy for ever the reputation she has acquired for exceptional virtue, she reserves confession for a posthumous letter. Whatever earthly damage may result from ending the illusion (the illusion that love has permanently been converted into friendship), Julie is not there to suffer its consequences.

The trials of 'amour' and 'amitié' are movingly portrayed in Rousseau's novel, if in terms which the modern reader is likely to find overstated. Although Rousseau's exhibitionism and that indulged in by his characters are of their age, other eighteenth-century works of fiction give subtler expression to the torments attendant on converting friendship into love, or love into friendship. So clinical is Wolmar's plan for effecting the latter that the frustration of his aim is a cause of some satisfaction for most readers. Valmont's scheme, later in the century, for carrying out the reverse procedure— persuading the Présidente to admit that what she feels for him is the intensity of passion, not the controlled warmth of amity—becomes touching only when it is tempered by real emotion on his part. Plots and subterfuges directed at the alchemical mutation of feeling themselves hold little appeal for feeling hearts. Far more poignant are the sufferings endured by lovers who respect their beloved's desire not to know them fully, and disdain the plans of a Wolmar or the machinations of a Valmont.

Mme de Grafigny depicts such a lover in *Lettres d'une Péruvienne*. There is unconscious cruelty in Zilia's explaining to Déterville the real meaning she attaches to the words 'Je t'aime' which she addresses to him. In keeping with her race's principle of never lying, she tells the man who loves her that the emotion she retains for her fiancé, the Inca emperor Aza, is 'amour', and that what she feels for Déterville is something less intense. When Déterville points out the many occasions he has had throughout their acquaintance to take advantage of her virtue,[24] Zilia pleads with him to retain his respect in the name of the sentiment he has awakened in her: 'Au nom de l'amitié, ne ternissez pas une générosité sans exemple . . . Ne condamnez pas en moi le même sentiment que vous ne pouvez surmonter.'[25] In the face of such resistance to his passion, there is little Déterville can do but resolve on a celibate existence, which he effectively does in becoming a Knight of Malta. But when Zilia

[24] p. 312.   [25] p. 318.

divines that Aza's feelings for her have changed, she sends an appeal to Déterville to return to help her in the name of friendship.[26] Despite her changed circumstances, however, he is not to win the bliss to which he aspires. For Zilia's love, albeit now frustrated, was exclusive, all-encompassing, and changeless: 'Le cruel Aza abandonne un bien qui lui fut cher; ses droits sur moi n'en sont pas moins sacrés: je puis guérir de ma passion, mais je n'en aurai jamais que pour lui.'[27] The only crumb of comfort for Déterville is the offer Zilia makes him of intensified friendship, a feeling strengthened by the impulses love familiarized her with:

tout ce que l'amitié inspire de sentiments est à vous, vous ne les partagerez avec personne, je vous les dois . . . Tout ce que l'amour a développé dans mon cœur de sentiments vifs et délicats tournera au profit de l'amitié.[28]

How Déterville reacts to this well-meant offer we are not informed.

In its emphasis on the capacity of passion to absorb an individual, making him or her unable to feel love for any person other than the original object of that emotion, Mme de Grafigny's novel has its place in a familiar tradition of thought about love and friendship. She anticipates Rousseau's treatment of the topic in *La Nouvelle Héloïse*, where Saint-Preux angrily meets Julie's suggestion that he should transfer his love from her to Claire with the observation that to do so would be a kind of sacrilege. Once a man has conceived a passion for Julie, he has fallen in love for life. Nothing can alter the absoluteness of such love or lessen its power to absorb the lover's being. Zilia's offer of platonic friendship to Déterville is in its way as insulting to him as Julie's proposal to Saint-Preux.

The collective which Wolmar establishes at Clarens founders for reasons anticipated by Du Puy in an early eighteenth-century treatise on friendship, the *Réflexions sur l'amitié* (1728). It is folly, Du Puy writes, to think that the sexes can live in constant proximity to each other without sustaining emotional damage. He echoes Senault's opinion that when men have become angels, but not before, they may safely form close friendships with women.[29] Ordinarily, as Déterville is to discover with Zilia, and Julie and Saint-Preux with

---

[26] p. 355.   [27] p. 361.   [28] Ibid.
[29] *Réflexions sur l'amitié* (Paris, 1728), p. 230.

each other, friendship will be translated into more intense feelings which are detrimental to the tranquillity of both parties (or, as in the case of Rousseau's hero and heroine, will only apparently be an 'amitié amoureuse'). This does not mean, Du Puy continues, that it is best for males to be cold by temperament, and remain at the level of friendship with the opposite sex simply because they are incapable of more intense relationships. What he deems truly virtuous and meritorious is to face temptation and rise above it, curb the imagination and have enough consideration for one's female friends to be repelled by the idea of abusing their virtue. Such Stoic fortitude will win men the ultimate reward of respect from the object of their affections. If, despite his best efforts to the contrary, a man is insufficiently master of himself to quell all feelings of love, and lacks the delicacy to find compensation in the pleasures of chaste friendship, then he should do the decent thing and withdraw from the relationship.[30]

For Du Puy's contemporary Mme de Lambert, the very difficulty of sustaining such a friendship is what gives it its peculiar charm and worth. Much virtue and restraint are required from both parties; women who know only the conventional sort of love are unfitted for it, as are men whose opinion of women's intellectual and spiritual qualities is damningly low. But when such unions are successfully achieved, they have much greater vivacity than does friendship between people of the same sex.[31] This notion is later confirmed in Lesbros de la Versane's *Les Caractères des femmes* of 1773, where a contrast is drawn between the unstable pleasures of love and the settled happiness of amity. The peculiar delight of friendship between a man and a woman derives from the hint of potential danger such a relationship contains:

il s'y mêle toujours un peu de ce feu que la nature a mis en nous, et qui est une étincelle d'un sentiment plus fort; mais il est si mitigé qu'on ne s'en aperçoit que pour s'aimer davantage.[32]

So, at least, runs the theory. And in the sentimental climate of the eighteenth century it is unsurprising that the successful transformation of love into friendship should sometimes be hailed as an admirable and meritorious achievement, at least where the

---

[30] pp. 240–1.   [31] *Traité de l'amitié* (n.p., n.d.), pp. 81 ff.
[32] *Caractères des femmes*, 2 vols (London, 1773), i. 99.

act of yielding to love would have had socially undesirable consequences. There is certainly something noble about many such acts of self-denial, normally performed (at least in fiction) for the sake of increasing someone else's happiness at the cost of one's own.

Ethical standards change, of course, and there is no guarantee that the self-mortification found socially admirable at one point in history will continue to be so regarded at another. But few modern readers will deny the name of virtuous to the abdication by a lover of his claim on a woman's heart which Diderot presents in his *conte Les Deux Amis de Bourbonne*. The two friends of the title, Félix and Olivier, fall in love with the same woman. Directly this fact has been realized by one of them, Félix, he withdraws from the contest for her affections, refusing to engage in love-rivalry with a man who is as dear to him as a brother. There is no suggestion by Diderot that the renunciation has been undertaken for any but the most selfless reasons.

Whether or not two human beings are erotically attracted to each other is not necessarily easy to determine, in literature or in life, and many eighteenth-century novels arouse uncertainties in their readers' minds in this respect. Richardson's Clarissa, for instance, was declared by her creator to have been, not in love, but 'in liking' with Lovelace. For the sake of example, he wrote, it was meant to be clear throughout the story that she could never have married the man who raped her, both because of his immorality and because he exploited to such dastardly ends the liking that, in part, induced her to run away with him. What is generally called love, Richardson remarks, should be designated by another term like cupidity or Paphian stimulus, to make clear how little of spiritual regard it contains.[33] Although Richardson does not say so explicitly, it is friendship that we should see as the tender emotion, not the love that so brutally asserts its rights over individual freedom.

The community at Clarens is portrayed by Rousseau as one over which the god of friendship presides; eros is kept under strict control. Even in marriage, as we have seen, erotic love is regarded as a bad thing; far better for husband and wife to show friendly and respectful attentiveness towards each other, as the Wolmars do, than to live

[33] *Clarissa*, 4 vols (London, 1962), iv. 558–9.

in each other's pocket and so prevent the serious business of life from being done. The dangers of passionate love are graphically illustrated in the novel, not merely by Julie's liaison with Saint-Preux, but also by Milord Edouard's threatened misalliance with an unsuitable Italian woman (from which he is saved by Saint-Preux), as well as by his inevitable penchant for Julie herself. The bracing climate of Clarens and its benign influence on those who, like Milord Edouard, temporarily absent themselves from it are calculated to snuff out passion and replace it by more wholesome emotional fare.

In returning to the house to become her own children's tutor, Julie tells Saint-Preux (without meaning to be malicious), he will find a substitute for 'le bonheur d'être père', which heaven in its wisdom has denied him.[34] To desire a closer bond with the woman who was once his mistress would be the height of irregularity. Indeed, so tightly knit has Wolmar's community become that all stand in a parental, filial, or fraternal relationship to one another. As Julie says to Saint-Preux,

Dans le nœud cher et sacré qui nous unira tous, nous ne serons plus entre nous que des sœurs et des frères [although Wolmar is father to them all]; vous ne serez plus votre propre ennemi ni le nôtre; les plus doux sentiments, devenus légitimes, ne seront plus dangereux; quand il ne faudra plus les étouffer, on n'aura plus à les craindre.[35]

That part of the homily over, Julie turns to what she perceives as Saint-Preux's next duty, namely the wooing and marrying of Claire. Her cousin, she reminds him, has an impeccable sense of wifely rectitude, and deserves a responsible husband. 'Elle aime comme Julie, elle doit être aimée comme elle.' If Saint-Preux feels worthy of Claire, he should speak; Julie's friendship will see to the rest. But if she, Julie, has placed exaggerated hopes in him, Saint-Preux, being a decent man, will do the proper thing and reject a happiness that might deprive Claire of her own.

The notion of marriage as a union of friends rather than passionate lovers which this novel extols contrasts with that encountered in other eighteenth-century novels. For the upper classes, as we have seen, marriage might be no more than a union of convenience, joining man and woman for the sake of social advantage and financial prestige. So, at least, Laclos describes it in the case of Cécile and Gercourt.

---

[34] p. 671.   [35] Ibid.

Mme de Tourvel's arranged marriage to the Président, if it inspires the former more with a sense of matrimonial duty than with 'amour-passion', none the less appears from the letters the absent Président reportedly sends his wife at least an affectionate relationship. And Louvet's *Faublas*, a novel of high society, shows some marriages being contracted for reasons of individual preference as well as family ambition. Unaware that the union his father proposes for him is with his beloved Sophie, Faublas objects that since the Baron de Faublas himself married for love he should allow his son to do so too. But in his acquaintance with society women, as we have seen, Faublas is struck by the reprehensible evidence of marriages settled by parents for dynastic reasons rather than the emotional gratification of their children. In Laclos's novel Mme de Volanges comes to judge that she has been an unreasonable mother in imposing the stranger Gercourt on her daughter rather than allowing her to marry the person she loves, Danceny. But she is overruled by Mme de Merteuil, who has private reasons for wanting Cécile to wed the man who had the temerity to desert the Marquise herself.

However, an ideal view of marriage as a relationship based on friendship as well as love is promoted by other novels of well-to-do life, albeit ones which show it being lived under somewhat unusual conditions. Restif's *Le Nouvel Abeilard* was written, according to its author, to inspire enthusiasm for the institution of matrimony, and to show that conjugal, paternal, and filial love cannot subsist without each other. Although the affection between the boy and girl destined to become spouses at the end of the book must not be too familiar from its origin, lest they come to feel mere sibling-love for one another, the marriage towards which the children tend is still portrayed as one based on amity. On the eve of their union, indeed, the new Abélard writes to his intended in terms underlining the moderate feeling, far removed from passion, with which she inspires him as they approach the wedding-ceremony. He informs her reassuringly that

mes sens se calment; le sentiment se concentre dans mon cœur, et je sens . . . que le respect va l'emporter sur l'amour . . . Un époux . . . ne doit pas aimer en amant . . . non, il ne le doit pas. Son amour doit être grave, circonspect, respectueux, avec dignité. Il ne doit jamais, je crois, exprimer le désir; sa femme est un autre lui-même, et la nature et la raison me crient qu'il doit respecter ce corps chaste comme le sien propre.[36]

[36] pp. 347–8.

This, we recall, is also the burden of the highly erotic intercalated story of the Prince Charming who, in order to ensure that he and his wife know each other's hearts before they become physical lovers, leaves her closeted with her soul-mate Isabelle for experiences of a different kind until she is ready to receive his marital advances in the right spirit.

The novel of deferment has a peculiarly rich eighteenth-century history. For obvious reasons, fiction based on the story of Abélard and Héloïse has a central part to play in this tradition, although a stricter interpretation of the historical material would have led its later adapters to write novels of renunciation rather than postponement. *L'Abailard supposé*, *Le Nouvel Abeilard*, and in a sense *La Nouvelle Héloïse* all climax in a sexual assertiveness which is fundamentally at odds with the tale of the real Abélard and Héloïse. In other works deferment has to do less with the need to know a partner spiritually before acquiring carnal knowledge than with learning how to cope responsibly with one's own sexual nature. This is the main theme of *Le Rideau levé* and, indirectly, of *Les Bijoux indiscrets*, where the reader has to encounter the hordes of sexually immoderate women before finally discovering the type of responsible moderation represented by Mirzoza. Like Crébillon's *Le Sopha*, Diderot's novel shows a reversal of the usual postponement theme: from multiple evidence of unchastity we move towards a final revelation of sexual purity.

Les Egarements du cœur et de l'esprit describes its hero as being taught the necessity to respect the other before consummating a physical relationship, and *Paul et Virginie* so firmly subordinates the latter to the former that it ends with Virginie's maidenhood secured for eternity. *Les Liaisons dangereuses*, or that part of it which is concerned with Valmont and the Présidente, depicts the gradual conversion of friendship into love, and its necessary accompaniment by physical union. At least, what the Présidente perceives as a gradual conversion; for Valmont it has long been clear that her feelings for him are stronger than she is prepared to admit. He knows that she is frightened by words as much as by feelings when she takes stock of her relationship with him:

Ce mot ['amour', which he has been contrasting with 'la sterile amitié'] vous intimide! et pourquoi? un attachement plus tendre, une union plus forte, une seule pensée, le même bonheur comme les mêmes peines, qu'y a-t-il donc là d'étranger à votre âme?[37]

[37] p. 180.

And later on he triumphantly proclaims to Mme de Merteuil his victory in a battle which has had to do with language as much as emotions:

Il y a quelques jours que nous sommes d'accord, Mme de Tourvel et moi, sur nos sentiments; nous ne disputons plus que sur les mots. C'était toujours, à la vérité, *son amitié* qui répondait à *mon amour*: mais ce langage de convention ne changeait pas le fond des choses.[38]

Les Liaisons dangereuses also describes another kind of fusion between love and friendship, or the former's radiative power. Cécile tries to capture this phenomenon in a letter to Sophie:

Je crois même que quand une fois on a de l'amour, cela se répand jusque sur l'amitié. Celle que j'ai pour toi n'a pourtant pas changé; c'est toujours comme au couvent: mais ce que je te dis, je l'éprouve avec Mme de Merteuil. Il me semble que je l'aime plus comme Danceny que comme toi, et quelquefois je voudrais qu'elle fût lui. Cela vient peut-être de ce que ce n'est pas une amitié d'enfant comme la nôtre.[39]

Wrong in its details as Cécile's diagnosis may be, the halo effect she describes is the very stuff of eighteenth-century sensibility (and probably explains why relationships which appear to a later age unconsciously homosexual were rarely perceived as such by those who enjoyed them). Intense feeling, once conceived, moves outwards, touching everything it encounters with its brilliance. Such contagion may, however, be hard to control. Spreading indiscriminately, it can lead to the kind of illusion or mistake which is a perennial theme of novels about love, and which is akin to Stendhal's 'crystallization'.

Rampant sensibility can be dangerous. It is incapable of making the fine distinctions which more reflective feeling can draw, and it is prone to gather together what is qualitatively different: hence its encouragement of errors like that unwittingly committed by Cécile, who confuses love with friendship in her relationships with Danceny and with the Marquise. Founded on excess, sensibility expends its superfluous force on objects outside its primary area of concern. Deluded by the energy it generates, it sees in its own rhythm the rhythm of the whole world. This is more or less what both Claire and Saint-Preux take Julie to task for in *La Nouvelle Héloïse*. Looking with the eyes of love, she finds love where only more temperate

[38] p. 221.   [39] p. 112.

emotion exists; and this is why she insensitively tries to engineer the pair's marriage.

Such a mistake is unlikely to be made by the rationalist, who is not duped by his feelings. Seeing clearly, he gauges emotional states at their true value. Where his vision is clouded, it may be because, like Valmont and Mme de Merteuil, he cannot exclude all traces of emotion from his inner life. Uncertainty may then arise as to the real nature of the relationships in which he is involved. Does Mme de Merteuil love, like, or just respect Valmont? The reader is never entirely sure, any more than he is about Valmont's feelings towards the Marquise.

In other novels marked by rational deliberation about emotions, conclusions may more readily be drawn. The middle-aged man who pays his suit to Marivaux's Marianne can give an exact analysis of the regard he feels for her. The marriage he proposes will be one of companionship rather than passionate love. No more than Wolmar is he blinded by distracting emotion: he esteems Marianne, and that is enough. Although, as Marivaux's contemporary Du Puy assures us, estimable qualities are not necessarily likeable ones,[40] this respect is presented by Marivaux as a solid base on which to build a union. Nevertheless, Marianne hesitates to answer her suitor's proposal affirmatively. Perhaps, like Julie after her, she realizes that the unimpassioned tranquillity offered by the marriage of regard may become dull. As the Présidente de Tourvel acknowledges, it is a safe kind of relationship, and to that extent reassuring; but it can leave the heart that has known extreme emotion dissatisfied.

If marriage bores and heterosexual acquaintance alarms, there is always the resort—not all would call it a refuge—of relationships with members of one's own sex. This is a fertile theme of the eighteenth-century novel, although there are wide divergencies in the nature of such friendships. Some appear Platonic, in the original rather than the dictionary definition of the word: that is, they represent a stage in what is seen as the individual's ascent towards definitive love, but one which is of the flesh rather than the *Symposium*'s spirit. Far from being disembodied and heterosexual, therefore, such relationships are characteristically carnal and homosexual, and they normally figure in the erotic-pornographic rather than the sentimental novel.

The sexually permissive father in Mirabeau's *Le Rideau levé*, despite enjoying the favours of a charming boy on one occasion, tells

---

[40] p. 51.

his daughter bluntly that men have no excuse for not preferring sexual intimacy with a member of the opposite sex to congress with their own. There are plenty of women available from whom to choose a partner; moreover, 'le chemin qu'ils [male homosexuals] recherchent n'est pas moins semé de dangers que celui qu'ils fuient dans les femmes.'[41] Perhaps as a general affirmation of his views, the eighteenth-century French novel tells us comparatively little about sexual relations between men. Where explicit acts are described, it is often with distaste, although the Sade who portrays Justine's horrified response to homosexual copulation in *Les Infortunes de la vertu* reports it far from disapprovingly in *La Philosophie dans le boudoir*. But the disgusted reaction of a Rousseau to such 'unnatural' activities, detailed on two occasions in the *Confessions*, is probably more typical of his age. There seems to be no literary reflection of the positive evaluation given to them in works like the *Symposium*, where they are seen as an essential part of the ascent towards the kind of love that really matters.

Physical attraction between women is generally viewed much more tolerantly, if not with positive approval, in novels of the period. Perhaps because of the physiologically less decisive nature of female sexual arousal, it provokes less antipathy in the (predominantly male) writers who describe it than does sexual activity between men. There is every reason to believe that Laure's father is speaking with Mirabeau's own voice when he explains to her that, while he regards a man's taste for his own sex as 'plus que bizarre',[42] the equivalent penchant in women is far from being so:

il ne me paraît pas extraordinaire, il tient même à leur essence, tout les y porte, quoiqu'il ne remplisse pas les vues générales; mais au moins il ne les distrait pas ordinairement de leur penchant pour les hommes. En effet, la contrainte presque générale où elles se trouvent, la clôture sous laquelle on les tient, les prisons dans lesquelles elles sont renfermées chez presque toutes les nations, leur présente l'idée illusoire du bonheur et du plaisir entre les bras d'une autre femme qui leur plaît; point de dangers à courir, point de jalousie à essuyer de la part des hommes, point de médisance à éprouver, une discrétion certaine, plus de beautés, de grâces, de fraîcheur et de mignardises. Que de raisons, chère enfant, pour les entraîner dans une tendre passion vis-à-vis d'une femme![43]

---

[41] p. 436.   [42] p. 435.   [43] pp. 435–6.

In some cases, it is true, a female's rehearsal of another woman's charms seems deliberately provocative, if not intended to shock the person to whom it is addressed by its downright impropriety. Mme de Merteuil's references to Cécile's physical beauty are probably meant to impress on Valmont her sexual independence from the male, as well as to carry a hint of naughtiness. In Mirabeau's *Le Rideau levé*, however, a woman's admiring observation of, and even sexual intimacy with, another female's body is presented only as a prelude to her greater enjoyment in congress with the male. On seeing her father's mistress Lucette (who is also her own governess) undressed for intercourse with him, the heroine Laure reflects to her correspondent Eugénie:

la beauté des femmes a donc un pouvoir bien singulier, un attrait bien puissant, puisqu'elle nous intéresse aussi! Oui, ma chère, elle est touchante, même pour notre sexe, par ses belles formes arrondies, le satiné et le coloris brillante d'une belle peau! Tu me l'as fait ressentir dans tes bras, et tu l'as éprouvé comme moi.[44]

In *Fanny Hill*, similarly, erotic encounters with members of her own sex are shown as helping woman to prepare for a fully realized sexual relationship with man or men. And in both works, notwithstanding the apparent liberalism of the authors, the surrogate satisfaction which the female may find with others of her kind is made to appear a second-best expedient. Mirabeau's reference to the 'illusory' happiness sought by women in homosexual relationships gives the measure of his true convictions.

Another story concerned with intense erotic friendship between women, the intercalated episode entitled 'Le Mariage à la chinoise' of Restif's *Le Nouvel Abeilard*, confirms the masculine proclivity which the author of the *Confessions* admits to in himself, namely a disposition to find the sight of women caressing one another sexually exciting. Isabelle, the bosom companion of the newly married Princess Louise, hearing that the wedded pair have deferred consummation of their union, resolves to console her friend for her willed self-deprivation with special marks of fondness. Having obtained permission to do so, she proceeds to display to the Prince, waiting frustratedly in an adjoining room,

[44] pp. 325–6.

un tableau si voluptueux qu'il fut prêt à succomber . . . Le jeune prince, témoin [des caresses] que se faisaient les deux plus belles personnes du monde, trouvait sa femme mille fois plus intéressante que si elle n'avait eu que le secours de sa propre beauté.[45]

(Apropos of this episode Restif quotes Rousseau's opinion that women instinctively know how adorable the caresses they give each other make them in the eyes of men.) After their initial fondling, the mischievous Isabelle, who is portrayed as a less delicate creature than Louise, takes it into her head to dress in the prince's clothes, telling her friend to imagine that she is really her husband and adding, 'Pour moi, au feu qui m'anime sous cet habit, je le crois déjà.' Louise defends herself against this attack, but without conviction; for Isabelle's gestures of endearment 'étaient si tendres, si touchantes, si vives! mais elles étaient pures; elles partaient de l'amitié.'[46]

The princess is then undressed by her friend, but modestly; at this spectacle the prince can hardly contain his excitement. Falling to her knees, Isabelle reverently takes off Louise's shoes, an experience Restif predictably describes with mounting enthusiasm:

Les appas qu'elle touche sont d'une perfection! le joli soulier qu'on venait de quitter conservait une empreinte si charmante . . . Une mule mignonne reçut à nu le pied des Grâces, et la céleste Louise fit quelques pas, soutenue par Isabelle, pour s'approcher du trône du repos. Elle y monta, en dérobant même les trésors que lui a prodigués la nature.[47]

In transports of erotic delight, the prince tiptoes into the bedroom and covers his wife's hands with kisses (which she believes to be those of Isabelle, installed in the adjoining bed). When she remarks that she has never felt her companion's friendship more strongly, Isabelle replies:

Prenez garde que ce ne soit de l'amour! il se glisse souvent au lieu de sa Sœur; et ce qu'on croit ne donner qu'à celle-ci, c'est à ce petit traître qu'on le donne.—Je ne m'en défends pas: au trouble de mon cœur . . . à l'émotion que me causent tes baisers . . . je crois reconnaître un sentiment . . . plus vif que celui qui nous unit . . . Ah! mon Isabelle! c'est à mon mari sans doute que je dois cette nouvelle étendue qu'a mon âme!—Oui, c'est à lui . . . Ah! si vous devez me rendre plus chère à la céleste Louise, aimable Prince! vous allez devenir le Dieu d'Isabelle.[48]

When Louise subsequently announces that she will give her husband a faithful report of all the pleasures which Isabelle's tender affection

---

[45] p. 142.   [46] p. 143.   [47] Ibid.   [48] p. 148.

affords her, she adds that the most impassioned caresses she received from him would not fill her heart with such delicious joy as she now experiences with her friend.[49]

The night passes with other such diversions, the result of which is to make Isabelle prostrate herself the following morning before her beloved companion and tell her that Louise is her goddess, and their friendship sacred:

que je suis heureuse! la divine amitié a pris un corps pour moi! Ce n'est pas la belle princesse de . . . que tu vois, heureuse Isabelle, c'est l'amitié même! Ah! chaste déesse, mon cœur sera ton temple. Jamais le parfum des plus tendres sentiments ne cessera dy brûler! Louise! divinité de mon cœur![50]

The fact that Restif should have had this story relayed to his innocent fiancée by a virtuous young man, who himself had it from his exceedingly upright father, perhaps merely confirms what Diderot's *La Religieuse* had already shown, that prurience often leads the imaginative writer to forget the principle of verisimilitude and become visibly involved in the fiction he constructs. 'Le Mariage à la chinoise', like *La Religieuse*, makes the portrayal of an incipient lesbian relationship into a matter of greater magnitude than the story's point of departure might have led the reader to expect. Both sacrifice plausibility—the likelihood of a particular narrator telling the story he or she does—in the interests of drama and an erotic *frisson*.

An obvious distinction between Diderot and Restif, however, lies in their respective attitudes towards lesbianism. Restif savours every moment of the encounter he describes, and is quite without antipathy towards his two charming girls. Diderot's novel reveals a more ambivalent authorial stance. If he does not deem the sexual love of one woman for another repellently abnormal (as Richardson does in *Pamela*, where Mrs Jewkes is made to appear thoroughly immoral in her efforts to 'educate' the innocent heroine in her own tastes), he still regards it as an unnatural and reprehensible product of an unhealthy environment. (He is much more liberal about homosexuality in *Le Rêve de d'Alembert*.) Like Suzanne herself, he seems to be erotically excited by the show of lesbian affection, but as a polemicist intent on attacking the institution of convents he paints a black picture of the sexual type which claustral life allegedly produces.

[49] p. 156.   [50] pp. 157–8.

The relativity of the canons governing appropriate behaviour between and within the sexes is illustrated by an episode in *Faublas* which turns on the degree of physical expressiveness permitted between individuals in society. The Baronne de *** sets about explaining to Faublas, whom she pretends to take for the woman he is dressed as, the meaning of the word 'advances'. These, she tells him, are forms of behaviour innocuous when directed at one's own sex, but suggestive if aimed at the opposite one. Thus, disingenuously,

toutes les caresses que je vous ai faites ce soir ne sont que des amitiés; mais si vous étiez effectivement un jeune homme déguisé, et que, le croyant, je vous eusse traité de la même manière, cela s'appellerait des avances, et des avances très fortes.[51]

To squeeze the hand of her husband (whom she encourages Faublas to flirt with) as she has been squeezing Faublas's own, or to kiss him as she kisses Faublas, would be neither proper nor 'vraisemblable'.

None of this has very grave implications in a novel which treats love and the traditional sexual roles with levity as well as occasional solemnity. But in *La Religieuse* the interpenetration of friendship and obsessional love between women is examined critically, if far from dispassionately. The potential confusion arising from gestures of affection has more far-reaching results than in Louvet's novel. Innocent as Diderot portrays her, Suzanne knows only one interpretation to give to the Mother Superior's caresses: the latter is an 'âme sensible' whose endearments are merely physical tokens of her sensibility. Early on in her stay at Sainte-Eutrope Suzanne notes the Superior's lapses from 'bienséance', but only in her general, non-private behaviour. There is nothing very compromising in the 'unseemly' lifting of her wimple to rub her skin, her way of crossing her legs, or her inability to sit still.[52] When it comes to more ambiguous motions, Suzanne still sees nothing amiss: the Superior's praise of a half-undressed nun's looks when she makes her stop flagellating herself simply demonstrates her extreme compassion and the tenderness of her feelings. And when, immediately after the same episode, Suzanne observes that '[on] est très mal avec ces femmes-là',[53] her reflection still has to do with the general rather than the particular case: things cannot

---

[51] p. 432.   [52] p. 329.   [53] p. 330.

function smoothly in an institution where there are no fixed rules governing procedure, and where one never knows where one is, or who is in charge. Despite her instinctive compassion, she remains convinced that her refusal of the woman's claims for attention is in everyone's best interest: 'c'est le mieux pour vous et pour moi, j'occupe trop de place dans votre âme, c'est autant de perdu pour Dieu à qui vous la devez tout entière.'[54] For the Superior, Suzanne's apparent altruism is too negative in origin, betraying that harshness in friendship whose necessity Rousseau's Claire bewails in a letter to Julie: focused on pain and misery, bitterness and suffering, it seems to feed distress rather than encourage positive states of mind.[55] But the Superior is unable to see that Suzanne's seeming severity (the more difficult for the girl to display in that she does not see why it is needful) springs from a sense of shared humanity. For the older woman, it appears calculated only to intensify her own affliction.

Of non-physical friendship between members of the same sex, whether male or female, eighteenth-century fiction offers us countless examples. The ancient model of Orestes and Pylades is expressly invoked in *Les Deux Amis de Bourbonne*, where the peculiarly selfless devotion of Félix and Olivier to each other is constantly in evidence. Diderot emphasizes the instinctual nature of the men's love, a love which, contrary to the theory elaborated by some eighteenth-century writers on friendship, continues unabated after one of them has married. Like others who have grown up together—Paul and Virginie, or Julie and Claire—the two friends share a sibling-love that belies their real, less direct relationship. They do benefits to each other unreflectingly, seeming to love each other, not as themselves, but better than themselves. Diderot distinguishes between the friendship of Félix and Olivier, which manifests itself spontaneously, and the reflective type (a distinction which we find in Aristotle).[56] Despite the acts of lawlessness the two men commit, which lead to the death of one of them as he saves the other from the gallows, their relationship is meant to appear admirable in its selflessness.

Few other masculine friendships in the eighteenth-century novel are depicted with such positive approval. Those which Rousseau describes in *La Nouvelle Héloïse*—Saint-Preux's with the father-figure

---

[54] p. 374.   [55] *La Nouvelle Héloïse*, p. 168.
[56] See the *Encyclopédie* article 'Péripétitien'.

Wolmar, and with the Edouard who alternately saves and is saved by Julie's lover—are, of course, intended to be seen as constructive. Wolmar's role resembles that of the older partner in Greek male friendships, but without the latter's erotic character.[57] However, many novelistic relationships between men are more ambiguous in their morality, if rarely so in the sexual sense. Restif's Gaudet d'Arras, as already noted, derives a perverse pleasure from making love with Laure when she has just spent the night with Edouard. But he is otherwise principally remarkable for the generally corrupting effect he has on his friend, overseeing nearly every aspect of the perversion to which the novel's title refers. His role is similar to that of Versac *vis-à-vis* Meilcour, who at the point we leave the unfinished *Egarements du cœur et de l'esprit* seems set on the path of moral turpitude to which Crébillon's preface alludes. Much later in the century, Laclos will show Valmont using Danceny as the means to an end while professing sincere friendship to him.

Equally disquieting, in a different way, is the relationship between Prévost's des Grieux and Tiberge. Des Grieux needs goods which Tiberge can supply, but accepts the homilies which accompany the provision of those goods with ill grace. However, he also knows how to turn the religiosity of his friend to advantage. Tiberge is presented by des Grieux as a zealot, but one whose zeal may readily be diverted to ends more secular than he supposes. Tiberge's eager concern for his friend's welfare is exploited by the younger man, who shamelessly uses him for purposes (such as freeing Manon from gaol) he does not suspect. What is more straightforwardly revealed to the pious Tiberge is des Grieux's need for money at periodic intervals during his liaison with Manon, and Tiberge's indispensable function as a reliable source of cash. This is wholly in keeping with what des Grieux tells Renoncour near the start of his narrative, that even after being subjected to his friend's deceptions and subterfuges Tiberge's ardent desire to serve him never abated.[58] As des Grieux hypocritically declares before one of his many appeals to Tiberge's charity, nothing is more wonderful and a greater credit to virtue than

la confiance avec laquelle on s'adresse aux personnes dont on connaît parfaitement la probité. On sent qu'il n'y a point de risque à courir. Si elles

---

[57] See Kenneth Dover, *Greek Homosexuality* (London, 1978).   [58] p. 24.

ne sont pas toujours en état d'offrir du secours, on est sûr qu'on en obtiendra au moins de la bonté et de la compassion.[59]

If this were all he could realistically hope for from Tiberge, needless to say, des Grieux would not waste time appealing to him. What he expects and receives—he, who lacks the probity he counts on in his friend, and happily obliges him to take many risks on his behalf—is material support. And in allowing himself to be exploited as he recurrently is, it need hardly be said, Tiberge departs from the strict code of Christian charity set out by St Paul and other church fathers. For Paul, charity often demands austerity and refusal rather than generosity and complicity. Almsgiving is one of the forms in which it may be legitimately expressed, but almsgiving is not always appropriate to the spiritual well-being of the beneficiary. What Tiberge offers des Grieux, in contrast, is another kind of agape, that which is unconcerned with merit. On a strict interpretation of propriety, Tiberge should not be supporting des Grieux's irregular love as he regularly does unless he can regard that love as justifying all the means employed to gratify it, which seems unlikely. (He may, however, reasonably hope that his financial help will keep des Grieux from dishonesty.)[60] On the other hand, if he sees it as no business of his to save des Grieux's immortal soul, but his plain duty as a friend to support him in his hour of need, then he is right to suspend judgement and simply furnish him with what he requests.

Give or take the odd bout of lecturing, this is what Tiberge chooses to do. And the end of the novel underlines the fact that his friendship is of the humanly disinterested rather than religiously motivated kind. When he learns of des Grieux's departure to America, and arrives at Le Havre too late to give his friend the sum of money he has asked for, Tiberge spends several months searching for a suitable ship on which to sail across the Atlantic, is taken prisoner during the voyage by pirates, succeeds, with untypical deftness, in giving them the slip, and finally discovers another vessel to take him to New Orleans. None of this, we learn, was undertaken out of reformist zeal, but as a pure act of humanity. When des Grieux tells Tiberge that the latter's old lessons in piety have at last borne fruit, and that the erstwhile sinner has now returned to the ways of virtue, he gives him 'une joie à laquelle il ne s'attendait pas'.[61]

---

[59] p. 57.   [60] p. 60.   [61] p. 204.

Tiberge's altruism shows him to be more imaginative than his stolid characterization in the novel leads the reader to expect. One of the reasons why it is hard to see *Manon Lescaut* as an intendedly exemplary work is the fact that Tiberge, the representative of moral righteousness, is a less compelling character than either des Grieux or Manon. But Tiberge's regard for his friend's good is so active, his attention to his needs so human, and his response to his plight at times so urgent, that he commands our respect if not our affection. Tiberge is careful to tell des Grieux at one point that he too has felt the promptings of sensuality; but such limited identification of himself with his friend as this may permit him is apparently not a necessary precondition of his fellow-feeling. Rather than needing to know exactly what des Grieux is undergoing, it is sufficient to Tiberge to be able to reconstruct something of his experience in imagination. There is nothing self-regarding in his friendship, but rather a concern that is purely directed at the other's plight, and which asks for no return.

Rousseau said in the *Discours sur l'origine de l'inégalité entre les hommes* that pity is the first form of friendship.[62] The borderline between compassion and pity is a narrow one: they seem at times indistinguishable. Pity, however, unlike compassion, can have an undermining force, a sense, perhaps, that its object has brought his sufferings on himself. Julie's pity, on which Saint-Preux comments in the first letter of Rousseau's novel, often has this aspect; it is tainted by the inclination to pass judgement which she regularly displays. There is a purer kind of sorrow in the truly compassionate person, like the Mme de Rosemonde of *Les Liaisons dangereuses*. All she can do in the face of her young friend's suffering, she writes to Mme de Tourvel, is to sympathize, not to attempt alleviation; for the illness of heart-break is incurable. Far from demanding freedom from mortal deficiency, which often seems the urge behind Julie's pity, the compassion of a Tiberge or a Mme de Rosemonde acknowledges imperfection as an essential part of humanness. Their forgiveness, tenderness, and respect for suffering are the truest forms of friendship. Julie's righteousness, by contrast, is often redolent of self-love.

This is not to say that she relentlessly asserts her superiority to others in feeling or action. Her friendship with Claire is of the type which the eighteenth century knew as soul-friendship, and in

[62] p. 155.

connection with which it often invoked the mysterious essence of the 'je ne sais quoi'. It resembles that instinctive affinity Montaigne felt with La Boétie, whom he loved 'parce que c'était lui; parce que c'était moi'.[63] But it has to do as well with the cousins' distance from the insincerities and distractions of society, where, as Rousseau claims in the *Discours sur les sciences et les arts*, one never truly knows whom one is dealing with, or who is a friend.[64] This, he contends, leads to a loss of real amity and confidence. People now have reduced the art of pleasing others to a set of principles (one remembers here Mme de Sévigné's reaction to de Sacy's treatise on friendship: 'Je hais les règles sur l'amitié'),[65] and so removed authenticity from their relationships. But if the French are, as Voltaire described them, the 'crème fouettée' of Europe,[66] the Swiss soul is made of more solid matter. Frenchmen, according to Du Puy, profess friendship perpetually, but practise it rarely.[67] Their Swiss neighbours, according to Rousseau, talk of it less readily, but cultivate it more assiduously. Lacking what Mme de Lambert calls the greatest threat to amity, a frivolous spirit,[68] they can devote their tranquil lives to tending their affections and watching them grow. Unlike French friendships, which Saint-Preux describes as being born in a moment and dying with it,[69] theirs will last for life.

The relationship of friendship to love is pinpointed in a letter from Claire to Julie before the latter's marriage, and during Saint-Preux's enforced absence with Milord Edouard. Female friendships, she writes, are commonly of short duration; passion, petty jealousy, and envy bedevil them.[70] But their own union, she tells her cousin, could be destroyed by nothing.[71] In her *Traité sur l'amitié* Mme de Lambert declared that friendship inevitably falls victim to the stronger emotion of love;[72] yet the bond between Julie and Claire withstood all the turbulence of the former's burgeoning passion for Saint-Preux. Not even Julie's native modesty could decide her to conceal her

---

[63] *Essais*, ed. Pierre Villey, 3 vols (Paris, 1922–3), i. 242.

[64] In *Œuvres complètes*, iii. 8.

[65] See Mme de Genlis, *De l'influence des femmes sur la littérature française*, p. 252.

[66] See Joseph Texte, *Jean-Jacques Rousseau et les origines du cosmopolitanisme litteraire* (Paris, 1895), p. 102.

[67] p. 3.   [68] p. 61.   [69] p. 228.   [70] p. 205.

[71] Julie's words contrasting her and Claire's friendship with that of other females exactly follow a remark by Richardson's Lovelace on the friendship of Clarissa and Anna Howe.   [72] pp. 59–60.

weakness from Claire, the latter writes, and this trust was proof of the unbreakable bond between them.

Discussion both inside and outside fiction showed the contemporary interest in the convergence and divergence of the sexes. In the seventeenth century the Cartesian tradition had encouraged the notion that physical difference was all that separated male from female. The rigid distinction between mind and body led to the kind of feminism evidenced in Poulain de la Barre's *De l'égalité des deux sexes*, whose pronouncement that 'l'âme n'a point de sexe' was the logical consequence of Descartes's dualism. In the eighteenth century Poulain's formulation was frequently echoed by novelists describing love-relationships between men and women. Indeed, it sometimes appeared as a thinly disguised contention that woman's spiritual essence was in fact elevated above man's. Rousseau's Claire, exalting her friendship with Julie above her proposed marriage to M. d'Orbe, clearly discounts the possibility that female might need male to realize her full potential.[73] Like Clarissa's friend Anna Howe, she finds complete fulfilment in a relationship with another woman (in this case her cousin), a friendship of sentiment which the flesh has nothing to do with. In keeping with the idealistic tenor of much, though not all, of Rousseau's novel, it sets soul above body, and in proposing that the former is of a distinction in which female is not inferior to male it effectively argues that she is actually superior to him.

This feminist non-conformism is stated in the strongest possible terms, but it fails, perhaps surprisingly, to draw any counter-argument from Julie. According to the conventional view, not only did marriage remove the need for amity by concentrating woman's energies within the confines of her immediate family, but in creating her new duties as mistress of the household it made the cultivation of friends positively improper. The opinion that woman was a weaker vessel than man, likely to crack or explode if subjected to the force of passion, is reflected in the frequency with which eighteenth-century females were counselled to turn their emotional energies inwards and focus them on their families.[74] Men, on the other hand, being more diverse creatures, were permitted to enjoy as many friendships as they liked; their varied passions, according to contemporary belief, needed all the available channels through which to be expressed.[75]

[73] p. 206.
[74] See, for example, Du Puy, op. cit. 214–15.
[75] This is Lovelace's view of male friendship.

None of this accords very well with the evidence the eighteenth-century novel provides of torrid female friendships, often described in terms of such intensity that to the modern reader they may appear sexually suggestive. Clarissa's friendship with Anna Howe is one such, and a likely model for the relationship between Julie and Claire in *La Nouvelle Héloïse*. Its extreme nature prompts Clarissa's cousin Morden to reflect that friendship is too fervent a flame for maidenly sensibilities.[76] From the first pages of the novel Anna Howe proclaims that she loves her friend as never woman loved another of her sex,[77] and that their love is so strong no third party could ever come between them. Like Rousseau's cousins, these two women demonstrate the attraction of opposites, Clarissa being gentle, thoughtful, and grave, and her friend hasty, impatient, and sometimes flippant.[78] Neither she nor her beloved, Anna writes to Clarissa, would marry if their male acquaintances would leave them alone.[79]

She later informs Clarissa that the latter is the joy, stay, and prop of her life, and it is Anna's glory and pride that she was capable of so deep a love for such a pure and matchless creature.[80] By contrast, her experiences at the hands of Lovelace inspire Clarissa with such repugnance for the opposite sex that she declares everything connected with the male of the species to be dangerous. Even in her wasting illness she refuses to admit an apothecary to her presence because he is a man.[81] Her last words, as she looks at her miniature portrait of Anna, are those of impassioned affection: 'Sweet and ever-amiable friend—companion—sister—lover!'[82] And Anna's frenzied state after Clarissa's death only confirms this image of all-encompassing mutual love.[83]

The similar scene at the end of *La Nouvelle Héloïse* does nothing to support the stereotyped view of the Swiss character as stolid and unexcitable. Like the triumvirate of women friends described in Valmont's letter to Mme de Merteuil about the rake Prévan, Julie and Claire have been inseparable from one another. Are they, then, justly subject to the same suspicions as the three bosom companions, whose 'scandalous constancy' leads their suitors to complain that they have, not 'rivaux', but 'rivales' for their intended's affections?

---

[76] iv. 408. [77] i. 3. [78] i. 128. [79] i. 341.
[80] iii. 419. [81] iii. 443. [82] iv. 309.
[83] Anna's friendship is more impassioned than Clarissa's, though not stronger; the same is true of the relationship between Claire and Julie.

Surely not. The climate of sensibility permitted and encouraged public expressions of fondness between members of the same sex far more extreme than would now be considered usual. The eighteenth-century world, which imposed a comparatively strict control on the open declaration of love between man and woman, relaxed it where female friendship was concerned.

Not all such displays of affection were necessarily innocent, but where they contained an obvious sexual element it was usually presented as titillating rather than base. The 'bonne amie' generally acquired in convents by girls being educated there, and from whom, if Mme de Merteuil is to be believed, they habitually learnt the theory of love and its pleasures,[84] might also provide practical instruction; so we infer from Rosambert's description to Faublas of the 'intimate liaisons' formed in that closed environment.[85] But such seems not to have been the case with Laclos's Cécile and Sophie, since the former is apparently completely unaware of the nature of Mme de Merteuil's advances to her.

The Marquise's friendship with the young girl, like Valmont's with Danceny, and unlike Anna Howe's with Clarissa or Claire's with Julie, is impure because it has its origin in self-interest. It is a utilitarian friendship, and as such properly subject to censure irrespective of the adverse effects it may or may not have on Cécile. Theorists of friendship often distinguished between the type motivated by egoism and that prompted by altruism. Cicero's observation in *De amicitia* that in friendship zeal should be ever-present, and hesitation always absent,[86] belongs to this tradition: if our concern is for the other rather than ourselves, reflections on the expediency, from our personal point of view, of helping him will never win the upper hand against our desire to serve him well. Tiberge's readiness to succour des Grieux is of this selfless order.

Doing good to others may not always be a selfless act, however. Philanthropy can be practised, not purely for the benefit of its object, but for the gratification it affords the benefactor. Valmont performs an act of charity towards the peasant family for the sake of the good he believes it will do him in the eyes of the Présidente. It also confirms his power and superiority in his own regard, and is gratifying to him in the evidence it affords of his ability to win submissive thanks for the least benefit conferred. (Lovelace enjoys this feeling too.) And

---

[84] pp. 171–2.   [85] p. 521.   [86] xiii. 44.

quite independently of that, as he reports to Mme de Merteuil, his charity has the unexpected effect of filling him with involuntary joy:

J'ai été étonné du plaisir qu'on éprouve en faisant le bien; et je serais tenté de croire que ce que nous appelons les gens vertueux n'ont pas tant de mérite qu'on se plaît à nous le dire.[87]

Virtue, then, may be not without its admixture of pride or self-interest. Even Mme de Tourvel, like Clarissa, chides herself for the self-pride she hoped to gratify by converting a man from his rakish ways. Extreme preoccupation with his own being, by definition, prevents a person from becoming involved in relationships that may lead to friendship or love, except in so far as these promise personal satisfaction at no cost to himself in terms of attention or effort. Such a person is a Narcissus, trapped within his being and capable of admiring himself alone. Valmont is a Narcissus until his involvement with the Présidente has been fully developed, and almost immediately that stage has been reached he destroys their relationship because he has been told that it reflects adversely on his reputation. For a brief period, however, he has dropped the self-concern that had hitherto motivated his affairs with women and allowed himself to become absorbed by a preoccupation with another.

Love, as a passion more intense than friendship, can correspondingly stimulate the greater degree of narcissism. A man may like himself for being a philanthropist, a friend to mankind; but the self-obsession which can be engendered in him if he fancies himself in love is likely to be of a different order. Rousseau's withering description of Grimm's symptoms after a disappointment in love well illustrates the narcissist's fascination with his own emotions. We read in book VIII of the *Confessions* of Grimm's reaction to being rejected by Mlle Fel, which was to sink into a state of torpor in which he neither slept, spoke, moved, nor ate (apart from a few candied cherries which Rousseauu solicitously placed on his tongue). Sénac, the author of a treatise on the passions, pronounced that Grimm's illness was nothing to worry about, and after several days of lethargy the sufferer suddenly got up and continued life as before.[88] True love, Rousseau is suggesting, leads to self-forgetfulness, not the self-concern displayed by Grimm on this occasion. It is a common enough experience in love to find the other's welfare of more urgent importance than

[87] pp. 46–7.   [88] p. 370.

one's own. On the other hand, the lover is constantly brought back to himself by the other's corresponding concern, for in a relationship of reciprocated affection the loved one will perpetually remind him of his own value.

The Jacob of *Le Paysan parvenu* displays both self-interest and altruism in his relationships with women, but his self-interest is never taken to the exaggerated lengths that would brand him an egoist. Yet it is sufficiently pronounced for Marivaux's reader to entertain some doubts as to the reality of his love for the women in his life. After marrying Mlle Habert, as we have seen, he flatters the vanity of other females both because it pleases them that he should do so and because he enjoys seeing the effect of his masculine charms on them. Marivaux reserves judgement as to the depth of feeling ever aroused in Jacob by the women around him, perhaps suggesting that what is usually called love is often a blend of temperate self-regard and diplomatic concern for others.

Some of those who have suffered a disappointment in love without becoming victims (like Grimm and the Superior of Sainte-Eutrope) of their wounded self-regard or obsessional emotion may readily find solace in social living. Like her forebear, Rousseau's Héloïse exchanges erotic fulfilment with a single person for the different rewards of communal friendship. Except in her last moments, she refuses to concede that the latter's satisfactions are any less intense than the former's. The ideal society which she and Wolmar construct at Clarens exhibits in microcosm the virtues of the collectivist moral philosophy preached by the Enlightenment. The Wolmars set themselves up as lovers of mankind, publicly caring for their fellows and, wherever possible, saving their souls from the torments that threaten ordinary human existence. Admittedly, there is a single being on whom the attentions of the community are focused—Julie herself—but the homage they pay her never prevents them from answering the call of duty towards others. If Julie's homiletics sometimes smack of self-love, none of those at whom she preaches ever seems aware of it. The heightened emotions awakened in them by their worship of her apparently also serve to stimulate feelings for others in their hearts.

At Clarens the soul worn out by passion finds reassurance in friendship. Milord Edouard writes to Wolmar that at the age he has reached there is no repairing the sentimental losses sustained in the affairs of life; he has consequently decided to devote the rest of his days to the cultivation of such feelings as remain with him in the peace of Clarens (delegating important business in the Upper House to a fellow peer).[89] Saint-Preux, the bourgeois upstart in Julie d'Etanges's household, persuades the lowly Italian mistress of his noble friend to renounce her claims on him: she obligingly withdraws from the world and becomes a nun.[90] Edouard is therefore free to become a friend of the world in a new setting, one which exerts a moral influence on its inhabitants out of all proportion to its small compass. There, as Mme de Wolmar writes to Saint-Preux, 'les plus doux sentiments, devenus légitimes, ne seront plus dangereux.'[91] In such an environment, she tells him, he will remain possessor of the better half of herself, that devoted to amity, and Julie will be able freely to give herself over to all her attachment to him. What stands before them, she tells her erstwhile lover in ringing tones, is the greatest kind of triumph, an exemplary catharsis. For '[on] étouffe de grandes passions; rarement on les épure.' And Clarens will be the living testimony to their achievement: 'O ami, quelle carrière d'honneur nous avons déjà parcourue! Osons nous en glorifier pour savoir nous y maintenir, et l'achever comme nous l'avons commencée.'[92]

At Clarens the cultivation of communal friendship does not preclude self-recollection in private. Saint-Preux describes to Milord Edouard the institution by Julie and Wolmar of the 'matinée à l'anglaise', a period during which the communion of hearts takes place in tranquil and reflective silence, and where the babbling that passes for amicable intercourse in France is replaced by the wordless expression of feeling. The raptures induced by this state, Saint-Preux avers, are unknown to their French neighbours:

l'amitié, milord, l'amitié! Sentiment vif et céleste, quels discours sont dignes de toi? Quelle langue ose être ton interprète? Jamais ce qu'on dit à son ami peut-il valoir ce qu'on sent à ses côtés? Mon Dieu! qu'une main serrée, qu'un regard animé, qu'une étreinte contre la poitrine, que le soupir qui la suit, disent des choses, et que le premier mot qu'on prononce est froid après tout cela![93]

This ecstatic immobility, he tells Edouard, possesses a charm that is reserved for 'sensible' hearts, and which importunate outsiders prevent the communicants from enjoying to its fullest extent:

[89] p. 654.   [90] pp. 652–3.   [91] p. 671.
[92] p. 664.   [93] p. 558.

j'ai toujours trouvé que . . . les amis ont besoin d'être sans témoin pour pouvoir ne se rien dire qu'à leur aise. On veut être recueillis, pour ainsi dire, l'un dans l'autre: les moindres distractions sont désolantes, la moindre contrainte est insupportable.[94]

Among the activities available to humans, such contemplation is perhaps the most settled and individually self-sufficient. So the ancient philosophers held, and so the experience of the Wolmars and their friends seems to confirm. But above all, and contrary to the assumptions of the professionally friendly, it is an essentially mutual state; the feeling of another person's presence, unstated but known, is a part of the fulfilment it brings. Nothing very important can subsist, Saint-Preux became convinced, when people are denied such deep communication. Although individuals need to cultivate self-sufficiency too (as Saint-Preux, banished for years from Julie's presence, has reason to know), the good life demands social relations. Aristotle remarks in the *Nicomachean Ethics* that the self-sufficiency which characterizes the highest kind of human life is communal, not solitary.[95] The best kind of existence, as the Wolmars invite Saint-Preux the tutor and Edouard the friend to discover, is that lived for others, whether they be children, partners, or other loved ones. The fostering of such an awareness is the purpose of the 'matinée à l'anglaise'.

Other eighteenth-century novels of love appear to reach this conclusion too. For the greater part of *Manon Lescaut* the lovers exist on the margin of society, either because they refuse to live according to its conventions and are therefore outlawed, or because society is inhospitable to them. Their love (as against the concupiscence that marks Manon's liaisons with other men) seems unable to flourish in an environment lacking a firm centre. Once in the New World, however, they adopt a different mode of living, changing their old acquisitive ways—in friendship as in other respects—for new, benevolent ones. They are greeted on landing as friends and companions come to lessen the solitude of colonial life. In return they let no occasion slip of assisting their new neighbours and doing small services to them. 'Cette disposition officieuse,' des Grieux tells Renoncour, 'et la douceur de nos manières nous attirèrent la confiance et l'affection de toute la colonie.'[96] The idyll ends when eros intrudes into the preserve of amity. The Governor's nephew

---

[94] Ibid.   [95] 1097^{b}1.   [96] p. 189.

Synnelet is apprised that Manon, whom he desires, is not married to des Grieux; the two men fight for possession of her, and des Grieux flees the town with his beloved after (as he believes) killing Synnelet in their duel. The desert wastes to which they retreat, in the knowledge that they have once again put themselves outside society's law, symbolically represent the loss of their new social identity. When he has laid Manon in her grave des Grieux can only determine to return to his homeland to repair the scandal of their past dissolute existence.

The end of earthly life also signals the disintegration of a small community in *Paul et Virginie*, but for different reasons. Within a single person, Virginie, the twin currents of love and friendship were combined. She was the force that held a private world together; she inspired erotic love in one, *philía* in others, and without her a guiding impulse in their life seems to have been lost. The pomp and circumstance and general disarray that attend Virginie's passing point to one certain conclusion: that the island mourns the loss of someone who mattered. But it is not simply Mauritius that grieves, not simply the girls who praise her sanctity, the mothers her filial obedience, the young men her womanly love and fidelity, the poor her friendship, and the slaves her compassion. Virginie's death is bewailed by the inhabitants of neighbouring islands and the entire Indian subcontinent, 'tant la perte d'un objet aimable intéresse toutes les nations.'[97] All sense that in her they have lost a benefactor and a friend.

It is of little use for Saint-Pierre's modern reader to ask what, precisely, all the fuss was about. We are given few details of the goods associated with Virginie and the positive benefits she bestowed, and this is perhaps appropriate. In the ample prose of eighteenth-century sensibility, the *feeling* of what constitutes virtuous and humane behaviour matters much more than the dry recounting of it. We know equally little about the goodness of Julie and Mme de Tourvel, but— especially in the former's case—we sense it from report, or from the awed and worshipping response of others to its manifestations. The recording of benefits, indeed, generally induces a reaction either of unease or of distaste (although it may not have done in the age of sensibility). Merely to suggest their reality, by contrast, as it is the special facility of literature to suggest, opens wide and hospitable areas for the play of the imagination.

[97] p. 210.

Persons who are loved like Virginie, it might be thought, must possess some special quality which is not possessed by ordinary people. Perhaps this is true where community love is in question. But no more in love than in friendship do one person's feelings for another seem necessarily to depend on the other's being endowed with qualities standardly found admirable, like moral goodness or loving-kindness. Many humans are liked or loved in spite of characteristics or proclivities which might be thought to render them unlovable. Manon is fickle, selfish, and materialistic, and yet des Grieux cannot help himself from adoring her. The reasons for this helplessness are probably half-perceived by him: her looks clearly have much to do with it, as does the indefinable 'je ne sais quoi', and no doubt other of the love-inducing properties already discussed. But moral worth does not seem to feature among them, at least until the couple arrive in New Orleans and Manon protests the tenderness and compassion she feels for the lover on whom she has brought so much suffering. Des Grieux, equally, does not seem to merit the devotion which the faithful Tiberge shows him: he repays his friend's trust and generosity as ill as Manon has repaid his own.

Cicero believed friendship to be rational, in the sense that it could only exist among good men.[98] We make friends of people who are virtuous, he declares, although their virtue is not necessarily the prime reason for our doing so. Given this prerequisite, he continues, a law may be established governing the procedures to be observed between friends: they should neither ask dishonourable things of each other, nor do them if asked.[99] Cicero moderates this view subsequently in *De amicitia*, observing that where the wishes of a friend are not wholly honourable, but need to be promoted in the interests of his life or reputation, a person may turn aside from the straight path and answer them, provided that utter disgrace does not ensue.[100] Such a case as Tiberge's fits the latter dispensation, for his succouring of des Grieux has more to do with the furthering of his friend's intemperate desires, however inadvisably, than with keeping him on the way to rectitude. But the very fact that Cicero could so qualify his prescription indicates the unease that may generally be felt at the unconditional demand of moral excellence in those we make our friends.

Most people recognize faults in themselves and in others they are close to, and would be made uncomfortable by the assumption that

[98] v. 18.   [99] xii. 40.   [100] xvii. 61.

deviation from the moral norm was a sufficient reason for ending a relationship enjoyed, not for its utility or exemplariness, but for the intangible pleasures of community and consolation it brought. A person may be liked for one set of qualities irrespective of other, less attractive ones that go with them. Indeed, many people are so irrationally biased towards the limitation of goodness that they may find qualified imperfection more reassuring, because more human, than absolute virtue. Our response to ethical values, too, changes over time. Julie, whose absolute goodness after her initial fall is proclaimed by every character in *La Nouvelle Héloïse*, is probably now found less appealing on that score than she was in the eighteenth century, although the crucial admission contained in her posthumous letter shows her in a different, more human light. Claire, on the other hand, burdened with minor mortal imperfections as she is, presents a much more likeable aspect than her saintly cousin.[101]

The ideal image of humanity which Rousseau presents in his new Héloïse is one far removed from our ordinary affection for fellow-beings. Liking and love for people involve concern, tolerance, and sympathy, and in a balanced relationship these qualities are reciprocated. Where someone is truly a masterpiece of perfection, he must appear as the deity appears in the Hebraic and Christian traditions. Such persons, like Julie or Saint-Pierre's Virginie, can love others irrespective of merit (although Julie often seems peculiarly bothered by human imperfections); the righteous win no more love than the wicked or sinful. But they are too close to the godhead for comfort. Julie and Virginie, although admirable, are barely human; Laclos's Présidente, by contrast, displays the weakness and pathos of humankind in her fall, but is moving by virtue of her imperfection. She is a flawed being who loves another flawed being, and who thereby fits into a pattern humans can recognize.

Reason may prompt us to see the goodness of a virtuous person, but it cannot make us like or love him. There is, after all, no particular reason why virtue should exercise a more compelling force than other human qualities and properties, like beauty, wit, or gentleness. Feelings are different from conditioned responses, and the patterns which theorists establish to determine what is lovable

---

[101] Virtue shown in practice is more attractive than virtue in the abstract, which is perhaps why Clarissa is more appealing than Julie (except in the letters where Anna Howe recounts all her friend's perfections to Belford).

or likeable are seldom followed in ordinary life. One may, like Mme de Sévigné, hate rules in amity precisely because of their artificial attempt to reduce the particular to the general. The human character is far more various than the schemes of thinkers like de Sacy would suggest. This, more than anything else, explains why the friendship-republic established by Wolmar at Clarens had to fail: in its attempt to organize the responses of different humans to each other, it ignored the characteristic diversity of such reactions.

To be a lover of mankind, as men of the Enlightenment professed themselves to be, is to widen the potential theatre of one's actions in a way that appears selflessly praiseworthy, and to live by the Terentian maxim that the eighteenth-century *philosophe* applied to himself: 'homo sum, humani nihil a me alienum puto.'[102] In practice, however, such extension may result in a dilution of sentiment. As we have seen, Aristotle considered that a man could not be a friend in the complete sense to more than a few people, largely because he believed that friendship required time and effort to bring to maturity.[103] And this, in the view of Saint-Preux, was what branded French friendships as insufficient: they were struck too quickly and unreflectingly, and failed to stand the test of time.

At the end of *Manon Lescaut* des Grieux determines to return to a mode of existence which, by helping to rebuild his regard for himself, will make him once more a useful member of society. He will embrace anew the ways and values of his caste, largely forgotten during his irregular liaison with Manon, and so win back the stability that belongs with adhering to a fixed order of things. He had attempted to achieve a similar settledness with Manon in the New World, but been pulled back despite himself to a state of irregularity by another man's concupiscence. Now that Manon, the involuntary cause of the latter, has been taken away from him and the world, he is at liberty to recall 'des idées dignes de [sa] naissance et de [son] éducation', and can give himself over entirely to the 'inspirations de l'honneur'.[104] His determination to live in accordance with the decent Christian principles which Tiberge tried to reawaken in him throughout the lost days of his relationship with Manon is a

---

[102] [Dumarsais?], 'Le Philosophe', in *Nouvelles Libertés de penser*, contained in *Le Philosophe*, ed. H. Dieckmann (Washington, 1948); Michel Delor, '"Homo sum . . .": Un Vers de Térence comme devise des Lumières', *Dix-huitième Siècle*, 16 (1984).
[103] *Nicomachean Ethics*, 1158ª1.   [104] p. 202.

confirmation of that new intent, and of his desire to become, through the repair of his moral character, a helpful member of society.

We should not, all the same, exaggerate the extent to which des Grieux abnegates eros for the sake of responsible friendship and the love of God. The memory of Manon, recalled in all its force by the account he is giving Renoncour of their life together, still exerts too strong an influence to be ignored. The man who has returned from America after burying his mistress still believes that erotic love procures ultimate felicity, marred only by the fact that it is transitory.

Dieux! pourquoi nommer le monde un lieu de misères, puisqu'on peut y goûter de si charmantes délices? . . . Quelle autre félicité voudrait-on se proposer, si elles étaient de nature à durer toujours?[105]

These reflections of the des Grieux who has supposedly emerged from his experiences with Manon a sadder and wiser man are no different from the ones he gave voice to in his triumphant debate with Tiberge at Saint-Lazare. They effectively qualify the affirmations of the novel's conclusion, which themselves strike a more temperate note than had been the case in the first edition of the book.

The self-esteem des Grieux intends to attempt regaining appears contingent on the permanent removal of Manon from his life, at least were it to be lived as it had been before her deportation. The regularizing of their relationship would perhaps have changed matters; but then, conceivably, habit would have transformed passion into something altogether more prosaic. Rousseau was to describe this phenomenon in *Emile*, where, anticipating the developments in *Emile et Sophie*, Emile's tutor remarks on the virtual impossibility of preserving love within wedlock. Bonds which are too tightly drawn eventually break, he declares; constraint and love are unhappy bedfellows, for pleasure will not be commanded.[106] Making caresses and marks of affection into duties renders them odious; the only law worth respecting is that of mutual desire.

In a way, this is reminiscent of Héloïse's argument against marriage in her letter to Abélard, except in the emphasis it places on pleasure. While for Héloïse it is the moral glory of unconstrained fidelity that argues for a non-marital relationship, for Rousseau hedonistic considerations are to the fore. In the ideal marriage, he continues, hearts are united but bodies unfettered. Emile should make Sophie

---

[105] p. 66.   [106] pp. 862–3.

his perpetual mistress, then, but also permit her to remain mistress of herself. The couple should be respectful lovers, appealing never to duty, but always to what each other wants. The least favours should be granted, not to right, but to inclination, and free will should determine the dispensing of affection. 'Souvenez-vous toujours que, même dans le mariage, le plaisir n'est légitime que quand le désir est partagé.'[107] If Emile and Sophie keep these principles always in mind, uncommon happiness will be their reward.

As the continuation to *Emile* reveals, they do not do so, and their reward is misery. But although their love ends, they preserve in marriage a vestigial esteem for each other that constitutes a low-grade kind of friendship, the sort of respect they do not see in the other loveless marriages of Paris. In so doing, they retain something of the proper self-regard which misfortune brought by humans on themselves always threatens to destroy, and which Emile finally loses when Sophie's pregnancy by another man is revealed: 'je n'étais plus; c'était ma propre mort que j'avais à pleurer.'[108] Now, ironically, he thinks back to the earlier days of their disunion as to a time of comparative happiness, when concern for his wife made him heed the advice his mentor had given him, and when he should have been content

à l'estimer, la respecter, la chérir, à gémir de sa tyrannie, à vouloir la fléchir sans y parvenir jamais, à demander, implorer, supplier, dési.. :r sans cesse, et jamais ne rien obtenir. Ces temps, ces temps charmants de retour attendu, d'espérance trompeuse valaient ceux même où je la possédais.[109]

What hurts him now is not the necessity of giving her up, but the need to despise her. There remains, none the less, the small comfort he can take in Sophie's honesty towards him, her refusal to pass off her love-child as Emile's own.

The loss of self-esteem which attends Emile's discovery of Sophie's adultery can occur outside marriage as well as within it, and the individual's sense of being voluntarily untrue to himself may be at its origin. Meilcour seems to suffer such a diminution of self-respect on consummating his relationship with Mme de Lursay, when the thought of his concurrent infidelity to Hortense is a source of torment. Try as he may to convince himself that his obligations to the girl are non-existent—for he scarcely knows her— he is yet

---

[107] p. 863.   [108] p. 894.   [109] p. 895.

distraught at the notion that he is breaking faith with her. On the face of it, his duty is towards the Mme de Lursay who has gone as far as a respectable woman may in making clear her readiness to be his lover. But whatever the niceties of etiquette seem to indicate, his real obligation is towards the Hortense to whom he has barely spoken, and whose response to him thus far has hardly been encouraging. Why should this be? Presumably because the imagination is at liberty to develop what is at most incipient friendship into full-blown love, and because failure to live up to the latter's demands brings with it a sense of self-disgust as powerful in the sensitive soul as would be its response to real inconstancy.

Although Meilcour is initially able to keep the thought of his infidelity at bay, it returns to plague him as the physical pleasure he is tasting with Mme de Lursay becomes blunted. An importunate sense of guilt wells up in him despite the best efforts of his rational mind to assuage it. 'Ce fut en vain que je le tentai [to dispel his feeling of guilt], et chaque instant me rendait plus criminel sans que je m'en trouvasse plus tranquille.'[110] His loss of quietude arises from the fact that, in being untrue to Hortense, he is being untrue to himself and to the moral conscience that his social mentors have failed to destroy. A person who cannot like himself will not be happy in love, because of the insistence with which the lover's self is recalled to him by the loved one's concern. In this sense 'amour de soi' is quintessentially a part of amatory experience. That is what Diderot meant in writing to Sophie Volland of the need for the other always to bring to the lover's attention ways in which he could improve himself, not in the self-conceited way of Rousseau's Julie, but unselfishly, in the interests of mutuality. By bettering oneself, as Aristotle remarks apropos of friendship, one benefits others by making oneself a better friend to them.[111]

This is one of the principal ways in which love and friendship can combine in an erotic relationship. It seems to be perfectly possible to feel friendship towards someone whom rational consideration leads us to regard as somehow deficient, but not towards someone for whom we have no respect at all. This is perhaps because we have faith in the capacity of emotion to change people, morally or otherwise. Equally, however, it may be because, as imperfect beings ourselves, we prefer mixing with others of our kind to being reminded

[110] p. 188.   [111] 1169[a]1.

of personal shortcomings by the perfection of our acquaintances. Some kinds of superiority may be easier to bear than others. To be with a person more physically beautiful than ourselves can be simply exalting, as Claire seems to find it in the presence of Julie; Marivaux's story about the reassurance ill-endowed women contrive to find on comparing themselves to better-favoured ones, however, shows the other side of the coin. But neither of these examples has to do with developed erotic relationships. Where the latter are at issue, genuine care seems to be precluded where all feeling of moral admiration for the other is absent. A merely sexual liaison can dispense with it altogether, but is not ordinarily thought to be the equivalent of a loving union.

Although we surely do not choose our friends as rationally as Aristotle and Cicero suggest, a mental component appears, no less surely, to belong to the process. Such was not the view of Du Puy, who asserted in his *Réflexions* that friendship is of the heart, and esteem of the mind.[112] But his opinion clearly rests on the assumption, often present in the 'sensible' confusion of the eighteenth century, that feeling and reason are mutually exclusive. Much of the period's literature suggests otherwise—not so much in describing the rational marriage enjoyed by the Wolmars, and proposed to Marianne by her middle-aged suitor, as in the depiction of the brief idyll experienced by Valmont and the Présidente de Tourvel, or the attraction Meilcour feels towards Hortense. The 'regular' relationship initiated by des Grieux and Manon in the New World is perhaps a third illustration of the type, and the thwarted union between Paul and Virginie a fourth.

To say this is possibly to do no more than draw attention to the familiar distinction between bodily and spiritual pleasures, frequently discussed in the eighteenth century in terms of the 'jouissance' or 'volupté' of the flesh and that of the soul. Put differently, it is close to the perception formulated by Rousseau's Claire and others, that 'l'âme n'a point de sexe': friendship so 'feeling' that it is barely distinguishable from erotic love exists independently of gender. And yet the relationship between Valmont and the Présidente, des Grieux and Manon, Paul and Virginie, manifestly have their share of heterosexual eros. If a spiritual essence infuses all of them, so too does a physical urge. 'Pure' (non-erotic) friendship like that between

[112] p. 35.

Julie and Claire, their prototypes Clarissa and Anna Howe, Saint-Preux and Milord Edouard, Félix and Olivier, is by contrast informed by emotion free from sensation. It is probably true that such friendship—perhaps all friendship— is more stable than love, because the emotions that constitute it are less intense than those of erotic passion. Love's flame, to borrow the old metaphor, flares more brightly and is sooner extinguished; that of friendship is a steady glow which outlasts its brilliant neighbour.

But to opt for friendship in preference to love, many eighteenth-century novels suggest, is to forfeit passionate intensity for controlled comfort in a way that diminishes the soul. The desire for safety may cripple the personality, preventing it from expanding under the benign influence of extreme, life-enhancing emotion. Yet in the happiest of instances, as novels like *La Nouvelle Héloïse* and *Les Liaisons dangereuses* tentatively suggest, eros may combine with the respect and regard for another person that constitute amity. A lover, in such circumstances, will also be a friend.

Whatever the case, it appears on the evidence of the period's prose fiction that female amity comes closer to erotic love than male. If the soul is androgynous, its linguistic gender—at least in French—nevertheless points to a propensity reflected in many friendships described by novels of the age. In other words, there is considerable literary dissent from the popular contemporary notion that woman needed little in the way of friendship, and none of it at all after marriage, while the diverse passions of man made it advisable for him to cultivate the amity of many fellow-males. Particularly at a time when arranged marriages were still common, the preservation of intense female friendship filled an obvious need. Claire's position in Julie's life after the latter becomes Mme de Wolmar is an instance of such a union, and one in which, it might be speculated, she becomes the legitimate object of emotions Julie cannot allow herself to direct at Saint-Preux, and of which Wolmar would be an altogether inappropriate recipient. Erotic relations, too, are more commonly depicted between women than men in eighteenth-century literature: there is no male equivalent of *La Religieuse*, perhaps because male homosexuality aroused greater public revulsion than female on the grounds of its apparently more direct carnality. Suzanne Simonin and Queen Victoria were not alone in wondering what could possibly constitute sexual activity between women.

All in all, it might seem that the line from Terence's *Heauton-timorumenos* needs reformulating in the light of eighteenth-century fiction. One should, as a man (not *vir* but *homo*, not *Mann* but *Mensch*), love all humanity; but as a woman one enjoys a bond with one's own kind more essential in its affective rewards than anything known between biological males. This runs counter to the Greek experience of masculine friendship, and to the one enjoyed by Montaigne and La Boétie. In that limited but crucial respect, the eighteenth-century novel reveals its age to have been a feminist one.

Conclusion

FOR the majority of leisured eighteenth-century women—keen novel-readers, as well as favourite subjects for novelistic treatment—love was probably perceived as the 'grande affaire de [la] vie' (the formulation of *La Nouvelle Héloïse*). If it was not the exclusive matter of female existence, as Byron declared it to be in *Don Juan*, it was certainly more than the 'thing apart' which the same poem held it to be for men. Even for the comparatively unleisured Julie de Wolmar, busy running the Clarens estate with her husband, love unquestionably bulked large; but more in memory and anticipation than in the well-occupied present, where the persistence of her passion for Saint-Preux could not be admitted, still less indulged. Without the spice of love, and marooned on an island of fairly tranquil ease, Julie is bored. In admitting to Claire that 'le bonheur m'ennuie' she is most honest with herself than the Présidente de Tourvel. For the latter, marriage and its concomitants also represent a safe harbour where she cannot be rocked or shaken in her moorings by the storm-tossed sea of passion; but its very safety is an implicit threat, which libertines as experienced as Valmont know how to turn to advantage.

Mme de Lambert saw the adverse consequences for women of their being raised in society for little but love. Social prejudice, she wrote, damns women for being concerned with frivolities, but the 'haut monde' conspires to prevent them from developing the capacity to live as variously as men:

Il faudrait prendre partie. Si nous ne les destinons qu'à plaire, ne leur défendons pas l'usage de leurs agréments. Si vous les voulez raisonnables et spirituelles, ne les abandonnez pas quand elles n'ont que cette sorte de mérite. Mais nous leur demandons un mélange et un ménagement de ces qualités qu'il est difficile d'attraper et de réduire à une mesure juste.[1]

She quotes Saint-Evremond's dictum that it is less difficult to find the solid reason of man in woman than the attractions of woman in man.

---

[1] Op. cit. 31-2.

But where woman is, like Laclos's Mme de Merteuil, both endowed with a powerful intellect and possessed of physical charms, society seems to conspire to quash, or at least to deprecate, the former. One of the reasons for the *succès de scandale* enjoyed by *Les Liaisons dangereuses* when it was first published was doubtless the fascinated indignation aroused, in female readers at least, by the spectacle of a woman's cleverness turned to thoroughly immoral ends. People were, and probably still are, more tolerant of sexual plotting in men. Richardson's readers, a predominantly female band, could not gainsay their attraction to Lovelace, and vainly begged the author to have his anti-hero reform at the end of the novel.

Is woman redeemer or damned? Mme Riccoboni, whose own novels exalted the pure ideal of womanhood, attacked Laclos for betraying her sex in the person of the Marquise. 'C'est en qualité de femme, Monsieur,' she wrote to him, 'de Française, de patriote zélée pour l'honneur de ma nation, que j'ai senti mon cœur blessé du caractère de Mme de Merteuil.'[2] Laclos attempted to mollify her by suggesting that the genuine image of female virtue was to be found in her own *Lettres de Fanny Butlerd* and *Lettres de Juliette Catesby*.[3] But he also upheld the novelist's right to create a composite character, however repellent in her multiple evils, as Molière had doubtless combined the hypocrisies of numerous real-life models in creating Tartuffe. He did not add, as he might have done, that the horrors perpetrated by the Marquise were palliated by their contrast with the angelic sweetness of the Présidente. That observation was left to one of Laclos's other critics, Tilly, who nevertheless manages still to damn woman with faint praise:

C'est un très grand art d'avoir fair Mme de Merteuil si corrompue, puisqu'elle en contraste mieux avec cette candeur angélique de Mme de Tourvel, et que Valmont est moins méchant qu'elle; l'auteur est en règle, puisque les femmes valent mieux que nous, mais vont plus vite et plus loin dans le chemin du vice, quand elles ont commencé à y marcher.[4]

But for moralists of Mme Riccoboni's persuasion, the charge stuck: Laclos had traduced the notion that 'l'esprit n'a point de sexe' by making Mme de Merteuil a woman of too much sex and too much 'esprit'.

---

[2] Laclos, *Œuvres complètes*, p. 759.   [3] Ibid.
[4] *Mémoires*, ed. Christian Melchoir-Bonnet (Paris, 1965), p. 176.

What *Les Liaisons dangereuses* most forcefully demonstrates is the virtual impossibility of beating the male at what he has deemed to be his own game by seeming to employ only those weapons he judges proper in a woman. The result, as the Marquise declares to Valmont, is what Mme de Lambert described: twice the difficulty and half the acknowledgement. The recognition Mme de Merteuil craves cannot come from the society which has endorsed the male-orientated rules of play. Only Valmont's approval can compensate for that; but Valmont, tragically for the Marquise, withdraws co-operation from their joint enterprise. There then remains little for his old accomplice to do but abandon caution and play recklessly, as a result of which she loses what her society considers most valuable in woman, her reputation. The fact that she coincidentally contracts smallpox and loses that other social asset, beauty, is an additional irony, and a crowning one for a woman so reliant on appearance. What she has not forfeited, however, is her mind. Laclos does not tell us what she achieves with her intellect after the *débâcle* described at the end of the novel, but it is hard to believe that she achieves nothing.[5]

What we observe in the Marquise throughout the book is different only in degree from what we see in other females who figure in the eighteenth-century French novel. Mme Merteuil becomes infamous by using emotion as a means to an end; but this, to a greater or lesser extent, is what her society ordains that woman shall do. In *Les Egarements du cœur et de l'esprit*, much earlier in the century, Crébillon had shown how the male is educated (by women as well as other men) to advance in the world by a tactical deployment of emotion and knowledge of people's accessibility to it. Mme de Merteuil fights a system which proscribes such open calculation in the female. The kind of sexual opportunism in which her contemporary Faublas happily indulges is still, shortly before the Revolution that proclaims the liberty and equality of all, denied to members of her sex. She knows her mental parity with (or superiority to) Valmont; she terminates their fraternity with a chilling 'Eh bien; la guerre!'[6] because he refuses to grant her right to the same liberty as he enjoys.

---

[5] However, a footnote of Laclos's to the final letter of the novel refers to the 'sinistres événements qui ont comblé les malheurs ou achevé la punition de Mme de Merteuil' (p. 386). [6] p. 351.

There are some forms of social deception more acceptable than others, endorsed by the world either because they seem relatively harmless to those who wield real power, or because they appear spontaneous. Diderot's Suzanne Simonin, who suffers much from the institutional inequities of her age, can justify her appeal to one of woman's oldest weapons in arguing her case against society, because she has been left with very little else to assist her. It is society that deems an unmarried female's proper place to be in the home or convent, and social prejudice that declares a mother's illegitimate offspring an embarrassment in the family. Where such inequity reigns, woman reaches for the tools closest at hand, and of whose efficacy she has an apparently inborn knowledge. To trade on one's attractions is called legitimate by the world that has treated Suzanne so ill, and she can thus admit with relatively little compunction to her imagined reader:

je me suis aperçue que, sans en avoir le moindre projet, je m'étais montrée à chaque ligne , . . beaucoup plus aimable que je ne le suis. Serait-ce que nous croyons les hommes moins sensibles à la peinture de nos peines qu'à l'image de nos charmes? et nous promettrions-nous encore plus de facilité à les séduire qu'à les toucher?[7]

Suzanne's authenticity, as she realizes, is not impaired by the resort to such an appeal, for it stems from 'un instinct propre à tout mon sexe'. Marivaux's Marianne draws a similar conclusion. Since the world is so fashioned that men and not women make the advances that are the usual prelude to a love-affair, women must use a degree of deception to ensure that their own interests are well served. Such deception either is, or becomes, second nature, so that women cannot justly be criticized for indulging in it. Besides, Marianne is careful to qualify what she means by coquettishness. Indecency is rigorously excluded from it:

[j'ai] toujours cherché l'honnête, et par sagesse naturelle, et par amour-propre . . . Je soutiens qu'une femme qui choque la pudeur perd tout le mérite des grâces qu'elle a: on ne les distingue plus à travers la grossièreté des moyens qu'elle emploie pour plaire; elle ne va plus au cœur, elle ne peut plus même se flatter de plaire, elle débauche; elle n'attire plus comme aimable, mais seulement comme libertine, et par là se met à peu près au niveau de la plus laide qui ne se ménagerait pas.[8]

[7] p. 392.   [8] pp. 208–9.

What Marianne is describing is the rationale behind a series of tactical moves, but where tactics are only half-knowingly used. Her activities contrast in their spontaneity with those the Marquise de Merteuil details in the eighty-first letter of *Les Liaisons dangereuses*. The difference between the two women's behaviour resides in their respective ambitions. Marianne aspires to no higher goal than that conventionally regarded as proper to females, namely obtaining a husband and a respectable social position (in her case, consonant with what she believes to have been her elevated origins). The Marquise, however, wants what convention does not readily accord her sex: recognition of an intellect equalling that of the most gifted male. But to achieve this end she must deploy far greater skill than is required of a man in such affairs; she cannot succeed without possessing both a sharp mind and feminine charms. This, she maintains, is the true and deplorable inequality.

Although Mme de Merteuil's cleverness compels admiration, and although it is hard to believe that Laclos did not feel strong sympathy for his creation, we can readily understand how this 'nouvelle Dalila'[9] was seen by many female contemporaries as having betrayed their own sex as well as the opposite one. Her activities have little or nothing to do with the quality that gave its name to the novel-genre in French, romance. But she is not alone among the female characters of the eighteenth-century novel in being strangely disabused with this quality, or in seeing how elusive it is. '[De] l'amour, en a-t-on quand on veut?' Mme de Merteuil asks the Vicomte, knowing what the answer must be. For many novelistic heroines of the day, love seems less of a desideratum than the Marquise wistfully implies: it may bring, not delight and exaltation, but disappointment and grief. The Hortense of *Les Egarements du cœur et de l'esprit* is certain that peril and disillusionment attend the loving heart; and both Marianne and Tervire, in Marivaux's unfinished novel, experience the pain of man's insincerity and inconstancy.

Another heroine, too, cautions against the romantic interpretation of eros. Yet are the professions of Julie to be believed when she speaks out against love as the basis for marriage? It is true that hers is the voice of experience, for she has known deepest passion. But does this mean that she should be heeded when she rates it below her

9 p. 175.

unimpassioned companionship with Wolmar? Given that she opts out of life when the force of her persistent love for Saint-Preux becomes too great to resist, it hardly seems so.

Other heroines may seem (in a different sense) ambiguous. In the midst of experiences which leave a part of their being curiously untouched, they retain a kind of transcendent naïvety about human emotions. One such is Manon Lescaut, and another Suzanne Simonin. Neither, for long stretches of the novel which recounts her story, seems fully aware of the nature of the emotion she arouses in other people. Indeed, in Manon's case it is the element of childlike incomprehension in her attitude which, more than anything else, permits her to retain the reader's indulgence. How well she understands the situations into which she propels des Grieux and herself it is impossible to say. The charitable interpretation would be that she grasps the truth of her attractiveness to men, but not its worldly implications; the uncharitable, that she knows she will never be without material male support, so that there is no urgent need to reflect on her moral conduct. If the latter was indeed her assumption, the fact that the law catches up with her in the end and has her deported to America shows how wrong it was.

Suzanne Simonin's situation is not really comparable to Manon's, for she is far more the victim of circumstances than the latter. But she displays the same incomprehension about the meaning of passion and its effect on those with whom she associates. Her naïvety is perhaps easier to accept than Manon's, given the unlikelihood that a girl of her age and background would have a developed understanding of homosexuality. But it is undeniable that she emphasizes her innocence rather too insistently for someone so naïve: the reader cannot entirely believe in it. Whatever the case, Suzanne's story is perhaps the least hopeful about woman and love of the many cautionary tales contained in the eighteenth-century novel. From other writings of Diderot's we know that he was not unvaryingly pessimistic in assessing the female's chances of gaining emotional happiness in the world. For the purposes of this novel, however, it suited him to describe the situation of women moulded by convent life as one without hope, for they were denied the possibility of redemption through feeling.

Whether or not love offers a kind of salvation to Laclos's Présidente de Tourvel or Saint-Pierre's Virginie is a more difficult question to answer. In the latter novel uncertainty seems to have been involuntarily imposed by the author on recalcitrant narrative material; in the former its presence appears intentional. Bernardin's theories about the role of women in the world may not have varied very much over his writing career—his general view was that they should keep their place in the household and leave the running of extra-domestic affairs to men—but his opinion about woman's chances of knowing and retaining happy love certainly did. In *Paul et Virginie*, contrary to what we find in his later *Chaumière indienne* and *Voyages de Codrus*, it is made plain that insuperable difficulties face the female who seeks the simple good of a happy married life. It eluded the mothers of both Virginie and Paul before them, and Saint-Pierre attempts to show why it must similarly elude the children. He never does so very clearly, but it is at least evident that he was torn between wanting to depict female felicity in a part-sensual, part-ascetic world, and to prove that all human bliss is chimerical. This is possibly Laclos's intention in *Les Liaisons dangereuses* as well, but it appears more likely that he links female suffering in love to a particular artificial and corrupting social milieu, where sincerity and trust are discouraged and sensation takes the place of sentiment.

A good number, if by no means all, of eighteenth-century novels dealing with love do end in its gratification: the protagonist becomes 'complete'. But the titles of these novels often sound a note of caution. Love leads to dangerous liaisons (Laclos), perverts the innocent (Restif), or causes moral and intellectual aberrations (Crébillon). Countless obstacles are put in the way of lovers, entailing frustration of their desires, postponement of their felicity, separation, suffering, and even death. They may be forced to marry where they do not want to, or hindered where they do. They may be afflicted with a passion that is unrequited, endure the pain of a partner's inconstancy, be prevented from avowing their feelings, or regret the avowals they make. Society may oblige them to abandon their love or to keep it a guilty secret. They may be told to await the coming of sexual and social maturity, or be pressed into a lifelong union before their natural time. They may be torn from their families by the demands of matrimony, or have no acquaintance with familial affection. They may be educated in love by the unscrupulous, or never given a sentimental education. They may, if they are women, be advised to suppress their longings in the name of decency, but observe that such constraints rarely apply to the opposite sex. They may reflect that love is unfair, but be informed that it creates its own justice. And when they believe that they have won mastery of it, they may be brought to nothing by its destructive power.

Most of the novels discussed in this book include one or more of these scenarios. It might be concluded from *Manon Lescaut, La Nouvelle Héloïse, Les Liaisons dangereuses*, and *Paul et Virginie* that great love is always thwarted. This notion has a certain appeal for the romantically minded. Perhaps because of the ancient link between tragedy and nobility, we attribute a kind of greatness to the terminally suffering lovers of the eighteenth-century novel, and a lesser stature to the gratified ones. It may be a case of the greater travail provoking the greater response: if this hypothesis is true, it would fully vindicate the eighteenth-century theory of sensibility, according to which the expression of emotion should be enlarged in order to guarantee a grand reaction to it. But at the same time the contemporary novel often sought to demonstrate that small as well as big objects, humble as well as great people, should have the capacity to move others, and that the old aristocracy of class was ripe for replacement by a new aristocracy of feeling. Saint-Preux the bourgeois upstart is no inferior to Milord Edouard where matters of emotion are concerned. Indeed, the sentimental revolution posited as a fundamental truth that all beings deserved an equal measure of human love.

Whatever their merit, however, feeling creatures had to acknowledge the precariousness of love and its seemingly inevitable association with human mutability. Sometimes, as with the libertines of the age, this uncertainty became a veritable *Lebensphilosophie*, the habitual demand for immediate gratification from emotion and sensation. It is rarely predicated of women in the narrative fiction of the day, perhaps because of the age-old conviction that woman's essence is to be settled and unified, and man's to be diverse and 'disponible'. There are exceptions, foremost among them Laclos's Mme de Merteuil. But the most memorable eighteenth-century example of addiction to this particular philosophy is not the Marquise, but Prévost's Manon. And even here we should not ever-emphasize her flightiness: she spends two years (if she is to be believed) with the wealthy man for whom she originally leaves des Grieux, and seems to find des Grieux a perpetually necessary point of return. Much more striking are the male inconstants of the novel, the Valvilles, the Faublases, and their like.

The argument against exclusive attachment to another person, for more delicate souls than the last-named, is that human changeability exposes people to the experience of pain and grief. But there is another solution to the problem of human love and its attendant hazards. The conviction that lovers may meet beyond the grave with those they have most cherished gives Julie strength, not to resist mortal extinction, but to court it. The hope of reunion with Virginie is translated into the swift demise of Paul in Saint-Pierre's novel. What Mme de Tourvel can hope for from her death, if hope is possible in her deranged state, we may only guess at. But the sentimental spirit of the eighteenth century found the affinity of death and love as compelling as it was terrible. Death unites innocents or near-innocents; death may cleanse the Lovelaces and Valmonts of the world and make them fit husbands for the chaste. The dust of those who cannot love on earth, like Abélard and Héloïse, Saint-Preux and Julie, or Diderot and Sophie, mingles restoratively in the grave.

Perhaps this conclusion carries compensation for the tribulations visited upon those who seek legitimate or illegitimate felicity through eros. The recalcitrance of love does not always assume tragic dimensions, but it is always a more or less provoking obstacle. Feeling for another cannot always be expressed as one would wish. Nor can reciprocity of emotion be guaranteed, in life or in literature: the *coup de foudre* does not necessarily strike two people at the same time. Some of the most moving eighteenth-century novels about love play on the uncertainty which the *coup de foudre* can generate, whether in the lover or the reader. Does Hortense love Meilcour, or is it another 'inconnu' who has won her heart? Does Manon love des Grieux in the same instant as he falls in love with her? The evidence we have about her character and her opportunism suggests otherwise, but we cannot be sure. When does Mme de Tourvel fall in love with Valmont? When, precisely (if these things can be pinpointed to a moment), does he fall in love with her? The reader wonders, and so, sometimes, does the fictional character.

As Marivaux's Jacob has reason to know, there are ways of persuading one's own heart which make it appear less certain that love cannot be summoned at will than Mme de Merteuil believes. Sentimentalists may concede little or no contribution by intent to the state of love (so powerful is the myth of love's thunderbolt); realists and pragmatists like Jacob, and perhaps Manon, know differently. As love may be willed, it may also appear wilful, a child of caprice which functions according to its own logic. Passion which is excited by obstacles is of this order. For Valmont, obstacles provoke determination if not constancy; when they are removed, the pursuer's love is in danger of dying. The risk of reciprocating love under these conditions is one which the Présidente de Tourvel tragically accepts; her tragedy might seem to prove, yet again, the danger of love in any of its guises. But all lovers are not Valmonts, whose loyalty is divided between his egoism and his non-narcissistic passion. Laclos leaves the question whether the consummation of eros leads necessarily to destruction unresolved.

For some if not all of his readers, none the less, it appears that Laclos rates the expression of a full-valued love in which sexual desire and tender care are combined as the most complete of human relationships. Notwithstanding conventional moral prohibitions against free sexual expression, and notwithstanding his own sexual oddities, Rousseau seems to arrive at the same conclusion in *La Nouvelle Héloïse*. Merely erotic love appears to both authors, and must generally appear to the novelist of sentiment, as an imperfect form of bonding between human beings, one that is properly refused consummation by the man or woman of modesty. There is nothing very surprising in the implication both Rousseau's novel and *Les Liaisons dangereuses* contain, that desire which does not aim at a lasting spiritual union is somehow second-rate: the appeal of such a notion is wide, and can be felt by people of rational as well as romantic disposition. Both love which is exclusively of the senses, and the respect-love (or amorous friendship) of couples like the Wolmars, is shown by major writers of the eighteenth century to be inferior to love that is of the soul and the body together. But, perhaps unexpectedly for the age of sensibility, many novelists of the period also reveal the latter kind of love as fragile, and in its instability dangerous to lovers of less than settled temperament.

In so far as they depict a society in which people are, to varying degrees, in the grip of sexual obsession, novelists like Crébillon, Laclos, and even Rousseau may be called unsentimental writers. In their different ways, *Les Egarements du cœur et de l'esprit*, *La Nouvelle Héloïse*, and *Les Liaisons dangereuses* all confirm the perception formulated by Diderot in his letter to Damilaville, that 'il y a un peu de testicule au fond de nos sentiments les plus sublimes et de notre tendresse la plus épurée.' Privileged societies have the means, in all the paraphernalia of refinement, to conceal the operations of brute sexuality. Those depicted by Crébillon, Rousseau, and Laclos can use the camouflage of language, the elaborate behavioural code of *mondanité*, and even the tangible resort of the 'petite maison' to hide their activities. Stendhal later described the operations of a similar world in *De l'amour*, under the heading 'amour-goût', and remarked on its absolute exclusion of the disagreeably frank in the interests of *bon ton* and delicacy. The well-born man knows in advance all the procedures he must observe through the successive phases of an 'amour-goût', excluding the passionate at all times and preferring delicacy in everything.[10] Marivaux's Jacob, although of peasant stock, learns through his exposure to high society that what it calls 'tendresse' is something trumped up, a thin layer of civility covering up primal instinct.

There is much irony in all of this. Indeed, the ability to analyse feeling, which is manifested in the fiction of Crébillon, Marivaux, and Laclos, may be a defining characteristic of the non-sentimentalist. (In the novels of Crébillon and Marivaux, at least, this ability often depends on the use of a 'double register' according to which narrators sometimes relate what they felt at the time of their actions, and sometimes reflect on their feelings from the vantage-point of age.)[11] People who think about feeling rarely emote: Rousseau proposed this view in the second preface to *La Nouvelle Héloïse*, and it seems to be borne out by the novels of his predecessors as well as by the author of *Les Liaisons dangereuses*. With their taste for lengthy reflection about and psychological analysis of feeling, Marivaux and Crébillon set the novel of love in a new direction. Such musings are seen by Marianne as excluding the manifestations of sensibility; as she says of herself, 'son esprit ne laisse rien passer à son cœur.'[12] Minds like this are rarely bemused by emotion, whatever the latter's complexity or multifariousness. Marianne's intellect is incapable of letting a 'mouvement' pass without attempting to dissect it.

There is a contrasting lack of subtlety in other eighteenth-century novels which deal with love, and Diderot's observation to Damilaville is enlightening about these works too. Louvet's *Faublas* has the rumbustiousness of a seventeenth- century realist novel like Sorel's

---

[10] p. 31.

[11] On the 'double registre' in Marivaux, see Jean Rousset, *Forme et signification* (Paris, 1970), pp. 45 ff.

[12] See Jean Fabre, *Idées sur le roman de Mme de Lafayette au Marquis de Sade* (Paris, 1979), p. 90.

Francion and the sentimentality of its own contemporary *Paul et Virginie*. It shows its hero loving like a stallion and simpering like a virgin, donating his seed to one woman and offering his purest adoration to another. Faublas's promiscuity catches up with him in the end; but until than he has seemed the epitome of male privilege, protesting pious and respectful love for one female while assuaging physical desire with another. His is perhaps an extreme case, but other eighteenth-century novels reveal the coexistence of sensibility and sensuality in the worlds they describe. Restif, the 'Rousseau du ruisseau', provides several examples of such duality. *Le Paysan perverti* and *La Paysanne pervertie*, to say nothing of *L'Anti-Justine*, constantly juxtapose or even mingle scenes of near-obscenity with passages of mawkish romanticizing, lyricism *à la* Rousseau with pornography *à la* Sade. Both can appear crude; both are products of the same historical moment and temper. Each is as true to an aspect of the eighteenth century as Marivaux's or Laclos's blend of irony and sensibility.

The notion that in giving free rein to one's sexual urges one is simply following nature is one often encountered in the eighteenth-century novel, and is part of the general interest of the age both in a purely physical 'science' of man and in his formation through society. The anti-civilization movement associated with Rousseau was paradoxically wedded to many of the beliefs about controlling human sexuality on which modern states had based their laws; and the 'civilized' world, in return, covertly availed itself of many sexual liberties associated—whether appropriately or not—with the 'natural' one. Neither Rousseau nor Saint-Pierre, both known for their attacks on the corruptions of civilization, wanted animal appetites to prevail in the ideal societies they constructed in their fiction; nothing could be more alien to the moral thought behind *La Nouvelle Héloïse* and *Paul et Virginie*. On the contrary, old-fashioned social virtue, with all that it implies of self-restraint in both the physical and ethical realm, is the cornerstone of the communities established at Clarens and on Mauritius.

In contrast to the 'primitivist' Diderot of the *Supplément au Voyage de Bougainville*, Rousseau sees stable marriage based on legal sanction and social approval as absolutely fundamental to right living (despite his own informal union with Thérèse Levasseur), and his ideal of marriage—in literature at least—is probably coloured by the Calvinism of his Genevan youth. Diderot is concerned with a different kind of stability, seen as equally beneficial to the community, but quite divorced from the Christian ideal of lifelong fidelity to a single sexual partner. Free sexual love, he argues in the *Supplément au Voyage de Bougainville*, is man's natural right (and woman's only if she is capable of conceiving a child). He finds the Christian assumption that desire and love should be located in a single being throughout adult life both a psychological absurdity and a matter for the populationist's regret. Both Diderot and Rousseau have their own account of love (and both offer different accounts in different works); neither, for the sake for the view he is currently espousing, is willing to accept the validity of his opponent's. Eighteenth-century fiction is host to their debate, and to many others, on the merits of different kinds of social living and loving. Diderot is less ready than Rousseau to set his up as a universal truth, although both believe that workable worlds can be constructed on the basis of their theories. Fiction, it appears, can make the most serious of contributions to the debate about sentiment and the human world. It is perhaps in this sense, rather than what Rousseau says in the *Confessions* about the educative effects of reading novels, that the imaginative literature of the age can be said to illuminate areas of emotional life.

As a more *gaulois* writer than Rousseau, Diderot approached sexuality less delicately than his one-time friend. While both men wrote with feeling about the natural life, the latter implied for Diderot an acceptance of 'un peu de testicule' that Rousseau would have found vulgar or bestial. Although both wrote rapturously and sentimentally about women (Diderot, notably, in the dithyrambic essay *Sur les femmes*), only Diderot could have composed *Les Bijoux indiscrets*—a work no less sexist than many of Rousseau's—and *Jacques le fataliste*. Diderot is more prurient than Rousseau, and is more consistently disposed to see erotic relations between the sexes (if not within them) as healthy and normal. Where Rousseau is uncertain, Diderot is clear: the passions, left in their natural state, are good and life-enhancing. What is deplorable is to pervert them by containing individuals within unnatural environments:

On déclame sans fin contre les passions; on leur impute toutes les peines de l'homme, et l'on oublie qu'elles sont aussi la source de tous les plaisirs. C'est dans sa constitution un élément dont on ne peut dire ni trop de bien ni trop de mal. Mais ce qui me donne de l'humeur, c'est qu'on ne les regarde jamais que du mauvais côté. On croirait faire injure à la raison si l'on disait un mot en faveur de ses rivales. Cependant il n'y a que les passions, et les grandes passions, qui puissent élever l'âme aux grandes choses. Sans elles, plus de sublime, soit dans les mœurs, soit dans les ouvrages; les beaux-arts retournent en enfance, et la vertu devient minutieuse.[13]

These are strong words. As Diderot remarks elsewhere, the grandeur that accompanies passion can be grandeur in immorality as well as goodness. But there is virtue (in the old sense of *virtus*, strength, as well as the more recent, ethical one) in the mere fact of feeling and being impressionable—a possibility which Stoic temperaments, devoted to *apátheia*, positively denied. Diderot's view that impassiveness is morally and aesthetically regrettable was shared by many in his age. If Rousseau bemoans his parental legacy of a feeling heart, and Saint-Preux exclaims to Julie that an 'âme sensible' is a fatal gift from heaven, condemning its possessor to pain and grief on earth, *La Nouvelle Héloïse* gives only limited support to their contention.

Des Grieux's view of the matter is straightforward. Man's nature directs him to search for happiness. The greatest happiness known to him consists in love. In the interests of bliss, therefore, he should allow the passion of love to direct him where it will—often towards darkness, as des Grieux concedes in his debate with Tiberge, but frequently too towards utter felicity. We know that Tiberge fears passion, and finds des Grieux's readiness to live for the present at whatever cost for the future reprehensible. So, presumably, would Julie's husband, for the economy of his estate functions on very different principles. But the religious ascetic and the Stoic find an unlikely partner in disapproval. This is the Don Juan, the far-from-ascetic, the man given to brief and multitudinous sexual encounters, but whose roving activities would be impeded by the upheaval of passion. Rakes like Versac and Valmont must necessarily be free: they must be agents, not patients, masters to act rather than matter to be acted upon. Passion, in the strict sense of the word, is therefore not for them. Their deprecation of it, unlike Tiberge's, has nothing to do with moral censure. It is closer to Wolmar's, which is based on business efficiency. But neither he nor they, one imagines, would acknowledge the likeness.

For certain members of society, then, the old, pejorative sense of passion remained. It might, as in its original meaning, be ethically

---

[13] *Pensées philosophiques*, pp. 9–10.

neutral, but it still implied a vulnerability to overpowering affection or seizure.[14] For Stoics, in antiquity as in the eighteenth century, its opposition to action was supplanted by a contrast with reason: the turbulence of passion was set against the calm order of mind, and firmly subordinated to it. It is certainly possible to argue that practical wisdom consists in dissociating oneself from the claims of emotion. This remains Julie de Wolmar's opinion through the larger part of *La Nouvelle Héloïse*. But she would not, one presumes, have denied what was often proposed by thinkers from Descartes to Rousseau, that strong emotion opens the individual to sympathetic awareness of other people's feelings, and so makes morality possible. Nevertheless, much of what the regenerate Julie proclaims in Rousseau's novel suggests that she also considers the passions to be brute facts of man's fallen condition, physical states to be fought against and, where practicable, suppressed. Although she sees the imagination, intimately connected with emotion, as a prerequisite for the 'feeling' morality preached throughout *La Nouvelle Héloïse*, she persists in regarding passion as a poor, and possibly destructive, relation of the mind.

Such a view of their antithetical nature, if not of sentiment's inferiority to reason, is supported by what Rousseau says in the second preface to *La Nouvelle Héloïse* about the non-rational quality of his characters' letters. When people are in love they are often incapable of ratiocination, and may indeed be accused of refusing to listen to reason. (This is Tiberge's grievance against des Grieux.) But if the eighteenth century saw many thinkers incline to this view, its converse found much support too. Self-possession, as Socrates argued in the *Phaedrus*, might lead to narrowness of vision. This is the burden of Diderot's first *Pensée philosophique*: starving and suppressing the emotions and appetites might fatally weaken the whole personality. The ascetic plan of Plato's *Republic*, which rests on depriving citizens of the emotional sustenance provided by the family, may cripple them and ultimately make them socially useless. The same could not be said of the scheme behind the Wolmars' little state, for it encourages familial relationships; but it founders through the starvation of other loves. Des Grieux tries to persuade Tiberge that forfeiting the joys of love goes against human nature, and is

---

[14] See Erich Auerbach, *Literatursprache und Publikum in der lateinischen Spätantike und im Mittelalter* (Bern, 1958), pp. 54 ff.

on that account reprehensible. And Valmont attempts to convince
the Présidente that passionless marital life, by firmly excluding the
ultimate bliss of eros, prevents those who live it from discovering
their fullest potential, from becoming complete.

Yet it would be mistaken to regard deep feeling as necessarily
incompatible with rationality. As Jane Austen suggests in *Sense and
Sensibility*, the head and the heart do not have to be antagonists.
Nor, as des Grieux tries to impress on Tiberge, is eros a purely
daemonic force; it is also a god, or as embodied in Manon a goddess.
But in the eighteenth-century novel we often find it subordinated to
another, more temperate emotion, that of friendship. In the *Phaedrus*
Plato says that the best lovers sometimes deny themselves sexual
intercourse because they fear that in enjoying its pleasures they may
do harm to other important elements of their relationship, like respect
and tender feeling. This is the fear that leads the newly married prince
and princess in Restif's *Le Nouvel Abeilard* to refrain from
consummating their marriage until they know each other's souls. It
doubtless explains something of Rousseau's distaste for and
disappointment in the act of intercourse, which he saw as cheapening
the women he adored. Valmont comes to feel that passion, far from
being compromised by a couple's desisting from carnal love, may
be enhanced by it. For in cultivating an emotion that is essentially
non-physical one is enjoying the deep communion the eighteenth
century called 'volupté de l'âme', an emotional plenitude which, like
the ultimate passion of Diotima's lovers, is great by virtue of its
detachment from the body.

In the eighteenth-century novel, however, this kind of love is still
of the fallibly human, and can therefore lead to the kind of error
on which the little republic of Clarens founders. Julie, Saint-Preux,
and the others play a dangerous game in confusing virtue with 'amitié
amoureuse'. At its idealistic best, it is true, such friendship gives a
new meaning to the notion of chastity. But we may find something
amiss in the total sublimation of physical desire by ordinary flesh-
and-blood humans—people who may wear the trappings of other-
worldliness, but in whose humanity we are also asked to believe.

The chastity of body to which both Julie and Virginie attach such
importance may appear to us, even if it did not appear to
contemporary readers, far less momentous than that inhering in
spiritual chastity, which sexual congress cannot compromise. This
is the purity preserved by Richardson's Clarissa and Laclos's

Présidente de Tourvel, and to many it will seem more impressive than the physical chastity to which several eighteenth-century heroines more or less ostentatiously lay claim. For it marks a condition in which the ideal and the real can combine, which denies nothing of the ordinarily mortal, and—at least in the Présidente's case—gives to the erotic its full value. Friendship does not reach the depths of erotic love, and when two people subordinate the latter to the former they may lose something valuable. They commune, but incompletely; and they surely deceive themselves in pretending that the best kind of friendship is superior to every kind of love.

It is perhaps a form of self-deception to which women are more prone than men, if only because convention has historically required the female to control her sexual desires more strictly. (Women rarely have independent sexual appetites in eighteenth-century fiction save in erotico-pornographic works like *La Philosophie dans le boudoir* and *Le Rideau levé*, although Crébillon's Mme de Senanges and Laclos's Marquise de Merteuil are obvious exceptions.) Not that the opposite sex is invariably shown as lacking 'pudeur' in the age of sensibility. Biological inhibition aside, it is not difficult to understand how young males may be at a loss to know how to declare their attraction to females. Both Meilcour and Danceny, as we have seen, feel the tension between the substantial pleasures of the flesh and the intangible ones of the spirit, and as idealists both are inclined to put the latter above the former in the scale of amorous values.

But the eighteenth-century novel shows the battle between erotic love and modesty as being essentially a female one. Woman's untaintedness has, for practical as well as mythic reasons, greater *cachet* than man's, and correspondingly receives fuller attention in the literature of love. Even where the female is as ambiguously portrayed as Prévost's Manon, there is a fascination about her notional purity which is rarely attendant on the male's. In separating physical love from love of the heart, Manon achieves a paradoxical kind of virginity. The compromising of her technical purity counts as nothing for her, when her tenderness for des Grieux remains intact. Her heart is faithful, even though her body is not. So she claims, at any rate; Prévost's reader may decide differently, remembering her thoughtlessness about her lover's feelings on the many occasions when she abandons him for richer pickings than he can provide. If Manon is not quite the mysterious creature many have claimed her to be, the narrative silence about various of her actions gives her a remoteness resembling that of more virtuous eighteenth-century heroines, but which is often more compelling than theirs. She possesses the quality of the unattainable, the not-quite-decipherable, which is identified in romantic literature as quintessentially feminine. In Manon the quality may be closer to the supposedly female one of caprice; in heroines like Suzanne Simonin, Mme de Tourvel, and Virginie, it is a sublime innocence.

Ostensible female virtue, needless to say, may not be quite what it appears. There is a provocativeness about refusal which can be tantamount to invitation. Mme de Lursay understands that truth; so, possibly, does the Présidente. Compliance is frequently less exciting to the pursuer than apparent resistance: in seeming to erect barriers, woman may be coaxing man into making advances. So she both flatters and goads him, intimating that there are prizes to be won while simultaneously hinting that he lacks the persistence needed to secure them. Embattled womanhood, then, is often less innocent than it looks. If the female is indeed a victim of male exploitation, as Sade suggests in *Les Infortunes de la vertu*, she may also exploit the male. Though she appears beset by masculine attack, she may in fact be wielding her own particular form of power.

The eighteenth-century French novel, indeed, shows many women who are manifestly stronger than the men with whom they consort. Hortense's mysterious remoteness bewitches Meilcour; Mme de Lursay's more tangible charms bend him to her will, as Manon's do des Grieux. Julie is infinitely more resourceful, imperious, and decisive than Saint-Preux, and is never worsted by him as he is by her. She governs her lover, chastens him, advises him to imitate her fortitude, and humiliates him into the bargain. Saint-Pierre's Virginie matures more quickly than Paul, inspires him, and ultimately seems to have power of life and death over him. And Laclos's Marquise senses her superiority of mind to Valmont, teasing him with his ineptitude, hinting that he has forfeited her respect through his subjection to romantic love, and contrasting the hindrances she faces as a woman with the easy path a male follows through amatory life. Whether as saviour, temptress, or senior partner, woman exerts a force in the eighteenth-century novel which man rarely matches. While seeming disadvantaged, she often carries off the moral or intellectual victory.

How much all this owes to the sex of the writers who created these characters is difficult to determine. There may be some hostility in

Laclos's creation of Mme de Merteuil, and in the depiction of certain female characters in Diderot's fiction—the Eastern women of *Les Bijoux indiscrets*, *Jacques le fataliste*'s Mme de la Pommeraye, the Mother Superior of Sainte-Eutrope. There is ambivalence in the portrayal of Manon Lescaut, possibly reflecting the Church's old mistrust of her sex, and in Crébillon's Mme de Lursay. However, there is nothing but authorial admiration behind the characters of Julie de Wolmar, the Présidente de Tourvel, and Virginie. Male novelists do not necessarily adopt a uniform view of the opposite sex. Diderot shows scant regard for women in *Les Bijoux indiscrets*, but a great deal in the ecstatic *Sur les femmes* and in the creation of Suzanne Simonin. And Laclos reveals how a single society can allow two very different types of women—the Machiavellian and the pure in heart—to coexist. (In the seventeenth century nothing was thought more decisively to prove the authenticity of Mariana Alcoforado's love-letters than the distinctive accents of female passion they contained, but the collection of *Lettres portugaises* has since been proved to be the work of a man, Guilleragues.) Men, it appears, may probe the female psyche as effectively as women, and the persuasiveness with which they present the latter seems less a function of their own gender—with or without an attendant range of prejudices for or against the opposite sex—than of their sensibility and rational understanding.[15]

Women readers may question the accuracy with which men describe certain female responses to matters erotic, but the work of male novelists in the eighteenth century gives us, on balance, little reason to complain about the way the opposite sex depicts women in fiction. On the contrary, there are some grounds for arguing that novels of the period portray women too uncritically, and that the eternal feminine is shown to be an implausibly virtuous and edifying quantity.

As with attitudes to women, so with attitudes to love generally. One period's style of love-making may not be to the taste of another age. 'C'est toujours l'homme,' Saint-Evremond remarked in a letter to a female friend, 'mais la nature se varie dans l'homme; et l'art, qui n'est qu'une imitation de la nature, se doit varier comme elle.'[16]

---

[15] See Roy Roussel, *The Conversation of the Sexes* (New York and Oxford, 1986).

[16] *Œuvres*, new ed., 9 vols (n.p., 1753), iv. 268.

Although love rests on eternal foundations, Sainte-Beuve wrote in his *Portraits littéraires*, it changes endlessly in form from one epoch to another, with the result that it notoriously deceives those who try to write about it imaginatively.[17] It is perhaps difficult for us, standing at a distance from the eighteenth century, to assess the accuracy of its portrayal of feeling: much of the historical evidence about the temper of the age, inevitably, is taken from its fiction. Where there is extra-literary testimony about people's affective lives—their proneness to weep or swoon, for example—the familiarity of fictional types which confirm the historical evidence will, at the very least, add authenticity to the historian's sketch. But even if novels are in this limited respect 'true', the portrait they paint will scarcely on that account alone appeal to a later readership, except of the type that regards documentary truth as artistically desirable.

The literature of love can probably teach its reader a number of things, among them the shifting perception of what is 'good' or 'bad' behaviour between lovers, what the relation between word and thought has been at different stages in history, what it means and has meant to be 'lover-like', what is commonly thought to be within the emotional scope of ordinary people, and what is thought to be attainable by a select few. It may also afford insight into what, in passional terms, seems scarcely believable, inhuman, hopelessly idealized, or straightforwardly wrong. Each reader takes and leaves what he will, often assuming his own emotional proclivities to be universal and judging the views and feelings of others accordingly.

Rousseau does not, in that memorable passage from the *Confessions*, tell us what were the novels he and his father read in the evenings after his mother's death: they may have been the bawdy tales of Rabelais, the realist fictions of Scarron and Furetière, the heroic romances of La Calprenède and the Scudérys, or the elevated narratives of Mme de Lafayette. Whatever the case, they are unlikely to have taught their reader, in Rousseau's household or elsewhere, the sum total of human knowledge about the passions. Their lessons might be partial, fanciful, outmoded, or otherwise irrelevant to the lives of real people in the real world. But they will, at the very least, have stimulated thought about what passion is, and done so in a form more palatable than that of a philosopher's treatise or a preacher's sermon.

[17] *Portraits littéraires*, 2 vols (Paris, 1852), i. 86.

If a great many eighteenth-century novels seem to confirm the general prejudice against French literature of the period—that it suffers by comparison with the previous century's and that of the following one—some fiction of the age still has a strong claim on our attention. Its appeal is not just as a moral portrait of the times, or an exercise in wish-fulfilment: bad novels can be those things as well as good. What many modern readers value in the eighteenth century's literature of love is its combination of controlled sensibility and sophisticated eroticism. This surely explains the continuing appeal of *Les Egarements du cœur et de l'esprit*, *Manon Lescaut*, and *Les Liaisons dangereuses*. If our admiration for other 'great' works is lesser—for *La Nouvelle Héloïse*, *La Religieuse*, or *Paul et Virginie*—the reason possibly lies in their excess: the self-indulgent amplitude of the first, the melodrama of the second, and the cloying sentimentality of the third. Novels which leave the reader in a state of teasing uncertainty, unsure which values are really being asserted, may hold greater attraction for our age than do wholeheartedly 'sensible' ones: the ironic hints of a Marivaux, a Crébillon, or a Laclos perhaps entice us more than the bombast of a Rousseau or the muddled virtue-mongering of a Bernardin.

After reading *La Nouvelle Héloïse* or *Paul et Virginie*, the present-day reader may be left with the suspicion that they belong to the literature of pretence; they are too direct in their efforts at winning our affective allegiance than seems to us proper. Later ages may well judge differently, however, and find the accusation of sentimentality and its related phenomena less of a slur than it currently appears. They may equally find the loveless universe of a Sade closer to their own ideas of what is true and important in the world of emotion than the elegantly sceptical one of Marivaux. But if today's reader gains nothing else from his acquaintance with these novels, he will probably emerge from it with his understanding of love at least partially clarified, whether in positive or in negative terms. The eighteenth-century reader of such fiction must have been impressed, as many modern readers remain, by its devotion to a subject about which it seems possible to say so much, so variously. The actions and perceptions associated with love are hard to describe, which makes the novelist's sustained attention to them foolhardy from one point of view, admirable from another. We may with some reason complain that a large number of eighteenth-century authors apparently found love all too easy to write about, and that greater restraint would have resulted in a more impressive body of work for future ages to savour. Chamfort declared that love is the one subject on which it is impossible to express an absurdity, and also claimèd that in love all is true and all is false. If the latter proposition holds, then there is every and no reason to write extensively about it—a paradox which the eighteenth-century French novel, still uncertain whether its proper goal lay in instruction or delight, was well justified in exploring.

Select Bibliography

Peter Ackroyd, *Dressing Up. Transvestism and Drag: The History of an Obsession* (London, 1979).

D. J. Adams, *Diderot, Dialogue and Debate* (Liverpool, 1986).

Joseph Addison, Richard Steele and Others, *The Spectator*, 4 vols (London, 1970–3).

Jean le Rond d'Alembert, *Essai sur les éléments de philosophie ou sur les principes des connaissances humaines, Œuvres philosophiques, historiques, et litteraires*, 18 vols (Paris, 1805), ii.

L'Ami des femmes, ou la philosophie du beau sexe (Paris, 1759).

Philippe Ariès, *L'Homme devant la mort* (Paris, 1977).

John Atkins, *Sex in Literature*, 3 vols (London, 1970–8).

Erich Auerbach, *Literatursprache und Publikum in der lateinischen Spätantike und im Mittelalter* (Bern, 1958).

Bailly, *Avis aux mères qui aiment leurs enfants, et aux dames qui aiment leur taille* (Paris, 1786).

Fernand Baldensperger, 'Gessner en France', *RHLF*, 10 (1903).

Roland Barthes, *Fragments d'un discours amoureux* (Paris, 1977).

Charles Batteux, *Principes de littérature*, 6 vols (Lyon, 1802).

Charles Baudelaire, 'Notes sur *Les Liaisons dangereuses*', *Œuvres complètes*, ed. Claude Pichois, 2 vols (Paris, 1975–6), ii.

Fanny de Beauharnais, *L'Abailard supposé ou Le Sentiment à l'épreuve* (Paris, n.d.).

Quentin Bell, *On Human Finery* (London, 1947).

Jean-Louis Bellenot, 'Les Formes de l'amour dans *La Nouvelle Héloïse*', *Annales de la Société Jean-Jacques Rousseau*, 33 (1953–5).

Lawrence Blum, 'Compassion', in Rorty (ed.), *Explaining Emotions* (q.v.).

Margaret A. Boden, 'Human Values in a Mechanistic Universe', in *Human Values*, ed. Godfrey Vesey (Hassocks, Sussex, and Atlantic Highlands, N.J., 1978).

Théophile de Bordeu, *Traité de médecine théorique et pratique extrait des ouvrages de M. de Bordeu par M. Minvielle* (Paris, 1774).

[Père Dominique Bouhours], *Les Entretiens d'Ariste et d'Eugène* (Amsterdam, 1671).

Bernard Bray, *L'Art de la lettre amoureuse* (The Hague, 1967).

R. F. Brissenden, *Virtue in Distress: Studies in the Novel of Sentiment from Richardson to Sade* (London, 1974).

Peter Brooks, *The Novel of Worldliness* (Princeton, 1969).

Mme de Brulart [Genlis], *Discours sur l'éducation publique du peuple* (Paris, 1791). *See also* Genlis.

A. Brun, 'Aux origines de la prose dramatique: le style haletant', in *Mélanges de linguistique français offerts à M. Charles Bruneau* (Geneva, 1954).

Ferdinand Brunot, *Histoire de la langue française des origines à nos jours*, new ed., 13 vols (Paris, 1966–72).

Georges-Louis Leclerc, comte de Buffon, *Histoire naturelle de l'homme, Œuvres complètes*, 25 vols (Paris, 1774–8), iv.

John Bugge, *Virginitas: An Essay in the History of a Medieval Ideal* (The Hague, 1975).

Edmund Burke, *Works*, 12 vols (London, 1887).

Robert Burton, *The Anatomy of Melancholy*, ed. Holbrook Jackson (London, 1972).

J. P. Bury, *The Idea of Progress: An Enquiry into its Growth and Origin* (New York, 1960).

Louis-Antoine de Caraccioli, *La Jouissance de soi-même* (Utrecht and Amsterdam, 1759).

Susan Lee Carrell, *Le Soliloque de la passion féminine ou le dialogue illusoire* (Tübingen and Paris, 1982).

John Carroll (ed.), *Samuel Richardson: A Collection of Critical Essays* (Englewood Cliffs, N.J., 1969).

H. H. O. Chalk, 'Eros and the Lesbian Pastorals of Longus', *JHS*, 80 (1960).

D. G. Charlton, *New Images of the Natural in France* (Cambridge, 1984).

François-René, vicomte de Chateaubriand, *Le Génie du christianisme, Œuvres complètes*, 22 vols (Paris, 1833–5), xi–xiii.

A. C. Chavannes, *Essai sur l'éducation intellectuelle, avec le projet d'une science nouvelle* (Lausanne, 1787).

Harvey Chisick, *The Limits of Reform in the Enlightenment* (Princeton, 1981).

Kenneth Clark, *The Nude* (London, 1956).

Jean-Antoine-Nicolas de Caritat, marquis de Condorcet, *Esquisse d'un tableau historique des progrès de l'esprit humain* (Paris, 1822).

Marie-Louise-Sophie de Grouchy, marquise de Condorcet, *Lettres à C*** [Cabanis] sur la théorie des sentiments moraux, ou Lettres sur la sympathie*, in Adam Smith, *Théorie des sentiments moraux*, trans. Mme [*sic*] Condorcet, 2 vols (Paris, 1798), ii.

Didier Coste and Michel Zératta (eds), *Le Récit amoureux (Colloque de Cerisy)* (Seyssel, 1984).

F. R. Cowell, *The Garden as a Fine Art from Antiquity to Modern Times* (London, 1978).

Claude Crébillon (*fils*), *L'Ecumoire ou Tanzaï et Néardané, histoire japonaise*, ed. Ernest Sturm (Paris, 1976).

——*Les Egarements du cœur et de l'esprit*, in Etiemble (ed.), *Romanciers du XVIIIᵉ siècle* (q.v.), ii.

Marin Cureau de la Chambre, *Les Caractères des passions* (Paris, 1640).

——*Discours sur l'amour d'inclination* (Paris, 1634).

Ernst Robert Curtius, *European Literature and the Latin Middle Ages*, trans. Willard Trask (London, 1953).

M. C. D'Arcy, *The Mind and Heart of Love* (London, 1945).

Robert Darnton, *Mesmerism and the End of the Enlightenment in France* (Cambridge, Mass., 1968).

P. Delarue, *Le Conte populaire français*, vol. i (Paris, 1957).

——and M.-L. Tenèze, vol. ii (Paris, 1964).

Frédéric Deloffre, *Une Préciosité nouvelle: Marivaux et le marivaudage* (Paris, 1955).

Michel Delor, ' "Homo sum . . .": Un Vers de Térence comme devise des Lumières', *Dix-huitième Siècle*, 16 (1984).

René Démoris, 'Les Fêtes galantes chez Watteau et dans le roman contemporain', *Dix-huitième Siècle*, 3 (1971).

René Descartes, *Des passions de l'âme*, ed. Geneviève Rodis-Lewis (Paris, 1955).

Denis Diderot, *Correspondance*, ed. Georges Roth, 16 vols (1955–70).

——*Eléments de physiologie*, ed. Jean Mayer (Paris, 1964).

——*Lettre sur les sourds et muets*, ed. Paul Hugo Meyer, *Diderot Studies*, 7 (1965).

——*Mémoires pour Catherine II*, ed. Paul Vernière (Paris, 1966).

——*Le Neveu de Rameau*, ed. Jean Fabre (Geneva and Lille, 1950).

——*Œuvres esthétiques*, ed. Paul Vernière (Paris, 1968).

——*Œuvres philosophiques*, ed. Paul Vernière (Paris, 1964).

——*Œuvres romanesques*, ed. Henri Bénac (Paris, 1962).

——*Salons*, ed. Jean Seznec and Jean Adhémar, 4 vols (Oxford, 1957–67).

——*Sur les femmes*, *Œuvres complètes*, ed. Jules Assézat and Maurice Tourneux, 20 vols (Paris, 1875–7), ii.

Heinrich Dörrie, *Der heroische Brief* (Berlin, 1968).

Amédée Doppet, *Le Médecin de l'amour* (Paris, 1787).

——*Le Médecin philosophe* (Turin, 1787).

Kenneth Dover, *Greek Homosexuality* (London, 1978).

[Abbé Jacques-Joseph Duguet], *Cas de conscience décidé par l'auteur de la 'Prière publique'. On demande s'il est permis de suivre les modes, et en particulier si l'usage des paniers peut être souffert? avec les réponses aux objections* (n.p., 1728).

[César Chesneau Dumarsais?], 'Le Philosophe', in *Nouvelles Libertés de penser*, contained in *Le Philosophe*, ed. Herbert Dieckmann (Washington, 1948).

Du Puy La Chapelle, *Dialogues sur les plaisirs, sur les passions, sur le mérite des femmes, et sur leur sensibilité pour l'amour* (Paris, 1717).

——*Réflexions sur l'amitié* (Paris, 1728).

Jean Ehrard, *L'Idée de nature en France dans la première moitié du XVIIIe siècle* (Paris, 1963).

Winfried Elliger, *Die Darstellung der Landschaft in der griechischen Dichtung* (Berlin and New York, 1975).

Winfried Engler, 'Jean-Pierre Claris de Florian, *Estelle*, roman pastoral?', *ZfSL*, 78 (1968).

Etiemble (ed.), *Romanciers du XVIIIᵉ Siècle*, 2 vols (Paris, 1965).

Andrew V. Ettin, *Literature and the Pastoral* (New Haven and London, 1984).

Jean Fabre, *Idées sur le roman de Mme de Lafayette au Marquis de Sade* (Paris, 1979).

——'Paul et Virginie pastorale', *Lumières et romantisme*, new ed. (Paris, 1980).

Pierre Fauchery, *La Destinée féminine dans le roman européen du dix-huitième siècle, 1715–1807* (Paris, 1972).

Robert Favre, *La Mort au siècle des Lumières* (Lyon, 1978).

François Salignac de La Mothe Fénelon, *Télémaque* (Paris, 1968).

Paul Feyerabend, 'Materialism and the Mind–Body Problem', in O'Connor (ed.), *Modern Materialism* (q.v.).

Leslie Fiedler, *Love and Death in the American Novel* (London, 1967).

Jean-Pierre Claris de Florian, *Fables* (Paris, 1858).

Bernadette Fort, *Le Langage de l'ambiguïté dans l'œuvre de Crébillon fils* (Paris, 1978).

W. W. Fortenbaugh, *Aristotle on Emotion* (London, 1975).

Michel Foucault, *Histoire de la folie à l'âge classique* (Paris, 1972).

——*Historie de la sexualité*, 3 vols (Paris, 1976–84).

Michel Fougères, *La Liebestod dans le roman français, anglais et allemand au dix-huitième siècle* (Ottawa, 1974).

R. A. Francis, 'Bernardin de Saint-Pierre's *Paul et Virginie* and the Failure of the Ideal State in the Eighteenth-Century Novel', *Nottingham French Studies*, 13 (1974).

Erich Fromm, *The Art of Loving* (London, 1975).

Marc Fumaroli, *L'Age de l'éloquence* (Geneva, 1980).

[Gautier], *Traité contre l'amour des parures et le luxe des habits*, 2nd ed. (Paris, 1780).

Stéphanie du Crest de Saint-Aubin, comtesse de Genlis, *Adèle et Théodore, ou Lettres sur l'éducation*, 3 vols (Paris, 1782).

——*De l'influence des femmes sur la littérature française* (Paris, 1811).

——*Discours sur la suppression des couvents de religieuses et sur l'éducation des femmes* (Paris, 1790). See also Brulart.

Jean Gillet, *Le Paradis perdu dans la littérature française de Voltaire à Chateaubriand* (Paris, 1975).

Etienne Gilson, *Héloïse et Abélard* (Paris, 1938).

René-Louis, marquis de Girardin, *De la composition des paysages*, ed. Michel H. Conan (Paris, 1979).

Edmond and Jules de Goncourt, *La Femme au dix-huitième siècle*, 2 vols in 1 (Paris, n. d.).

Monique Gosselin, 'Bonheur et plaisir dans *Les Liaisons dangereuses*', *RSH*, 35 (1970).

Thérèse Goyet, *L'Humanisme de Bossuet*, 2 vols (Paris, 1965).

Françoise de Grafigny, *Lettres d'une Péruvienne*, in *Lettres portugaises, Lettres d'une Péruvienne et autres romans d'amour par lettres*, ed. Bernard Bray and Isabelle Lany-Houillon (Paris, 1983).

Joseph Grasset, *Le Médecin de l'amour au temps de Marivaux: Etude sur Boissier de Sauvages* (Montpellier and Paris, 1896).

Grimm, Diderot, Raynal, Meister etc., *Correspondance littéraire, philosophique et critique*, ed. Maurice Tourneux, 16 vols (Paris, 1877–82).

J. Guillaume (ed.), *Procès-verbaux du Comité d'Instruction publique de l'Assemblée législative* (Paris, 1889).

Georges Gusdorf, *Naissance de la conscience romantique au siècle des lumières* (Paris, 1976).

Jean H. Hagstrum, *Sex and Sensibility: Ideal and Erotic Love from Milton to Mozart* (Chicago, 1980).

Richmond Y. Hathorn, 'The Ritual Origins of Pastoral', *TAPA*, 92 (1961).

Hoffman Reynolds Hays, *The Dangerous Sex. The Myth of Feminine Evil* (London, 1966).

Paul Hazard, 'Les Origines philosophiques de l'homme de sentiment', *RR*, 28 (1937).

Robert Hazo, *The Idea of Love* (New York, Washington, and London, 1967).

Stephen Heath, *The Sexual Fix* (London, 1982).

Lorenz Heister, *De anatomia in genere* (Altdorf, 1716) (= *L'Anatomie d'Heister*, trans. J.-B. Sénac (Paris, 1724).

Claude-Adrien Helvétius, *De l'esprit* (Paris, 1758).

——*De l'homme*, 2 vols (London, 1773).

John Hibberd, *Salomon Gessner* (Cambridge, 1976).

Christopher Hill, 'Clarissa Harlowe and her Times', in Carroll (ed.), *Samuel Richardson* (q.v.).

Paul Hoffmann, 'Le Discours médical sur les passions de l'amour de Boissier de Sauvages à Pinel', in Viallaneix and Ehrard (eds), *Aimer en France* (q.v.), i.

——*La Femme dans la pensée des lumières* (Paris, 1977).

R. L. Hunter, *A Study of Daphnis and Chloë* (Cambridge, 1983).

P. Jacquinot, *Les Prédicateurs du XVIIe siècle avant Bossuet* (Paris, 1863).

Werner Jaeger, 'Aristotle's Use of Medicine as Model of Method in his Ethics', *JHS*, 77 (1957).

Robert James, *Medicinal Dictionary* (London, 1743–5) (= *Dictionnaire universel de médecine*, trans. Diderot, Eidous, and Toussaint, 6 vols (Paris, 1746–8).

Ruth Kirby Jamieson, *Marivaux: A Study in Sensibility* (New York, 1941).

Dominique Julia, *Les Trois Couleurs du tableau noir: La Révolution* (Paris, 1982).

Eve Katz, 'Ambiguity in *Les Liaisons dangereuses*', *FMLS*, 10 (1974).

Anthony Kenny, 'Mental Health in Plato's *Republic*', *The Anatomy of the Soul* (Oxford, 1973).

Annemarie Kleinert, *Die frühen Modejournale in Frankreich* (Berlin, 1980).

Paul Kluckhohn, *Die Auffassung der Liebe in der Literatur des 18. Jahrhunderts und in der deutschen Romantik* (Halle, 1922).

Erich Köhler, '*Je ne sais quoi*: Ein Kapitel aus der Begriffsgeschichte des Unbegreiflichen', *Romanistisches Jahrbuch*, 6 (1954).

Julia Kristeva, *Histoires d'amour* (Paris, 1983).

David Kunzle, 'The Corset as Erotic Alchemy: From Rococo Galanterie to Montaut's Physiologies', in *Woman as Sex Object*, ed. Thomas B. Hess and Linda Nochlin (London, 1973).

Jean de La Bruyère, *Les Caractères*, ed. Georges Mongrédien (Paris, 1954).

Louis-René Caradeuc de La Chalotais, *Essai sur l'éducation nationale, ou plan d'études pour la jeunesse* (Paris, 1763).

Pierre Choderlos de Laclos, *Œuvres complètes*, ed. Laurent Versini (Paris, 1979).

Jean de La Fontaine, *Fables, contes et nouvelles*, ed. E. Pilon, R. Groos, and J. Schiffrin (Paris, 1948).

Anne-Thérèse de Marguenat de Courcelles, marquise de Lambert, *Réflexions nouvelles sur les femmes* (London, 1730).

——*Traité de l'amitié* (n.p., n.d.).

Denis Lambin, 'Ermenonville et le jardin paysager en France', in Paireaux and Plaisant (eds), *Jardins et paysages* (q.v.).

Julien Offroy de La Mettrie, *Anti-Sénèque, ou le souverain bien* (Potsdam, 1750).

——*De la propagation du genre humain* (Paris, an VII).

——*L'Ecole de volupté* (Cologne, 1747).

——*L'Homme-machine*, ed. Aram Vartanian (Princeton, 1960).

François, duc de La Rochefoucauld, *La Justification de l'amour*, ed. J. D. Hubert (Paris, 1971).

——*Maximes, suivies de réflexions diverses*, ed. Jacques Truchet (Paris, 1967).

James Laver, *Modesty in Dress* (London, 1969).

Stephen Leacock, *Nonsense Novels* (London and New York, 1921).

[Antoine Le Camus], *Abdeker ou l'art de conserver la beauté*, 2 vols (Paris, 1754).

——*La Médecine de l'esprit*, 2 vols (Paris, 1753).

Daniel Leclerc, *Histoire de la médecine* (Geneva, 1696).

Alphonse Leroy, *Recherches sur les habillements des femmes et des enfants* (Paris, 1772).

Louis Lesbros de La Versane, *Caractères des femmes*, 2 vols (London, 1773).

Julie de Lespinasse, *Correspondance entre Julie de Lespinasse et le comte de Guibert*, ed. comte de Villeneuve-Guibert (Paris, 1906).

Jean-Pierre Letort-Trégaro, *Pierre Abélard* (Paris, 1981).

A. H. T. Levi, *French Moralists and the Theory of the Passions* (Oxford, 1964).

de Lignac, *L'Homme et la femme considérés physiquement dans l'état de mariage*, 2 vols (Lille, 1772).

Longus, *Daphnis and Chloë*, ed. Otto Schönberger (Berlin, 1960).

Jean-Baptiste Louvet de Couvray, *Les Amours du chevalier de Faublas*, in Etiemble (ed.), *Romanciers du XVIIIe siècle* (q.v.), ii.

Arthur Lovejoy, ' "Nature" as Aesthetic Norm', *Essays in the History of Ideas* (New York, 1955).

Niklas Luhmann, *Liebe als Passion: zur Codierung der Intimität* (Frankfurt am Main, 1982).

E. Allen McCormick, '*Poema pictura loquens*: Literary Pictorialism and the Psychology of Landscape', *Comparative Literature Studies*, 13 (1976).

Alasdair MacIntyre, *After Virtue* (London, 1981).

Mary Martin McLoughlin, 'Abelard and the Dignity of Women', in *Pierre Abélard—Pierre le Vénérable: Les Courants philosophiques, littéraires et artistiques en occident au milieu du XIIe siècle (Colloque international du C. N. R. S., Abbaye de Cluny, 2 au 9 juillet 1972)* (Paris, 1975).

John McManners, *Death and the Enlightenment* (Oxford, 1981).

Carol McMillan, *Women, Reason and Nature* (Oxford, 1982).

M. Magendie, *Le Roman français au XVIIe siècle* (Paris, 1932).

Pierre Carlet de Chamblain de Marivaux, *Les Aventures de *** ou les effets surprenants de la sympathie, Œuvres de jeunesse*, ed. Frédéric Deloffre (Paris, 1972).

——*Journaux et œuvres diverses*, ed. Frédéric Deloffre and Michel Gilot (Paris, 1969).

——*Le Paysan parvenu*, ed. Frédéric Deloffre (Paris, 1959).

——*La Vie de Marianne*, ed. Frédéric Deloffre (Paris, 1963).

Pierre-Louis Moreau de Maupertuis, *Vénus physique, Œuvres* (Dresden, 1752).

Fernand Maury, *Etude sur la vie et les œuvres de Bernardin de Saint-Pierre* (Paris, 1892).

Robert Mauzi, *L'Idée du bonheur au XVIIIe siècle* (Paris, 1960).

Gita May, 'Diderot and Burke: A Study in Aesthetic Affinity', *PMLA*, 75 (1960).

Roger Mercier, *La Réhabilitation de la nature humaine (1700–1750)* (Villemonble, 1960).

Alain Michel, *Rhétorique et philosophie chez Cicéron* (Paris, 1960).

Mary Midgley, *Heart and Mind: The Varieties of Moral Experience* (London, 1983).

Honoré Gabriel Riqueti, comte de Mirabeau, *Le Rideau levé, Œuvres érotiques de Mirabeau* (Paris, 1984).

Mistelet, *De la sensibilité par rapport aux drames, aux romans et à l'éducation* (Amsterdam, 1777).

Ashley Montagu (ed.), *The Practice of Love* (Englewood Cliffs, N.J., 1975).

Michel de Montaigne, *Essais*, ed. Pierre Villey, 3 vols (Paris, 1922–3).

Charles de Secondat, baron de la Brède et de Montesquieu, *Lettres persanes*, ed. Paul Vernière (Paris, 1960).

Roland Mortier, 'Unité ou scission du siècle des lumières?', in *Clarté et ombres du siècle des lumières* (Geneva, 1969).

François Moureau and Alain-Marc Rieu (eds), *Eros philosophe: Discours libertins des Lumières* (Paris, 1984).

B. Munteano, 'Survivances antiques: l'abbé Du Bos esthéticien de la persuasion passionnelle', *RLC*, 30 (1956).

Vivienne Mylne, *The Eighteenth-Century French Novel: Techniques of Illusion* (Manchester, 1965).

Martha Nussbaum, *The Fragility of Goodness: Luck and Ethics in Greek Tragedy and Philosophy* (Cambridge, 1986).

Anders Nygren, *Agape and Eros*, trans. P. S. Watson (London, 1953).

John O'Connor (ed.), *Modern Materialism: Readings on Mind–Body Identity* (New York, 1969).

A. Paireaux and M. Plaisant (eds), *Jardins et paysages: le style anglais*, 2 vols (Lille, 1977).

Erwin Panofsky, 'Et in Arcadia ego', in *Philosophy and History: Essays Presented to Ernst Cassirer*, ed. Raymond Klibansky and H. J. Paton (Oxford, 1936).

Blaise Pascal, *Pensées*, ed. Louis Lafuma, 3 vols (Paris, 1952).

John Passmore, *The Perfectibility of Man*, 2nd ed. (London, 1972).

Ronald Paulson, *Emblem and Expression* (London, 1975).

D. F. Pears (ed.), *Freedom and the Will* (London, 1963).

Jérôme Peignot, *Les Jeux de l'amour et du langage* (Paris, 1974).

Régine Pernoud, *Héloïse et Abélard* (Paris, 1970).

Josef Pieper, *Love and Inspiration: A Study of Plato's 'Phaedrus'* (London, 1965).

Michèle Plaisant, 'Poésie et jardins anglais (1700–1740)', in Paireaux and Plaisant (eds), *Jardins et paysages* (q.v.), i.

Roy Porter, 'Spreading Carnal Knowledge or Selling Dirt Cheap? Nicolas Venette's *Tableau de l'amour conjugal* in Eighteenth-Century England', *JES*, 14 (1984).

Mario Praz, *The Romantic Agony* (London, 1951).

John Preston, '*Les Liaisons dangereuses*: Epistolary Narrative and Moral Discovery', *French Studies*, 24 (1970).

Abbé Antoine-François Prévost, *Histoire du chevalier des Grieux et de Manon Lescaut*, ed. Frédéric Deloffre and Raymond Picard (Paris, 1965).

B. P. Reardon, *Courants littéraires grecs des IIe et IIIe siècles après Jésus-Christ* (Paris, 1971).

[Rémond le Grec], *Agathon. Dialogue sur la volupté*, in *Recueil de divers écrits sur l'amour et l'amitié etc.* (Brussels, 1736).

Nicolas Restif de la Bretonne, *L'Anti-Justine, Œuvres érotiques de Restif de le Bretonne* (Paris, 1985).

——*Le Ménage parisien*, 2 vols (The Hague, 1773).

——*Le Nouvel Abeilard, ou Lettres de deux amants qui ne sont jamais vus*, 4 vols (Neufchâtel and Paris, 1778).

——*Le Paysan perverti*, ed. François Jost, 2 vols (Lausanne, 1977).

——*La Paysanne pervertie, ou les dangers de la ville* (Paris, 1972).

Aileen Ribeiro, *Dress in Eighteenth-Century Europe 1715–1789* (London, 1984).

Samuel Richardson, *Clarissa*, 4 vols (London, 1962).

——*Pamela*, 2 vols (London, 1962).

John M. Rist, 'The Stoic Concept of Detachment', in *The Stoics*, ed. John M. Rist (Berkeley, Los Angeles, and London, 1978).

Jacques Roger, *Les Sciences de la vie dans la pensée française du XVIIIe siècle* (Paris, 1963).

K. L. M. Rogers, *The Troublesome Helpmate: A History of Misogyny in Literature* (Seattle, 1966).

Erwin Rohde, *Der griechische Roman und seine Vorläufer*, 3rd ed. (Leipzig, 1914).

G. Rohde, 'Longus und die Bukolik', *RhM*, 86 (1937).

Charles Rollin, *De la manière d'enseigner et d'étudier les belles-lettres par rapport à l'esprit et au cœur*, 2 vols (Paris, 1740).

Amélie Oksenberg Rorty (ed.), *Explaining Emotions* (Berkeley, Los Angeles, and London, 1980).

Jeannette Geffriand Rosso, '*Jacques le fataliste*': *L'Amour et son image* (Pisa, 1981).

Denis de Rougemont, *L'Amour et l'occident*, revised ed. (Paris, 1939).

Jean-Jacques Rousseau, *Essai sur l'origine des langues*, ed. Angèle Kremer-Marietti (Paris, 1974).

——*Lettre à d'Alembert sur les spectacles*, ed. M. Fuchs (Lille and Geneva, 1948).

——*Œuvres complètes*, ed. Bernard Gagnebin and Marcel Raymond, 4 vols (Paris, 1959–69).

Pierre Roussel, *De la femme considérée au physique et au moral* (Paris, 1788).

Roy Roussel, *The Conversation of the Sexes: Seduction and Equality in Selected Seventeenth and Eighteenth Century Texts* (New York and Oxford, 1986).

Jean Rousset, *Forme et signification* (Paris, 1970).

——*Leurs Yeux se rencontrèrent* (Paris, 1981).

Roseann Runte, '*La Chaumière indienne*: Counterpart and Complement to *Paul et Virginie*', *MLR*, 75 (1980).

Donatien-Alphonse-François, marquis de Sade, *Idée sur les romans*, ed. Octave Uzanne (Paris, 1878).

——*Les Infortunes de la vertu* (Paris, 1969).

——*La Philosophie dans le boudoir* (Paris, 1972).

Charles-Augustin Sainte-Beuve, *Portraits littéraires*, 2 vols (Paris, 1852).

Charles de Marguetel de Saint-Denis de Saint-Evremond, *Œuvres*, new ed., 9 vols (n.p., 1753).

Jacques-Henri Bernardin de Saint-Pierre, *Œuvres complètes*, ed. L. Aimé-Martin, 12 vols (Paris, 1818).

——*Œuvres complètes*, ed. L. Aimé-Martin, 18 vols (Paris, 1820).

——*Œuvres posthumes*, ed. L. Aimé-Martin (Paris, 1840).

——*Paul et Virginie*, ed. Pierre Trahard (Paris, 1958).

——*Voyage à l'Ile de France*, ed. Yves Benot (Paris, 1983).

Gerhard Sauder, *Empfindsamkeit*, 3 vols (Stuttgart, 1974–80).

Tilo Schabert, *Natur und Revolution* (Munich, 1969).

Friedrich von Schiller, *Über naive und sentimentalische Dichtung*, ed. W. F. Maitland (Oxford, 1951).

Roger Scruton, 'Emotion, Practical Knowledge and Common Culture', in Rorty (ed.), *Explaining Emotions* (q.v.).

——*Sexual Desire* (London, 1986).

Naomi Segal, *The Unintended Reader: Feminism and 'Manon Lescaut'* (Cambridge, 1986).

François Senault, *De l'usage des passions* (Paris, 1661).

George Sherburn, ' "Writing to the Moment": One Aspect', in Carroll (ed.), *Samuel Richardson* (q.v.).

Georg Simmel, 'Female Culture', *On Women, Sexuality, and Love*, trans. Guy Oakes (New Haven and London, 1984).

——'Flirtation', ibid.

Irving Singer, *The Nature of Love*, new ed., 2 vols (Chicago and London, 1984).

I. H. Smith, 'The Concept "Sensibilité" and the Enlightenment', *AUMLA*, 27 (1967).

Bruno Snell, 'Arcadia. The Discovery of a Spiritual Landscape', *The Discovery of the Mind*, trans. T. G. Rosenmeyer (Oxford, 1953).

Maurice Souriau, *Bernardin de Saint-Pierre d'après ses manuscrits* (Paris, 1905).

Ronald de Sousa, 'The Rationality of Emotions', in Rorty (ed.), *Explaining Emotions* (q.v.).

Leo Spitzer, 'The Style of Diderot', *Linguistics and Literary History* (Princeton, 1948).

Germaine Necker, baronne de Staël-Holstein, *De l'influence des passions sur le bonheur des individus et des nations* (Lausanne, 1796).

Stendhal, *De l'amour* (Paris, 1965).

Jacques Sternberg, Maurice Toesca, and Alex Grall (eds), *Chefs-d'œuvre de l'amour sensuel* (Paris, 1966).

Philip Stewart, 'Décence et dessein', in Viallaneix and Ehrard (eds), *Aimer en France* (q.v.).

——*Le Masque et la parole: Le Langage de l'amour au XVIIIᵉ siècle* (Paris, 1973).

M. E. Storer, *La Mode des contes de fées, 1685–1700* (Paris, 1928).

P. F. Strawson, 'Freedom and Resentment', *Proceedings of the British Academy*, 48 (1962).

Théodore Tarczylo, *Sexe et liberté au siècle des Lumières* (Paris, 1983).

Tertullian, *De cultu feminarum*, ed. and trans. Marie Turcan (Paris, 1971).

Joseph Texte, *Jean-Jacques Rousseau et les origines du cosmopolitanisme littéraire* (Paris, 1895).

Keith Thomas, *Man and the Natural World* (London, 1983).

Paul van Tieghem, *Le Sentiment de la nature dans le préromantisme européen* (Paris, 1960).

Philippe van Tieghem, 'Les Idylles de Gessner et le rêve pastoral dans le préromantisme européen', *RLC*, 4 (1924).

Alexandre Tilly, *Mémoires*, ed. Christian Melchoir-Bonnet (Paris, 1965).

Charles-François Tiphaigne de la Roche, *L'Amour dévoilé, ou Le Système des sympathistes* (n.p., 1749).

C. J. Tissot, *De l'influence des passions de l'âme dans les maladies, et des moyens d'en corriger les mauvais effets* (Paris, 1798).

Janet Todd, *Women's Friendship in Literature* (Columbia, 1980).

Robert Tomlinson, *La Fête galante: Watteau et Marivaux* (Geneva, 1981).

[François-Vincent Toussaint], *Les Mœurs* (Amsterdam, 1762).

Theodore James Tracy, S. J., *Physiological Theory and the Doctrine of the Mean in Plato and Aristotle* (Chicago, 1969).

[Abbé Nicolas-Charles-Joseph Trublet], *Essais sur divers sujets de littérature et de morale* (Paris, 1735).

——*Panégyriques des saints, suivis de réflexions sur l'éloquence en général, et sur celle de la chaire en particulier*, 2nd ed., 2 vols (Paris, 1764).

Graeme Tytler, *Physiognomy in the European Novel: Faces and Fortunes* (Princeton, 1982).

[Charles-Augustin Vandermonde], *Dictionnaire portatif de santé*, 2 vols (Paris, 1759).

Ilza Veith, *Hysteria: The History of a Disease* (Chicago and London, 1965).

Nicolas Venette, *De la génération de l'homme, ou tableau de l'amour conjugal* (Cologne, 1726)

Laurent Versini, *Laclos et la tradition* (Paris, 1968).

Paul Viallaneix and Jean Ehrard (eds), *Aimer en France 1760–1860 (Actes du Colloque international de Clermont-Ferrand)*, 2 vols (Clermont-Ferrand, 1980).

Gregory Vlastos, 'Plato: The Individual as Object of Love', in *Philosophy through its Past*, ed. Ted Honderich (Harmondsworth, 1984).

Michel Vovelle, *Ville et campagne au dix-huitième siècle* (Paris, 1980).

David Watkin, *The English Vision: The Picturesque in Architecture, Landscape and Garden Design* (London, 1982).

Ian Watt, *The Rise of the Novel* (London, 1957).

H. Wendel, 'Arkadien in Umkreis bukolischer Dichtung in der Antike und in der französischen Literatur', *Giessener Beiträge zur romanischen Philologie*, 26 (1933).

The Whore's Rhetoric (London, 1683).

Huntington Williams, *Rousseau and Romantic Autobiography* (Oxford, 1983).

Elizabeth Wilson, *Adorned in Dreams: Fashion and Modernity* (London, 1985).

S. L. Wolff, *The Greek Romances in Elizabethan Prose Fiction* (New York, 1912).

Johann Georg Zimmermann, *Von der Erfahrung in der Arzneikunst*, 2 vols (Zurich, 1763–4) (= *Traité de l'expérience en général, et en particulier dans l'art de guérir*, trans. Le Febvre de Villebrune, 3 vols (Paris, 1774).

Index